Lecture Notes in Morphogenesis

Series editor

Alessandro Sarti, CAMS Center for Mathematics, CNRS-EHESS, Paris, France
e-mail: alessandro.sarti@ehess.fr

More information about this series at http://www.springer.com/series/11247

Serguey Kashchenko

Models of Wave Memory

 Springer

Serguey Kashchenko
P.G. Demidov Yaroslavl State University
Yaroslavl
Russia

The author V.V. Mayorov is deceased.

ISSN 2195-1934   ISSN 2195-1942 (electronic)
Lecture Notes in Morphogenesis
ISBN 978-3-319-36808-5   ISBN 978-3-319-19866-8 (eBook)
DOI 10.1007/978-3-319-19866-8

Springer Cham Heidelberg New York Dordrecht London

Printed on acid-free paper

Springer International Publishing AG Switzerland is part of Springer Science+Business Media
(www.springer.com)

Preface

This book examines the models of neural mediums described by a system of equations with delay. Each element of the medium (neuron) is the oscillator, which generates short pulses (spikes) in stand-alone mode. We discuss the models of synaptic interactions between neurons, which lead to complex oscillatory modes in the system. We study the structure of these modes and the ways of controlling their structure, that is, we solve the problem of choosing the weights of interaction to obtain attractors with predetermined structures. Such attractors are interpreted as images encoded in the form of autowaves (wave memory), thereby solving the problem of identifying attractors or image comparison. The system of equations defining the neural network is obtained from the biological premises, which include many parameters within the meaning of the task. The goal of the book is to develop methods for the asymptotic study of the neural system. These methods allow transfer to other types of equations.

This book discusses two physiological facts arising from theoretical studies: (1) the amount of short-term human memory correlates with the dimension (complexity) of electroencephalogram (EEG) signals; and (2) it is possible to identify visual stimuli on the evoked potentials (forced electric oscillations of the primary visual cortex). This book can be useful to both specialists in oscillatory neural networks as well as experts in differential equations. It is intended for senior students, graduate students, and young researchers engaged in oscillation theory.

This edition is published with financial support from the Russian Foundation for Fundamental Research on Project No. 08-01-07130. Work on the book was supported by the Yaroslavsky Scientific and Educational Center called *Nonlinear Dynamics*.

Contents

Introduction

Investigating the structure and principles behind the functioning of the human and animal nervous systems, one of the most perfect natural systems, is an interesting and significant task. Typically, study of the nervous system involves either the top-down approach, which is used by most psychologists, or the bottom-up approach, which is more common to neurophysiologists. Whereas the former begins with understanding of the high-level phenomena of higher nervous activity, the latter starts with study of the neurons in the elementary components of the nervous system. The top-down approach explains the outward manifestations of the nervous system and its underlying mechanisms. The bottom-up method provides an overall understanding of the functioning of the nervous system through a study of the properties of neurons and their synapses. Although significant advances have been made along both of these paths, neither has yet provided a comprehensive understanding of the nervous system as a whole. The breakthrough in interpreting brain phenomena may occur when these two directions come closer to merging, although doing so does not mean that all psychic phenomena can be explained at the level of neural functioning and the interaction of individual neuronal populations.

Psychologists and neuroscientists speak different "languages," which can be unified through mathematical modeling. Neurophysiology's greatest achievement was the understanding of the fact that even the simplest principles of neuronal activities and operations could be explained through a simple mathematical model to exhibit complex behaviour (e.g., memory, the ability to pay attention, and the ability to classify), which is characteristic of higher animals and humans. This observation has given rise to a new research direction called "neuroinformatics." Such systems (the so-called "artificial neural networks") are right now being successfully applied for solving practical problems and to demonstrate particularly notable advances in those areas, for which mathematical methods for obtaining the exact answer do not exist.

Artificial neural networks are based on our understanding of the mechanisms of their biological prototype and are inextricably linked with the study of the principles of the nervous system. The nervous system is formed externally by relatively simple structural elements, and all of its essential properties are caused by the interaction of these elements ("connectionism"). According to current estimates, there are approximately 10^{10}–10^{12} nerve cells or neurons in the brain. However, there is a general dearth of research into understanding the mechanics of neuronal interaction (including the morphology of the connections between neurons) and the effects caused by the collective behavior of neuronal populations. We understand some of the most general principles. For example, neural systems are able to adapt to changing conditions, i.e., they do not need strict programs. At the same time, collective neuronal populations, at least in the form of reflexes, exist in the nervous system. An experimental study of the effects of the collective behaviour of neural systems is intrinsically difficult because these systems have complex structures. In the human brain, each neuron is exposed to thousands of other neurons and consequently affects thousands of neurons.

An added challenge is to understand how information is transmitted through the nervous system, i.e., understanding the "language" of the brain. By determining the individual functions of the nervous system, neuroinformatics is creating workable models to further this understanding. The main characteristics of the neural network include the following:

- the structural element of the network, i.e., the neuron
- the connections between individual neurons
- the topology of the connections between neurons

Early neural network models were based on the simplest representations of the nervous system. Neurons could exist in one of two states. Neuronal communications were characterized by numerical weight using the simplest network topology, which consisted of a layer of neurons, the states of which comprised the vector output of the network. A neuron passed into one of these states depending on the sign of the weighted sum of its inputs. All of these assumptions are highly idealized.

In its brief excited state, a biological neuron forms a pulse output or "spike." Spikes are characterized by a relatively constant amplitude and shape; therefore, information can be encoded only by interpreting certain properties of the spike sequence. Frequency encoding is a commonly proposed hypothesis to understand the transmission of information based on the frequency of spike generation. Accordingly, the neuron is arbitrarily attributed to an analogue state, thus characterizing the current frequency of spike generation. The action of spike sequences on the neuro-receiver is determined by the state of the synapse (i.e., the mode of contact through which the neurotransmitter operates on the receiver). The numerical weight of the synaptic connections between neurons is an idealized model for determining the process of the action of the spike sequence on the neuro-receiver. Advances in artificial neural networks have been accomplished through the use of more realistic (from a biological perspective) ideas about the structure of the nervous system in which binary neurons were replaced by analogue neurons and the state of each analog neuron assigned a real number. Topological connections between neurons also became more complex: First appeared the layered networks (multilayer perceptrons) followed by the recurrent networks or the networks with feedback (e.g., Hopfield networks).

To a large extent, the use of the above-mentioned hypothesis regarding the frequency encoding of information determines the devices of the artificial neuron and the synapse. Simultaneously, frequency encoding of information contradicts current experimental data on the relationship between the temporal properties of the spike sequence and the speed of neural information processing: Frequency encoding is poorly adapted for rapid information processing. A substantial amount of research has been conducted to emphasize the importance of the temporal properties of the spike sequences. Recently, the wave hypothesis proposed that neural information is represented by certain temporal (phase) properties of spike sequences. In comparison with the frequency hypothesis, the wave hypothesis imposes much stricter requirements on biological neuron modeling leading to the active development of biologically plausible neuron models. Choosing a convenient, biologically plausible neuron model is challenging. The Hodgkin–Huxley model, obtained from the first principles of the laws of chemical kinetics, is poorly adapted for use in composing neural networks owing to its complexity. However, a new range of phenomenological models of biological neurons is being developed. These newer models more or less accurately reflect the dynamics of biological neurons and are simple enough for analytical or numerical investigations. Among such models, special mention must be made of the Zeeman relaxation models and those models based on the use of harmonic oscillators. Another challenge is to represent the relationship between static and dynamic information in the neural network. In the biological nervous system, the boundary between static and dynamic information is less distinct because learning takes place simultaneously when part of the information changes from dynamic to static. Static information represents the properties of the neural network as a system and transforms the input signals into output signals. Dynamic information corresponds to the signal at the network input at any given time. In classical neural networks, static information is usually integrated into the network structure in the form of the weights of the connections and introduced into it from the outside at the learning stage (i.e., the configuration process when solving a specific problem); dynamic information is presented by the inputs in the network. Thus, the framework of classical neural networks represents static and dynamic information in different ways, which is contradictory to the framework of the biological nervous system.

This book is dedicated to the development of neural networks wherein information is presented in spike sequences, which is in line with the wave hypothesis. Spike sequences generate biologically plausible model neurons described by phenomenological equations with delay. The model neurons reflect the dynamics of biological neurons well while avoiding the use of analytical methods. In the following chapters, we discuss various methods to represent the relationship between static and dynamic information through stable spike bursts.

Review of the Literature

Researchers have long known that neurons communicate through the exchange of pulses (spikes). However, there is no single or specific approach that demonstrates how information can be presented and stored in the brain. Moreover, the well-known monograph by Pribram (1975) convinces us that there is not enough evidence of homogeneity. All models of neural networks are based on various representations of the principles of information processing in the brain. Adrian (1934) (see also Adrian and Moruzzi 1939; Green et al. 1993, p. 272) demonstrated that spikes have a sufficiently stable amplitude, so researchers often say that neurons respond to pulse action according to the law of "all or nothing." In one of the approaches, incoming information is encoded by the spike frequency. Because the spikes can be combined into bunches, or "bursts," variations of this approach then become possible, i.e., information can be transmitted by a number of pulses in a burst or in highly variable interpulse intervals (Burns 1968), which can be measured by probabilistic coding or the coding of random flow (Frolov 1993). Timofeev (1997) proposed and investigated a model of the neural network, the state of which could be described by the spatial distribution of random binary stream densities generated by the neurons. At the entry level of every neuron, the flows are mixed with random flows that perform the role of synaptic weights at the inputs of each neuron. The output stream of the neuron is formed depending on the total input stream.

The detector neuron model is based on the frequency approach, in which the notion of the intensity of action on the neuron is introduced into the analysis. The intensity of action is determined by the spike frequency. In other words, the state of the neuron is characterized by the frequency of pulsation. The state of the neural network reflects the distribution of frequencies of pulsation of all its neurons minus the phase relationships. According to the synaptic hypothesis proposed by Hebb (1949) and later developed by Eccles (1966), neurons respond to external action depending on the current synaptic permeabilities of the distribution of synaptic weights, which characterize the efficiency of inputs. Synaptic permeabilities may change with time (referred to as the "plasticity of synapses"), thus providing a synthesis of incoming information and previous experience. Therefore, according to the synaptic hypothesis, information is reflected as the specific structure of synaptic permeabilities. However, it should be noted that the neuron response is dependent not only on the condition of its inputs (signals and synaptic permeabilities) but also on the current state of the neuron itself. For example, after pulsing, the neuron enters a state of refractoriness in relation to the stimulus (Eckert et al. 1991).

The detection approach, pioneered by McCulloch and Pitts (1943) and further developed by Hebb (1949), is a highly influential and widely accepted approach to measure the state of neural networks. Other scholars, such as Frolov and Muravyev (1987, 1988) (see also Frolov 1993); Dunin-Barkovsky (1978); Minsky and Papert (1971); Rosenblatt (1965), have expanded on and compared various aspects of the detection approach. Edelman and Mountcastle (1981) detected the neural modules also known as the "mini-columns." From synaptic theory, it follows that in the process of perceiving, storing, and reproducing information, the brain's main purpose is to ensure the collective functioning of all its elemental structures. This principle of collectivism is laid in all models of neural networks beginning with the Kohonen model (1980, 1984). Hopfield (1982, 1984); Hopfield and Tank (1985); Chua and Yang (1988a) further expanded this research to include network models consisting of neuron-like detector elements of both discrete and analog operational amplifiers. In such networks, a set of stable states of equilibrium is formed during the process of memorization (learning) on the corresponding system phase portrait. These stable states are called "finishing states" and act as storehouses of information. In the process of reproduction under external signals, the network lapses into one of the many finishing states.

Networks with finishing states have long been used for solving applied problems such as the use of neural networks for computational purposes (see Galushkin 2000; Gorban and Rossiev 1996; Neural networks for instrumentation 2003). For instance, three-layer cellular networks can be successfully used to distinguish the trajectories of a moving object on a background of intense interference (Korotkin and Pankratov 1997). Furthermore, perceptrons combined in micro-columns successfully cope with the allocation of the edge of the image (Sergin 2005). Nevertheless, networks with the finishing state, which simulates the detection approach, are static and act as storehouses of information. Regarding classical neural networks, we refer the reader to the meaningful review of Potapov and Ali (2002), which gives an overview of the different types of neural networks with finishing states, considers the rules for choosing the weights of connections (learning), and discusses the dynamics of a wide class of networks without finishing states.

Generally, the system of equations describing a network has complex attractors (due to feedback) including chaotic ones. Until the mid-90s, researchers avoided the interpretation of complex modes because of the difficulties involved. For example, the assumption about the symmetry of links allows us to prove that all paths in the Hopfield network tend to finishing states. At that time, we considered that the interpretation of periodic modes suggested itself. For instance, it is a well-known fact that by taking into account the signs of the phase variables of any analog network (without finishing states), we can obtain the sequence of binary vectors and view them as a sequence of images.

Sokolov and Izmailov (1984, 1989) conducted research with regard to the perception of visual information using the detector direction and the theory of information coding by channel number. Their framework is based on the concept of the reflex arc (Sokolov 1981), according to which the channel or chain of neurons responds to a stimulus. The inputs formed are best tuned to signal parameters (the vectors of synaptic weights and of stimulus are collinear or almost collinear). Hubel and Wiesel (1977) (see also Hubel 1990) described in detail the analogous mechanism for coding of the visual analyzer.

Networks with final states have already long been used for making decisions about application-oriented tasks. To get information about the domain's application, see Gorban and Rossiev (1996) and Neural networks for instrumentation (2003).

The monograph of Galushkin (2000) is dedicated to the application of neural networks. Even three-layered cell networks can be successfully used to highlight the trajectories of a moving object on a background of intensive interruptions (see Korotkin and Pankratov 1997). Moreover, perceptrons united in micro-columns can successfully manage to highlight the edge of the image (see Sergin 2005). However, networks with final state, modeling the detector approach, are static as is the information stored in them. Concerning classical neural networks, we can refer to the review by Potapov and Ali (2002), in which different types of such networks are introduced. This review also considers the rules of how to choose (training) the weights—coefficients of links and discusses the dynamics of a wide range of networks that have no final state.

The phase approach to coding information in the nervous system presents an alternative to the frequency approach. According to the phase-frequency approach to storing information in the brain, two types of coordinated work of the neurons should be taken into account with respect both to space and time. Neurons have the property of temporary selectivity (see Elul and Adey 1966; Elul 1967). Depending on the state of the neural network, an impact may or may not generate a pulse. Bekhtereva (1980, 1988) (see also Bekhtereva et al. 1977, 1985) stated that information coding is not a static but rather a dynamic process associated with the formation of ensembles of coherently working neurons. In one of the latest approaches, information coding is carried out in the form of waves of neural activity while also accounting for phase oscillations. This is called the "wave approach."

A hypothesis of a phase-frequency signal coding in the nervous system was also developed by Lebedev (see Zabrodin and Lebedev 1977; Lebedev 1985, 1992, 2002). Ideas about wave codes have been useful in explaining a number of psychological phenomena manifested in the search speed and rapidity of decision-making in a situation where there is a choice. Lebedev's hypothesis attributes a predominant role to waves of slow activity. Other researchers (Elul and Adey 1966; Adey 1967; Creutzfeldt 1966; Shulgina 1978; Danilova 1985) have also proven that there exists a close relationship between slow activity waves and pulse potentials. As Lebedev's hypothesis explains, the phenomenon of memory from a position of self-oscillation does not use possible mechanisms of fixing data based on the modification of synaptic conductivities. Although Lebedev's research on wave coding and information storing in the brain are based on experimental material, in many respects these experiments have been performed at the descriptive and empirical levels (see Lebedev et al. 1991).

The wave approach is modeled by neural networks with nonstationary behavior, which can be divided into two classes. In the first case, although individual elements in the neural network may not have their own auto-rhythmicity, the whole system can function in an oscillatory mode. A good example of this is Wiener's network (Wiener and Rosenbluth 1961), which is used to describe the process of the propagation of excitation in cardiac muscle. Oscillatory modes in Wiener's network have been studied by Kashchenko and Mayorov (1995). Mayorov and Shabarshina (1997, 2001) proposed a construction called the "W-net neurons," which are a type of integrative-threshold element to store and reproduce a sequence of binary vectors.

Researchers have developed a model of the neural system with stationary behavior (see Frolov and Shulgina 1977, 1983; Frolov et al. 1984; Frolov and Muravyov 1987, 1988; Shulgina et al. 1983, 1988; Shulgina 1990a, b, 1993). Their research has also demonstrated that periodic action on the overall input of the associative network of the system is an indispensable condition for memorizing and reproducing information. For detailed information about oscillatory networks consisting of nonoscillatory elements, see Potapov and Ali (2002).

In the second case, we consider nonstationary networks consisting of neural oscillators. Networks arise naturally in simulation of the olfactory and visual systems of humans and animals. Oscillatory neural networks can be divided into two classes based on the types of elements that are either close to or far from the harmonic oscillators. In the former scenario, the population of interconnected neural oscillators is situated close to the harmonic oscillators; these include the Van der Pol oscillators (see Ohsusa et al. 1989; Omata et al. 1988; Osipov et al. 1992). In this arrangement, the distribution of phases and amplitudes of oscillations can be interpreted as the image code (see Freeman et al. 1988; Kryukov 1988; Malsbyrg and Schneider 1986). Many scholars—such as Borisyuk et al. (1992, 2002)—have written detailed reviews about the functioning of neural oscillator networks situated close to the harmonic oscillators. Despite the variety of methods for studying these networks, it seems to us that they all fit into the following scheme. The transition to slow variables is carried out in what is called the "amplitude phase." Various assumptions about the structure of neural interactions are set forth in terms of these variables. In some cases, the nature of the interaction is also specified in the framework of the original variables. Next, all of the amplitudes are often regarded as the same, and the corresponding equations are omitted. The resulting outcome is called the "system-of-phase oscillator" (Ermentrout and Kopell 1984). We propose that by controlling the synaptic weights, the system-of-phase oscillator would have a stable mode with a predetermined phase distribution (stationary or more complex). In some cases, this problem can be solved analytically. For example, researchers (see Kazanovich et al. 1991; Baird 1986, 1989) showed that we can specify the rule for choosing the weights such that a predetermined portion of oscillators operates in phase and the rest in antiphase.

In our opinion, we can strictly justify this method only in the vicinity of the bifurcation point when partial neuron generators give rise to harmonic oscillations. In this case, the normal form is constructed (Shilnikov et al. 2001, 2004; Guckenheimer and Holmes 2002; Wiggins 1990), which is a system of approximate equations for the amplitudes and phase differences of the partial oscillations of generators. If the normal form has a stable state of equilibrium, then we can prove that the original system has a stable periodic or almost periodic solution called a "torus." Periodic solutions and tori correspond to the tori of the reference system. If the normal form contains chaotic modes, then it is not possible to connect them with possible chaotic modes of the original system. Nevertheless, the transition to the system for the amplitudes and phase differences is very useful. It allows us to hypothesize about the structure of the oscillation, which can then be verified numerically. To not be categorical, here we note that normalization is also possible for singularly perturbed systems (as demonstrated later in this book).

Correlating the above-mentioned information with biological data, we provide the following interpretations. First, in our opinion, it is not typical in the nervous system for neural oscillations to be close to harmonic oscillations. In the normal state, the neural network's separate elements, neurons, generate short pulses called "spikes." The connection between neurons is carried out through spike transfers. After removing the pulse component, oscillations averaged over neural associations are far from harmonic, and this is reflected in the asymmetry of the human α-rhythm wave. Second, the possibility of the transition to phase oscillators assumes uniformity of the interaction times of the neural oscillators. In the nervous system, both individual neurons and neural associations possess the property of absolute and relative refractoriness (i.e., immunity or weak susceptibility to external impacts at certain intervals of time). For instance, a biological neuron is unable (or almost unable) to perceive any impact during spike generation and for a short interval after aspike has been generated. Thus, the process of interaction is strongly nonhomogeneous in time where refractivity must be accounted for (Wiener and Rosenbluth 1961; Malsbyrg and Schneider 1986; Kryukov et al. 1986) unlike within the framework of harmonic approximations.

Researchers are studying neural networks from types of oscillators other than harmonic ones. A systematic description of the used formal neurons and some results on the organization of the oscillations in networks is available in the works of Rabinovich et al. (2006) and Borisyuk et al. (1992, 2002). Hodgkin–Huxley's classical model of representing a neuron by four ordinary differential equations is often used. Modifications of this model first occurred in the direction of simplification with a decrease in the number of equations and later toward complication wherein the number of equations increased. The latter is associated with the simulation of complex and chaotic modes; for a simpler model of the network, see Rabinovich (1996). Other ideologically related models include the Wilson–Cowan, Hindmarsh–Rose, and Shillien–Köenig models as well. Ivanitsky et al. (1994) discussed autowave processes in nonlinear excitable media. These investigators are specifically interested in studying networks, in which the elements are described by Fitz–Hugh–Nagumo equations and the closely related Zeeman (1973) model of differential equations.

Due to the complexity of nonlinear models, small neural networks are usually considered instead. However, the model can be simplified by passing to the phase variables using the interaction function, which is also known as the "Kuramoto H-function" (Kuramoto 1984) (see also Abarbanel et al. 1996). Using this function, it is difficult to prove the relationship of the solutions of phase equations and the original system. It is important to note that we cannot take any oscillator as a formal neuron in the model. We argue that (see Kashchenko and Mayorov 1995a) in addition to the network, we must also consider the elements described by Hutchinson, i.e., a simple equation for the system with a renewable resource taking into account the delay in recovery. In our model, we located neurons at regular intervals of a rectangular lattice lying on the plane with local connections. We analytically showed the existence of interesting oscillatory modes, except during the transition to a triangular lattice where the number of modes disappeared due to a disturbance in the balance between the time of wave propagation in triangles and the resource recovery time. We did not observe this phenomenon in phase oscillators.

We also note the importance of attention problems in neural systems. When information is being processed, part of the static information gets actualized and becomes embedded in the structure of the neural network. Thus, in Hopfield networks, one of the binary patterns memorized at the learning stage becomes actualized. Simultaneously, the effects of false memory can also be observed when the image absent in the learning sample also becomes actualized. To combat the problem of false memory and the construction of auto-associative memory systems, different approaches have been proposed based on the use of chaotic and stochastic modes and their synchronizations. For instance, scholars such as Izhikevich and Malinetskii (1992, 1993) have explained the workings of neural networks comprising neurons with stochastic dynamics. They demonstrated that such networks favorably differ from classic networks in combating false images. In addition, in such a network exists the state "I do not know," which corresponds to an unidentified input pattern. Dmitriev (1993, 2002) (see also Dmitriev et al. 2000) also constructed chaotic systems that store information in the form of cyclic unstable trajectories. In these networks, the procures of recording, storing, and reproducing information are based on the control and synchronization of oscillatory stochastic processes. Kryukov (1991, 2004) discussed other aspects of attention problems associated with the release of arrays of incoming sensory information.

Thus, at present, the prospects of developing new approaches to information processing are inseparably associated with the ideas of nonlinear dynamics (see Haken 1990; Nicolis 1989). In this approach, the information carrier is not a stationary distribution-bit value but rather a complex and, in general, nonperiodic mode of the dynamic system (see Malinetskii). According to some biologists (for example, Crick and Koch 1990), the future belongs to oscillatory networks, which better explain the processes of information processing in the brain.

Networks of Pulse Neurons Described by Differential Equations with Delay

Mayorov and Myshkin (1990a, b) proposed a differential equation with delay for oscillator biologically plausible neural elements, the oscillations of which are quite far from the harmonic type. The network also comprised diffusion-related neurons that were numerically investigated and conventionally referred to as the "network of pulse neurons" (Kashchenko and Mayorov 1992, 1993, 1995a, b, 1998; Kashchenko et al. 1993, 1995c, d, 1997, 1998). In our research, we have shown that the system of differential equations with delay allows for the possibility of not only computer-aided but also analytical research through asymptotic methods. We succeeded in solving problems of the structure, planning, and synchronization of oscillatory modes (see also Paramonov (2002) for an understanding of the pulse neuron model with variable delay).

Relying on the asymptotic formulae, Shabarshina (1994, 1999) proposed an axiomatic model network of neuro-cellular automata. There is one-to-one correspondence between oscillatory modes in the axiomatic model and the original system described by equations with delay. For example, Shabarshina (2000) compared the self-organization of a slightly nonhomogeneous network of neuro-cellular automata with the periodic sequence of synchronously functioning subsets of automata. Lagutina (2002) also found such self-organization within the original model.

As Mayorov and Myshkin (1990a, b) showed, the model of the neuron is universal to a certain extent and describes the workings of both pacemaker neurons (oscillators) and neuron detectors. Mayorov et al. (2003) also studied the planning of waves in annular patterns of associations of neurons detectors. Anufrienko et al. (2004) and Anufrienko and Mats (2007) used variant detectors for simulating the process of saltatory conduction of pulses by way of myelinated nerve fibers.

Based on the asymptotic formulae obtained for neuron detectors and on Shabarshina's model, Mayorov et al. (2006) and Mayorov and Konovalov (2007) proposed the axiomatic model of generalized neuronal automata (type of integrative-threshold element), which operates in continuous time and can function as an oscillator and detector. A medium constructed of these elements forms the foundation on which the wave memory is modeled.

To model nonstationary networks consisting of elements that are not oscillators, scholars have proposed the W-neuron network (see Mayorov and Shabarshina 1997, 2001). W-neurons operate in discrete time. Each of them sequentially passes through the various states of refractoriness and expectation. If a neuron receives a sufficiently strong impact in its pending state, it generates a pulse by itself and then passes into the refractory state as has been demonstrated by biological data. The network can memorize and reproduce sequences of binary vectors or a sequence of images. In playback mode, part of the sequence (possibly distorted) is presented to the network. After initialization, the network reproduces the entire sequence. In a certain way of initializing, the network can combine parts of several records into a single sequence.

A W-neuron can be upgraded so that depending on the force of impact, it will respond with a finite sequence of single pulses (burst). Thai significantly increases the possibility of the neural network for storing and reproducing information. The network is a ring of *associations of* W-neurons or units. For each neuron of every module, all the neurons of the two previous modules will operate and respond differently to the same vectors contained in the memorized sequence of binary vectors (Mayorov and Shabarshina 1999).

A network of W-neurons can also be successfully used as a solution to the problem of planning the safe paths of point robots. We have considered the problem of planning optimal paths for two point robots on the regular triangular lattice, in which part of the vertices play the role of barriers (Mayorov et al. 2005). When moving, the robots must bend around obstacles without mutually colliding.

The basic idea of finding the shortest safe paths consists of replacing the motion of two robots on the movement of one generalized robot using a specially constructed set in four dimensions. To solve this problem, we used the neural network with W-neuron elements (see Lebedev 2000). Anufrienko (2005) proposed a model of continuous neurons of Wiener based on differential equation with delay, and considered the propagation of waves in cellular neural networks from such neurons. Similarly, Korotkin et al. (2007) proposed a neural network of pulse neurons (see Mayorov and Konovalov 2007) for forming the textured attributes of images.

Review of the Contents of the Book

In Chap. 1, based on biological premises, we propose a differential-difference equation to describe the dynamics of the membrane potential of a neuron (i.e., the quantitative characteristics of the state). We analyzed the equation using special asymptotic methods and arrived at two types of solutions. One type is characterized by short-term and high-amplitude pulses that are separated from each other by the intervals of slow change. The spike formations, and the sections of the slow evolution of the membrane potential between them, are in reasonable agreement with experimental data. The second type of solution corresponds to slower development of processes in the neuron that have also been observed in experiments. This model is suitable for describing two types of neurons: pacemakers (oscillators) and detectors. For the latter, the pulse is a response to external impact.

In Chap. 2, we discuss the results of external actions on the neurons. We consider two types of electrical and synaptic actions (electrical and chemical synapses). We analyze the relevant equations to show that periodic pulse actions impose their frequencies on the neurons. In response to each external pulse, a neuron generates its own spike. If the external pulses are sufficiently continuous, then the action leads to oscillations of a new type, wherein the response spikes of the neuron are combined into bursts called "groupings." We observed both types of neuron activity occurring in response to the external action in biological experiments.

In Chap. 3, we consider ring neuron systems where each element is related to two neighboring elements. We use this model to describe the systems of differential-difference equations. Using asymptotic methods, we analytically obtained the values of synaptic weights (i.e., coefficients characterizing the interaction force of neurons), which guarantee the existence in the ring of waves of predetermined structures. The analogous problem is solved for the ring of neurons where each neuron acts on two consecutive elements.

In Chap. 4, we study focused uniform ring structures of neural modules containing excitatory and inhibitory elements. In each module, excitatory elements have equal access not only to all the neurons of the given module but also to the excitatory neurons of the next module. Inhibitory neurons could act only on the excitatory elements of the same module. We demonstrated that in the population within each module, a part of the excitatory elements synchronize spike generation, whereas the other elements suppress impulsation. Among groupings of neuron spikes of neighboring modules, temporary mismatches arise that are unequivocally determined by the number of unbraked excitatory neurons. In modules, the composition and size of the groups of synchronously functioning excitatory neurons depend on the initial conditions. If we interpret the spatial distribution of spikes (patterns) in the module as the image, then the network can store infinitely longer sequences of images. Both the patterns and their temporal distribution prove to be extremely informative. In Chap. 5, we propose the model of adapting synaptic weights in the neural ring structure. Under the influence of an external periodic sequence of pulses, synaptic weights are modified. On completing the adaptation and after being disconnected from the external signal, the neuron ring continues to generate a copy of the external pulse sequence.

In Chap. 6, we consider the mechanism of interaction of two neural rings. Any neuron of one ring can potentially act on any neuron of the other ring (communications do not have an address). The functioning of the neural structure for identical rings results in synchronization of their oscillations, especially if the initial oscillations are not too different in phase.

Chapter 7 deals with the organization of the oscillations in homogeneous populations of diffusion-related neurons. We consider the network topology of the segment, circle, and plane. We establish the existence of vibrational modes of the type of the leading centers. Traveling waves can exist on the circle, whereas any curve can serve cyclically as the repeated wave-front at the beginning of the phase on the plane. For fully connected neural networks by asymptotic methods, we prove the existence of the following two types of oscillations:

- The set of neurons can be arbitrarily split into two subsets. Neurons belonging to both subsets generate pulses almost synchronously. Neuron pulses belonging to different subsets are interspersed over time. We offer a simplified interpretation of this phenomenon. A pair of subsets taken in one order is seen as the image and in a different order as the negative of the image. The neural network can infinitely present to the observer the image as well as its negative for an infinitely long period of time.
- One can arbitrarily split the set of neurons into three subsets. The pulses of neurons in each of the subsets occur synchronously through the same intervals of time. Through self-organization, the start pulses of the first, second, and third subsets occur sequentially and in a cyclical manner resulting in three fully functioning neuron modules.

In the same chapter, we also discuss computer analyses of the system of equations of a fully connected homogeneous network of diffusionally interacting neurons. We establish the existence of complex modes. If the external pulse sequence acts on a system that is in a complex mode, then there appears the wave packet (groupings of pulses) bearing the features of the external action. These groupings of pulses break down over time. The phenomenon can be interpreted as short-term storage or memorization of wave packets, which is followed by their subsequently being forgotten.

In Chap. 8, in relation to an individual, we experimentally verify one of the phenomenological consequences of the model: The more complex the mode of oscillation in the neuron population, the more adequate is its response to the external action. We investigate an electroencephalogram (EEG) of a human being and estimate its complexity by assigning it a number called the "pseudo-correlation dimension." The formalism of computation of the pseudo-correlation dimension almost completely coincides with the procedure of the Grassberger–Prokaccia (1983a, b) computation of the correlation dimension. Furthermore, for the individual, using a special psychological test, we determine the amount of short-term memory required to remember decimal numbers. We demonstrate that a statistically valid amount of short-term memory positively correlates with the pseudo-correlation dimension.

In the concluding chapter, we investigate the so-called "evoked potentials" (EPs) of the brain, which are induced in an individual by way of color stimuli. The results of the analysis show that there are cyclically observed zones of difference in EPs for stimuli of different colors separated by intervals with no differences in EPs. We confirm this fact through two independent methods. The cyclic recurrence of the presence of zones of difference can be interpreted as an indirect confirmation of the wave-coding hypothesis of information in the human brain.The proposed methods provide an opportunity with high probability to identify the stimulus with respect to the appropriate EP.

Appendix A outlines the steps of integration of differential-difference equations with delay. Appendix B specifies the duration of the solution of the equation of neuron. We choose the term of the first order of smallness of magnitude inverse of the large parameter. In Appendix C, we consider the model of excitation of a myelinated nerve fiber. Appendix D introduces the model of neuro-cellular automata, axiomatizing the dynamics of the neuron, as described in Sects. 1.6 and 2.3. Appendix E proposes the construction called the "W-neuron network," which can store and reproduce the sequence of binary patterns. Here, based on the network of W-neurons, we solve the problem of planning optimal paths of motion for point robots on a regular triangular lattice.

References

Abarbanel, G. D. I., Rabinovich, M. I., & Selverston, A., et al. (1996). Synchronization in neural ensembles. *UFMN, 166*(4), 363–389.

Adey, V. R. (1967). Organization of brain structures in terms of communication and storage of information. *Modern problems of electrophysiology of the central nervous system* (pp. 324–340). Moscow: Nauka.

Adrian, E. D. (1934). Electrical activity of the nervous system. *Archives of Neurology and Psychiatry, 34*(6), 1125–1136.

Adrian, E. D., & Moruzzi, B. (1939). Impulses in the pyramidal tract. *Journal of Physiology, 97*(2), 153–159.

Anufrienko, S. E. (2005). Waves in cellular neural networks of continuous neurons of Wiener. In *Proceedings of the 7th All-Russian Conference "Neuroinformatics 2005"*, Part I (pp. 75–80). Moscow: Moscow Engineering and Physics Institute.

Anufrienko, S. E., & Mats, A. S. (2007). Model of saltatory conduction of excitation along the branching nerve fiber. *Modelling and Analysis of Information Systems, 14*(2), 17–19.

Anufrienko, S. E., Mayorov, V. V., Myshkin, I. Y., & Gromov, S. A. (2004). Investigation of the system of equations with delay, modelling saltatory conduction of excitation. *Modelling and Analysis of Information Systems, 11*(1), 3–7.

Baird, B. (1986). Nonlinear dynamics of pattern formation and pattern recognition in the rabbit olfactory bulb. *Physica D 22, 150,* 150–175.

Baird, B. (1989). A bifurcation theory approach to the programming of periodic attractors in network models of olfactory cortex. In *Advances in Neural Information Processing Systems* (pp. 459–467). Morgan Kaufmann Publishers Inc., CA.

Bekhtereva, N. P. (1980). *Healthy and diseased human brain*. Moscow: Nauka.

Bekhtereva, N. P. (1988). Mechanisms of activity of human brain. In N. P. Bekhtereva (Ed.), *Neurophysiology of man*. Leningrad: Nauka.

Bekhtereva, N. P., Bundzen, P. V., & Gogolitsin, Y. L. (1977). *Brain codes of mental activity*. Leningrad: Nauka.

Bekhtereva, N. P., Gogolitsin, Y. L., Kropotov, Y. D., & Medvedev, S. V. (1985). *Neuro-physiological mechanisms of thinking*. Leningrad: Nauka.

Borisyuk, G. N., Borisyuk, R. M., & Kazanovich, J. B., et al. (1992). Oscillatory neural networks. Mathematical results and applications. *Mathematical Modelling, 4*(1), 3–43.

Borisyuk, G. N., Borisyuk, R. M., Kazanovich, Y. B., & Ivanitsky, G. R. (2002). Models of neural dynamics in brain information processing—results of "decade". *UFMN, 172*(10), 1189–1214.

Burns, B. N. (1968). *Uncertainty in the nervous system*. Moscow: Mir.

Chua, L., & Yang, L. (1988a). Cellular neural networks: Theory. *IEEE Transactions on Circuits and Systems, 35*(10), 1257–1272.

Creutzfeldt, O. D., Watanabe, S., & Lux, H. D. (1966). Relations between EEG phenomena and potentials of single cortical cells. Spontaneous and convulsoid activity. *Electroencephalography and Clinical Neurophysiology, 20*(1), 19–37.

Crick, F. & Koch, C. (1990). Towards a neurobiological theory of consciousness. *Seminars in Neuroscience, 2,* 163–191.

Danilova, N. N. (1985). *Functional states: mechanisms and diagnostics*. Moscow: Moscow State University Press.

Dmitriev, A. S. (1993). Chaos and information processing in nonlinear dynamic systems. *RE, 38* (1), 1–24.

Dmitriev, A. S. (2002). Dynamic chaos as information carrier. In *New in synergetics: a look into the third millennium* (pp. 82–122). Moscow: Nauka.

Dmitriev, A. S., Andreyev, Y. V., & Bulushev, A. G. (2000). Information coding on the basis of dynamic chaos. In *Foreign Radioelectronics. Advances of modern electronics*. Moscow: IPRZHR, 11, 27–33.

Dunin-Barkowski, V. L. (1978). *Information processes in neural structures*. Moscow: Nauka.

Eccles, J. (1966). *The physiology of synapses*. Moscow: Mir.

Eckert, R., Randall, D., & Augustine, G. (1991). *Animal Physiology* (vol. 1). Moscow: Mir.

Edelman, G. M., & Mountcastle, V. B. (1981). *The mindful brain*. Moscow: Mir.

Elul, R. (1967). Amplitude histogram of the EEG as an indicator of the cooperative behaviour of neurons. *Electroencephalography and Clinical Neurophysiology, 23*(1), 87.

Elul, R., & Adey, W. R. (1966). Nonlinear relationship of spike and waves in cortical neurons. *The Physiologist, 8,* 98–104.

Ermentrout, G. B., & Kopell, N. (1984). Frequency plateaus in a chain of weakly coupled oscillators I. *SIAM Journal on Mathematical Analysis, 15*(2), 215–237.

Freeman, W. J., Yao, Y., & Burke, B. (1988). Central pattern generating and recording in olfactory bulb: a correlation learning rule. *Neural Networks, 1*(4)

Frolov A. A., & Muravyov I. P. (1987). *Neural models of associative memory*. Moscow: Nauka.

Frolov A. A., & Muravyov I. P. (1988). *Information characteristics of neural networks*. Moscow: Nauka.

Frolov, A. A. (1993). Structure and functions of learning neural networks. *Neurocomputer as the basis of thinking computers* (pp. 92–100). Moscow: Nauka.

Frolov, A. A., & Shulgina, G. I. (1977). Modelling of conditions of excitation at different modes of operation of the nervous network. In *The functional significance of the electrical processes in the brain* (pp. 190–197). Moscow: Nauka.

Frolov, A. A., & Shulgina, G. I. (1983). Memory model based on the plasticity of inhibitory neurons. *Biophysics, 28*(3), 445–480.

Frolov, A. A., Medvedev, A. V., Dolina, S. A., Kuznetsova, G. D., & Shulgina, G. I. (1984). Modelling of different modes of brain bioelectrical activity in norm and at increase of "convulsive readiness" on the network of neural elements. *ZHVND, 34*(3), 527–536.

Galushkin, A. I. (2000). *The theory of neural networks. Neurocomputers and their application*. Moscow: Publishing enterprise of editorial board of the journal "Radio engineering".

Gorban, A. N., & Rossiev, D. A. (1996). *Neural networks on the PC*. Novosibirsk: Nauka.

Grassberger, P., & Procaccia, I. (1983a). Characterization of strange attractors. *Physical Review Letters, 50*(5), 346–349.

Grassberger, P., & Procaccia, I. (1983b). Measuring the strageness of strage attractors. *Physica D, 9*(1–2), 189–208.

Green, H., Stout, W., & Taylor, D. (1993). *Biology* (Vol. 2). Moscow: Mir.

Guckenheimer, J., & Holmes, F. (2002). *Nonlinear oscillations, dynamic systems and bifurcations of vector fields*. Moscow-Izhevsk: Institute of Computing Research.

Haken, H. (1990). *Information and self-organization: A macroscopic approach to complex systems*. Moscow: Mir.

Hebb, D. O. (1949). *The organization of behaviour: neuropsychological theory* (p. 335). New York: Wiley.

Hopfield, J. J., & Tank, D. W. (1985). Neural computation of decisions in optimization problems. *Biological Cybernetics, 52*, 141–152.

Hopfield, J. J. (1982). Neural networks and physical systems with emergent collective computational abilities. *Proceedings of the National Academy of Sciences, 79*(8), 2554–2558.

Hopfield, J. J. (1984). Neurons with gradual response have collective computational properties like those of two-state Neurons. *Ibid, 81*, 3088–3092.

Hubel, D. 1990. *Eye, brain, vision*. Moscow: Mir.

Hubel, D. H., & Wiesel, T. N. (1977). Functional architecture of macaque monkey cortex. *Proceedings of the Royal Society*. London, *198*, 1–59.

Ivanitsky, G. R., Medvinsky, A. B., & Tsyganov, M. A. (1994). From the dynamics of population waves generated by living cells to neuroinformatics. *UFN, 164*(10), 1041–1071.

Izhikevich, E. M., & Malinetskii, G. G. (1992). On possible role of chaos in neural systems. In *Reports of the Academy of Sciences of Russia, 326*(4), 626–632.

Izhikevich, E. M., & Malinetskii, G. G. (1993). *A neural network model with chaotic behaviour*. Moscow: Russian Academy of Sciences, Institute of Applied Mathematics named after Keldysh, Preprint, 17.

Kashchenko, S. A., & Mayorov, V. V. (1992). *Investigation of difference-differential equations, modelling the dynamics of neuron*. Moscow: Russian Academy of Sciences, Institute of Applied Mathematics named after Keldysh, Preprint, 8.

Kashchenko, S. A., & Mayorov, V. V. (1993). On a differential-difference equation, modelling neuron impulse activity. *Mathematical modelling, 5*(12), 13–25.

Kashchenko, S. A., & Mayorov, V. V. (1995a). Wave structures in ring systems from homogeneous neural modules. *Reports of the Russian Academy of Sciences, 342*(3), 318–321.

Kashchenko, S. A., & Mayorov, V. V. (1995b). Wave structures in cellular network from formal neurons of Hutchinson. *Radio Engineering and Electronics, 40*(6), 925–936.

Kashchenko, S. A., & Mayorov, V. V. (1998). Model of adaptation of ring neural ensembles. *Radio Engineering and Electronics, 43*(11), 1–7.

Kashchenko, S. A., Mayorov, V. V., & Myachin, M. L. (1995c). Oscillations in systems of equations with delay and differential diffusion, modelling local neural networks. *Reports of the Russian Academy of Sciences, 344*(3), 137–140.

Kashchenko, S. A., Mayorov, V. V., & Myshkin, I. Y. (1993). Investigation of oscillations in ring neural systems. *Reports of the Russian Academy of Sciences,333*(5), 594–597.

Kashchenko, S. A., Mayorov, V. V., & Myshkin, I. Y. (1995d). Wave propagation in the simplest ring neural structures. *Mathematical modelling, 7*(12), 3–18.

Kashchenko, S. A., Mayorov, V. V., & Myshkin, I. Y. (1997). Wave structures in ring neural systems. *Mathematical Modelling, 9*(3), 29–39.

Kashchenko, S. A., Mayorov, V. V., & Myshkin, I. Y. (1998). Model of the neural system, synchronizing wave packets. *Reports of the Russian Academy of Sciences, 360*(4), 463–466.

Kazanovich, Y. B., Kryukov, V. I., & Luzianina, T. B. (1991). Synchronization via phase-locking in oscillatory models of neural networks. In *Neurocomputers and attention. Neurobiology, Synchronization and Chaos* (pp. 269–284). UK: Manchester University Press.

Kohonen, T. (1980). *Associative memory*. Moscow: Mir.

Kohonen, T. (1984). *Self-organization and associative memory*. New York: Springer.

Korotkin, A. A., & Pankratov, V. A. (1997). Selection of moving objects by three-layer cellular neural network. *Journal of Computational Mathematics and Mathematical Physics, 37*(5), 630–636.

Korotkin, A. A., & Mayorov, V. V., & Myachin, M. L. (2007). Application of pulsed neural network to generate texture features. In *Proceedings of the IX All-Russian Conference "Neuroinformatics-2007"*. Moscow: Moscow Engineering Physics Institute.

Kryukov, V. I. (1988). Short term memory as a metastable state. V. "Neurolocator, a model of attention". In: *Cybernetics Systems-88* (pp. 999–1006), Kluwer Academic Publisher.

Kryukov, V. I. (1991). An attention model based on the principle of dominant. In A. V. Holden & V. I. Kryukov (Eds.), *Neurocomputers and Attention. I. Neurobiology, Synchronization and Chaos* (pp. 319–351). Manchester: Manchester University Press.

Kryukov, V. I. (Abbot Theophan). (2004). The model of attention and memory, based on the principle of the dominant and comparator function of the hippocampus. *ZHVND, 54*(1), 10–29.

Kryukov, V. I., & Borisyuk, G. N., et al. (1986). *Metastable or unstable states in the brain*. Pushchino.

Kuramoto, Y. (1984). *Chemical Oscillations, waves, and turbulence*. New York: Springer.

Lagutina, N. S. (2002). Self-organization in slightly inhomogeneous fully connected network of pulsed neural oscillators. In *Neuroinformatics*, Part 2 (pp. 60–66). Moscow: Moscow Engineering Physics Institute.

Lebedev, A. N. (1990). Human memory, its mechanisms and limits, in *Research of memory* (pp. 104–118). Moscow: Nauka.

Lebedev, A. N. (1992). On the physiological basis of perception and memory. *Psychological Journal, 2*, 30–41.

Lebedev, A. N. (2002). Cognitive Psychology at the turn of the century. *Psychological Journal, 23, 85–92*.

Lebedev, A. N., Mayorov, V. V., & Myshkin, I. Y. (1991). The wave model of memory. In A. V. Holden & V. I. Kryukov (Eds.), *Neurocomputers and attention. neurobiology, synchronisation and Chaos* (Vol. 1, pp. 53–59). Manchester: Manchester University Press.

Lebedev, D. V. (2000). Planning of optimal paths of planar robots using the wave neural network architecture. In *Modelling and Analysis of Information Systems*. Yaroslavl, 7(1), 30–38.

Malsbyrg, C., & Schneider, W. (1986). A neural cocktail-party processor. *Biological Cybernetics, 54,* 29–40.

Mayorov, V. V., & Konovalov, E. V. (2007). Generalized neural automaton in the problem of propagation of the wave of excitation through the neural network. *Neurocomputer, 7*, 3–8.

Mayorov, V. V., & Myshkin, I. Y. (1990a). On one model of functioning of the neural network. In *Modelling of dynamic populations*. Nizhny Novgorod, 70–78.

Mayorov, V. V., & Myshkin, I. Y. (1990b). Mathematical modelling of neurons of the networks based on equations with delay. *Mathematical Modelling*, 2(11), 64–76.

Mayorov, V. V., & Shabarshina, G. V. (1997). Report on networks of W-neurons. *Modelling and Analysis of Information Systems*, 4, 37–50. (Yaroslavl).

Mayorov, V. V., & Shabarshina, G. V. (1999). Simplest modes of burst wave activity in the networks of W-neurons. In *Modelling and Analysis of Information Systems*. Yaroslavl, 6(1), 36–39.

Mayorov, V. V., & Shabarshina, G. V. (2001). Networks of W-neurons in the associative memory problem. *Journal of Computational Mathematics and Mathematical Physics*, 41(8), 1289–1299.

Mayorov, V. V., Myshkin, I. Y., Kuksov, A. G. & Myachin, M. L. (2003). Planning of waves in ring structures of associations of pulse neurons as detectors. In *Modelling and Analysis of Information Systems*. Yaroslavl, 10(2), 30–34.

Mayorov, V. V., Shabarshina, G. V., & Anisimova, I. M. (2005). Neuronet solution to the problem of planning of optimal paths for point robots. In *Proceedings of the VII All-Russian Conference "Neuroinformatics-2005"*, Part I (pp. 197–202). Moscow: Moscow Engineering Physics Institute.

Mayorov, V. V., Shabarshina, G. V., & Konovalov, E. V. (2006). Self-organization in a fully homogeneous network of neural cellular automata of excitatory type. In *Proceedings of the VIII All-Russian Conference "Neuroinformatics-2006"*, Part I (pp. 67–72). Moscow: Moscow Engineering Physics Institute.

McCulloch, W. A., & Pitts, W. (1943). A logical calculus of ideas immanent in nervous activity. *The Bulletin of Mathematical Biophysics*, 5, 115–133.

Minsky, M., & Papert, S. (1971). *Perceptrons*. Moscow: Mir.

Neural Networks for Instrumentation, Measurement and Related Industrial Applications (2003). In *NATO Science Series, Series III: Computer and Systems Sciences*, 185. IOS Press.

Nicolis, G. (1989) *Dynamics of hierarchical systems*. Moscow: Mir.

Ohsusa, M., Yamaguchi, Y., & Shimizi, H. (1989). Entrainment of two coupled van der Pol oscillators by an external oscillation. *Biological Cybernetics*, 51, 325–333.

Omata, S., Yamaguchi, Y., & Shimizi, H. (1988). Entrainment among coupled limit cycle oscillators with frustration. *Physica D*, 31, 397–408.

Osipov, G. V., Rabinovich M. I., & Shalfeev V. D. (1992). Dynamics of nonlinear synchronization networks (Vol. 2, p. 88). In *Proceedings of International Seminar "Nonlinear circuits and systems"*, June 16–18, Russia, Moscow.

Paramonov, I. V. (2002). Modelling of impulse activity of a neuron by differential equation with delayed argument with a variable value of delay. *Modelling and Analysis of Information Systems*, 9(2), 12–15.

Potapov, A. B., & Ali, M. K. (2002). Nonlinear dynamics of information processing in neural networks. In *New in synergetics: a look into the third millennium* (pp. 467–426). Moscow: Nauka.

Pribram, K. (1975). *Languages of the brain*. Moscow: Mir.

Rabinovich, M. I. (1996). Chaos and neurodynamics. *Proceedings of the universities. Radiophysics, XXXIX*(6), 168–757.

Rabinovich, M. I., Varona, P., Selverston, A. I., & Abarbanel, H. D. I. (2006). Dynamical principles in neuroscience. *Reviews of Modern Physics*, 78, October–December, 2–53.

Rosenblatt, F. (1965). *Principles of neurodynamics*. Moscow: Mir.

Sergin, A. B. (2005). Biologically plausible neural network edge detector, *Proceedings of the VII All-Russian* Scientific and Technical Conference *"Neuroinformatics-2005"*, Part I (pp. 249–256). Moscow: Moscow Engineering Physics Institute.

Shabarshina, G. V. (1994). Conduction of excitation along the ring structure of neural cellular automata. In *Modelling and Analysis of Information Systems*. Yaroslavl, 2, 116–121.

Shabarshina, G. V. (1999). Self-organization in homogeneous fully connected network of neural cellular automata of excitatory type. *Automation and Remote Control*, 2, 112–119.

Shabarshina, G. V. (2000). Self-organization in slightly inhomogeneous fully connected network. In *Mathematical Modelling and Analysis of Information Systems*. Yaroslavl, *7*(1), 44–49.

Shilnikov, L. P., Shilnikov, A. L., Turaev, D. V., & Chua, L. (2004). *Methods of qualitative theory in nonlinear dynamics, Part I*. Moscow-Izhevsk: Institute of Computing Research.

Shilnikov, L. P., Shilnikov, A. L., Turaev, D. V., & Chua, L. O. (2001). Methods of qualitative theory in nonlinear dynamics. Part II, in *World Scientific Series on Nonlinear Science. Series A: Monographs and Treatises, 5*. River Edge, NJ: World Scientific Publishing Co., Inc.

Shulgina, G. I. (1978). *Bioelectrical activity of the brain and the conditioned reflex*. Moscow: Nauka.

Shulgina, G. I. (1990a). Investigation of the role of rhythmic activity in the processes of recording and reproducing information on the model of the neural-like cells. *Reports of the Academy of Sciences of the USSR, 312*(5), 1275–1279.

Shulgina, G. I. (1990b). Neurophysiological mechanisms of reinforcement and internal inhibition. in *Neurobiology of Learning and Memory* (pp. 89–114). Moscow: Nauka.

Shulgina, G. I. (1993). Investigation of the conditions of conducting of excitation, processes of learning and formation of associations on the model of the network of excitatory and inhibitory neural-like elements. in *Neurocomputer as the basis of thinking computers* (pp. 110–128). Moscow: Nauka.

Shulgina, G. I., Ponomarev, V. N., Murzina, G. V., & Frolov, A. A. (1983). Model of neural network training based on changes of the efficiency of excitatory and inhibitory synapses. *ZHVND, 38*(5), 926–935.

Shulgina, G. I., Ponomarev, V. N., Rezvova, I. R., & Frolov, A. A. (1988). Effect of the background activity of a network of excitatory neural-like elements on the conduction of excitation. *ZHVND, 38*, 715–724.

Sokolov, E. N. (1981) *Neural mechanisms of learning and memory*. Moscow: Nauka.

Sokolov, E. N., & Izmailov, Ch. A. (1984). *Colour vision*. Moscow: Moscow State University Press.

Sokolov, E. N., & Vaitkevičius, G. G. (1989). *Neuro intellect: from neuron to neurocomputer*. Moscow: Nauka.

Timofeev, E. A. (1997). Modelling of neuron, transmitting information by density of the flow of pulses. *Automation and Remote Control, 3*, 190–199.

Wiener, N., & Rosenbluth, A. (1961). Conduction of impulses in the heart muscle. *Cybernetic Collection, 3*, 7–56 (Moscow: IL).

Wiggins, S. (1990). *Introduction to applied nonlinear dynamical systems and chaos*. New York: Springer.

Zabrodin, J. M., & Lebedev, A. N. (1977). *Psychophysiology and psychophysics*. Leningrad: Nauka.

Zeeman, E. C. (1973). Differential equations for the heartbeat and nerve impulse. in *Salvador Symposium on Dynamical Systems* (pp. 683–741). London: Academic Press.

Chapter 1
The Model of a Single Neuron

Lebedev and Lutsky (1972) pioneered the concept of modeling the dynamics of neurons on the basis of equations with delay. Generally, researchers agree that it is not necessary to introduce delay in explicit form into equations that describe the neuron. In examining the fundamental role of delays in neuron modeling, Hodgkin and Huxley (1939; see also Hodgkin 1965) achieved the effect of delay through the use of a chain of ordinary differential equations. A model always reflects the interpretation of a phenomenon; simultaneously, some aspects are taken for granted while some others are typically ignored. The purposes of modeling also differ. Some models are used to study processes taking place inside the object under examination; others are used to explain the relationship of objects to each other. The neuron is a complex formation, and it is difficult to expect the monotony of representations. In our view, a variety of mathematical models of neurons can be divided into two classes: static and dynamic.

In static neuron models, the output is completely determined by the current states of inputs or the state of inputs taken into account along with their prehistory. An example is the threshold element of McCulloch and Pitts (1943) and its numerous generalizations. Hopfield's (1982, 1984) formal neuron also belongs to this class on the basis of operational amplifiers. The relationships in neural associations may be explained through static models. The collective behavior of neurons in associations is studied by neurodynamics, the principles of which were laid by researchers such as Hebb (1949), Rosenblatt (1965), Minsky and Papert (1971).

In the dynamic models, the output of neurons is fundamentally unstable. It is determined not only by signals arriving at the neuron but also by the current internal state of the neuron. Wiener and Rosenbluth's (1961) axiomatic model and the Hodgkin–Huxley model of axon belong to the class of dynamic models of neurons. The model of Wiener and Rosenbluth explains the processes in which excitation waves propagate, i.e., it also serves as the apparatus to study neuron interactions. The Hodgkin–Huxley model is dual. On one hand, the model describes the generation of pulses by neurons, i.e., the internal processes. On the other hand, it reflects pulse propagation along the axon, which is one of the factors involved in neuron interaction.

© Springer International Publishing Switzerland 2015
S. Kashchenko, *Models of Wave Memory*,
Lecture Notes in Morphogenesis, DOI 10.1007/978-3-319-19866-8_1

1.1 Informal Description of the Processes of Electrical Activity of the Neuron and the Variety of Models of the Phenomenon

The structural unit of the brain is the single nerve cell or neuron. Each neuron consists of the central part of the body, within which is the nucleus, as well as tree-like appendages diverging from the body in different directions (Green et al. 1993). Relatively short appendages are called "dendrites," and the only long one is called the "axon." Regardless of their location and function, any neuron, like other cells of the body, is limited by the lipoprotein membrane, which determines its individual limits while functioning as a good electrical insulator. On both sides of the membrane, between the contents of the cell and the extracellular fluid, there exists the difference of electric potentials (Khodorov 1975). This difference is called the "membrane potential." The origin of the membrane potential is due to an uneven distribution of ions on both sides of the membrane. Most of the time, the external surface of the membrane is electropositive with respect to the internal surface. The membrane potential is negative and is large in absolute value, i.e., the membrane is strongly polarized. To avoid ambiguous interpretation, when we refer to the increase and decrease of value, we do not mean its absolute value.

At some point in the process of the neuron's functioning, the membrane potential rapidly grows, changes sign, and then decreases rapidly. As a result, the membrane potential becomes negative again and even greater in magnitude than it was at the beginning, i.e., the membrane reaches a state of hyperpolarization. This is the process of the generation of a neuron pulse, which is called a "spike" or "action potential."

The neuron is of a distributed formation. Conventionally, we can say that the spike is born in the neuron's central part, i.e., in the body. While propagating along the axon—a single centrifugal outgrowth of the body of the neuron and its branches —the action potential influences the other neural formations and the executive organs, i.e., the glands and muscles. The branches of the axon end in close proximity to the bodies and dendrites of neurons or receivers. The junctions of the neurons are called "synapses." Under the action of the incoming spike, the synapse goes into an excited state and exhibits either an excitatory or inhibitory effect, depending on its type, on the neuron-receiver.

The start of the spike is associated with the membrane potential achievement or the threshold value of generation. A relatively slow increase of the membrane potential from the state of hyperpolarization (after the spike) up to the threshold value of generation is caused due to a number of reasons. Of these reasons, many scholars (Hodgkin and Huxley 1939; Katz 1968; Khodorov 1975; Tasaki 1971) have identified the excitement incoming from other neurons as the main and unique one. At the same time, an important role is played here by the neural mechanisms, which also make a significant contribution to depolarization. It is of principal significance because the membrane potential of a neuron—which possesses such properties—sooner or later, even without external action, will reach the threshold generation of the pulse. Such a neuron will have the property of autorhythmicity, i.e., spontaneous activity in the generation of pulses. The ability to generate spikes with certain periodicity is inherent to many neurons and, in particular, to pacemaker neurons.

The problem of the periodicity of neural activity has long attracted the attention of researchers Deutsch (1970), Guselnikov and Suping (1968), Lebedev and Lutsky (1972), Livanov (1972), Singer et al. (1975). Here we can distinguish two main directions. First, neural activity is periodic activity as a manifestation of the normal functioning of nervous structures. Second, neural activity is the role of periodic activity in the mechanism of the processing and storage of information in the nervous system. From this perspective, the simulation and explanation of period-icity at the level of individual nerve cells is of particular importance.

Simultaneously, the internal mechanisms that ensure the slow increase of the membrane potential may be active only in a state of strong polarization and, later, atrophy. As a result, the membrane potential reaches a level that is smaller than the threshold value of spike generation. This level of the membrane potential is called the "resting potential." For the neuron that has resting potential, the spike is a response to external stimulation, which is characteristic of neurons as detectors (Hubel 1990). It is believed that the force of the stimulus needed to generate a spike by the neuron as detector must be larger than some critical level.

A review of the literature (Hodgkin 1965; Khodorov 1975; Tasaki 1971; Plonsi and Barr 1992) shows experiments about the artificial stimulation of the neuron by electrical pulses, which demonstrate that the amplitude of the spike does not depend on the force of the action. At the same time, the state of the neuron is important for generating the required force of the stimulus. The stimulus has no effect on the neuron during the spike. Immediately after the spike, when the membrane is hyperpolarized, we can observe the zone of refractoriness, i.e., full or relative immunity to the stimulus. According to different investigators Katz (1968), Plonsi and Barr (1992), Eckert et al. (1991), when caught in this zone a stimulus will either have no effect on the membrane potential or its sufficiently strong action will lead to spike generation. However, the amplitude of the spike is lower than usual, and increasing the stimulus will not increase the amplitude. The zone of refractoriness is followed by an interval of normal excitability, i.e., the stimulus leads to the spike of the standard for the given neuron amplitude.

The described peculiarities of the generation of a spike served as a starting point for a number of studies on the mathematical simulation of neurons and their associations. The following examples show the directions in which individual aspects of the phenomenon were absolutized. Historically, McCulloch and Pitts (1943) were the first to explore and discuss this phenomenon. It is based on the idea of the threshold element, which corresponds to the neuron as detector. The threshold element is subject to several laws, two of which are outlined below:

- The functioning of the neuron is subject to the "all or nothing" law (i.e., the spike is generated, or the membrane potential is at rest).

- For excitation of the neuron (spike) at a certain time, it is necessary to initiate a certain number of synapses (inputs) during some preceding period.

Later, the signals arriving at the neuron have come to be weighed. The initial rules for the adaptation of weights are connected with Hebb's (1949) research. Another important observation was made. There are detector neurons, which, in response to stimulation, generate a group of pulses. The number of pulses (or the frequency of pulsation) can be taken for the output characteristic of a neuron. The output characteristic is the nonlinear function of the weighted characteristics of the input similar to Rosenblatt's (1965) perception. Currently researchers are exploring different architectures of neural networks improving on and creating new rules for choosing the weights of interaction, and studying the possible fields of application of neural networks. We investigate neurons with different types of nonlinearities, and this has led, for example, to the emergence of the neuron model based on classical principles, but one that functions stochastically (Izhikevich and Malinetskii 1992, 1993). Another direction of simulation is associated with Wiener and Rosenbluth's research (1961). The Wiener-neuron operates in accordance with ticks. It may be in the excitation state (pulse generation), which is replaced by the state of refractoriness (full immunity to the action). This is followed by the pending state (i.e., when the membrane potential is at rest). When in a pending state, if any excitation arrives along even one synapse of the neuron, then the neuron will pass into an excited state. From a biological point of view, the Wiener representations seem to be more sophisticated than that of McCulloch–Pitts. The Wiener model is dynamic and has worked well, for example, in simulation of the cardiac muscle. The direction of study is being actively developed (Borisyuk et al. 2002; Mayorov and Shabarshina 1997, 1999, 2001; Mayorov et al. 2005).

Here is another example of absolutization of a certain part of the phenomenon of neuron. Experimental investigations of the neuron in living tissue show that the distribution of its pulses has a very complex nature. This is natural because the neuron is under the influence of thousands of other neurons. At the same time, it could be hypothesized that the impulsation has a random nature (Burns 1968). The probability of pulse generation depends on the density of random streams arriving at the neuron through the synapses. This is also the direction of study in the simulation of neural networks (Murray 1989; Timofeev 1997).

1.2   Nature of the Membrane Potential

At the beginning of the last century, Bernstein (1990) suggested an ionic hypothesis about the causes of polarization of the membrane. Further follows a discussion of the nature of the membrane potential. The membrane separates the cytoplasm of the neuron from the intercellular medium, which is mainly composed of lipids (fat-like substances) and into which protein molecules are embedded. However, some molecules permeate the lipid layer, whereas others get immersed only up to the core (peculiar villi). Some proteins play the role of channels and carry out the transport of ions through the membrane. Other molecules are the receptors, which allow the neuron to respond to action potentials incoming from other neurons (Nicholls et al. 2003).

The mechanisms of ion transport through the membrane can be divided into two classes: passive and active. Passive transport requires no energy input. It is determined by two factors. Diffusion helps equalize the concentration of solutions, and the ions are moved under the electric field. Active transport occurs with the help of chemical energy, which is exempt during the oxidation of amino acids. The chemical energy helps transfer ions both in domain with their higher concentration and against electrical forces.

Regularities of passive transport lean on some provisions of electrochemistry. Membranes of biological origin have selective permeability where some ions easily diffuse through them, whereas for others the membrane proves to be impermeable. Consider using the example of the assumption regarding electrochemical equilibrium. A cavity N is let into a vessel with a solution and bounded by the membrane in which the KCl solution is also present. The concentrations of the solution in the vessel and the cavity are, respectively, denoted by C_0 and C_i. Assume that the membrane is permeable for K^+ ions and impermeable for Cl^- ions. The concentration of the solution in the vessel is lower than that in the cavity, i.e., $C_i > C_0$. Then the potassium ions will pass from the cavity into the vessel because the diffusion tends to equalize concentrations. In the cavity, due to a scarcity of positively charged ions, the cavity is charged negatively. More precisely, the negative charge is concentrated on the inner surface of the membrane, whereas the outer surface remains positively charged. The electric field is directed such that the diffusion of positive K^+ ions will be difficult. As a result, at certain difference of potentials $E_K < 0$ between the inner and the outer surface of the membrane, the diffusion of ions will cease altogether. The value E_K is called the "equilibrium electrochemical potential." It is calculated by the Nernst formula:

$$E_K = (RT/FZ)\ln(C_0/C_i), \tag{1.2.1}$$

where R is the universal gas constant, T is the absolute temperature, F is Faraday's number, and Z is the ion charge (for the potassium ion it is equal to one).

Let $u(t)$ be the current value of the membrane potential, and let us consider some of the assumptions of the theory of electrical circuits. The membrane is the dielectric (isolator) and therefore has a capacity. Current i_c, flowing through the capacity and voltage u (it is also the membrane potential) are related as follows —$i_c = c\dot{u}$,—where the proportionality coefficient c is called the "capacity." Note that in some sense, the current flowing through the membrane is fictitious. It is not accompanied by charge transfer. According to the law of Nernst, the membrane is the source tension E_K. Current i_K, flowing through the voltage source E_K is related with the voltage u and on its poles by the formula $i_K = \bar{g}_K(u - E_K)$ where coefficient $\bar{g}_K$ is called "conductivity." According to Kirchhoff's law, the sum of the currents is zero: $i_c + i_K = 0$. Thus, we obtain the differential equation:

$$c\dot{u} + \bar{g}_K(u - E_K) = 0. \tag{1.2.2}$$

Trivial analysis of Eq. (1.2.2) shows that

$$u(t) = E_K + (u(t_0) - E_K) \exp(-(\bar{g}_K/c)(t - t_0)).$$

Thus, exponentially, the membrane potential is $u(t) \to E_K$ at $t \to \infty$.

Let us make a number of observations. In the analysis of Eq. (1.2.2), we assumed that the coefficients c and $\bar{g}_K$ do not depend on u. In actuality, for the membrane, the capacity is almost constant. However, conductivity $\bar{g}_K$ is a function of u. This analysis is valid only at small deviations $u(t_0)$ from the equilibrium value E_K.

Assume now that in addition to KCl in the cavity and in the vessel, there is NaCl in varying concentrations in the cavity and in the vessel. Its own equilibrium value E_{Na} corresponds to Na^+ ions. Sodium current i_{Na} is associated with the membrane potential u by the relation: $i_{Na} = \bar{g}_{Na}(u - E_{Na})$. Again, involving Kirchhoff's law, we obtain:

$$c\dot{u} + \bar{g}_{Na}(u - E_{Na}) + \bar{g}_K(u - E_K) = 0. \tag{1.2.3}$$

Now, according to (1.2.3), the equilibrium value of the membrane potential is the essence:

$$u_* = \frac{\bar{g}_{Na}E_{Na} + \bar{g}_K E_K}{\bar{g}_{Na} + \bar{g}_K}. \tag{1.2.4}$$

The sense of Bernstein's argument about the nature of the membrane potential is sufficiently transparent. He assumed that the concentration of K^+ ions in the cell is greater than that in the extracellular medium. Consequently, the K^+ ions diffuse from the cell. The relative scarcity of positive ions is formed inside the cell membrane, and the inner surface of the membrane is charged negatively to the Nernst Eq. (1.2.1). Thus, the resting potential is determined by the equilibrium electrochemical potential for K^+ ions.

Furthermore, Bernstein suggested that in the life of the neuron there sometimes comes a special moment—a state of excitement—when for a short time the membrane becomes permeable to other ions, particularly those of sodium and chlorine. As a result of the flow of positive ions into the cell, the membrane potential decreases in value; it vanishes and then, after the restoration of selective properties of the membrane, again it enters the normal range. Thus, spike generation was explained for the first time.

1.3 Sodium–Potassium Cycle

Bernstein's hypothesis well explained many facts known to physiologists of his time. In particular, the dependence of the resting potential on the concentration of potassium in the extracellular medium, as well as on the temperature, was a well-known concept. However, at that time the hypothesis could not be subjected to experimental verification. The researchers had no direct methods of measuring the membrane potential and ion concentrations at their disposal. Bernstein's hypothesis caused many disputes.

The situation changed after the announcement made by the English zoologist, George Jung, in 1936. He found that the long strands on squid and cuttlefish are not blood vessels as was previously thought, but rather they were unusually thick axons (appendages of nerve cells). They were termed "giant axons" and became a natural object for studying membranes. The diameter of axons reaches 1 mm, and thus they provide researchers an opportunity to introduce electrodes and capillaries into them. As a result, we can measure both membrane potential and the concentration of the intracellular solution.

It was found out that the membrane theory of Bernstein as a whole was true, although it needed the introduction of some significant additions and changes. In particular, the statement that the resting potential is caused mainly by the electro-chemical potential of K^+ ions was proven to be true. On the contrary, according to the theory of Bernstein, at the peak of the action potential (spike), the membrane potential must be zero (the ion current is stopped when the voltage is zero). However, British physiologists Hodgkin and Huxley (1939) showed that the maximum capacity of the membrane potential is not destroyed but rather changes sign. The outer surface of the membrane becomes negatively charged in relation to the internal membrane. The phenomenon was called "overshoot", and it had no explanation within Bernstein's theory. It is connected with the uneven distribution of sodium ions. It was found that the concentration of Na^+ ions inside the cell was approximately 10 times lower than that in the intercellular fluid. At the start of the spike, sodium ions, under the influence of the electric field and concentration gradient, rush from the extracellular medium into the cell. In this sense, Bernstein's hypothesis has proved to be true. After the change of sign of the membrane potential, sodium ions continue to move into the cell only under the influence of the concentration gradient. The membrane potential at the peak of overshoot approaches the equilibrium electrochemical potential for Na^+ ions. Then, the membrane permeability for sodium ions decreases and activates the conductivity for potassium ions, which move out of the cell where their concentration is greater. Potassium ions carry the positive charge from the cell. The inner surface of the membrane becomes negatively charged. This sodium–potassium cycle will be discussed in greater detail later in the text.

During the sodium–potassium cycle, sodium ions enter the cell, and potassium ions leave the cell. This substitution equalizes ion concentration. It would seem that after a number of cycles, the nerve cell should not generate spikes, but this is not so.

The uneven distribution of ions inside and outside of the cell is maintained by the active transport of ions. Mechanisms that transport ions across the membrane in the direction of greater concentration are called "membrane pumps" (Hodgkin 1965; Katz 1968; Cole 1968; Khodorov 1975; Blum et al. 1988; Nicholls et al. 2003). They use energy released by the oxidation of amino acids.

In this chapter, we study the sodium pump, which outputs Na^+ ions from the cell and simultaneously transports K^+ ions into the cell. This process includes a mandatory exchange of two K^+ ions in the external medium into three Na^+ ions in the cell (if we remove the ions of K^+ from the medium, then the ions of Na^+ will not be output from the cell). If the membrane potential is negative (i.e., the inner surface of the membrane is negatively charged), then the sodium pump performs electrical work outputting one positive charge from the cell. The sodium pump acts selectively. For example, it cannot transfer lithium ions. The molecular mechanism of active transport currently is not completely clear.

The sodium pump is one of the main mechanisms due to which a concentration of K^+ ions is maintained within the cell compared with that in the external medium along with an insufficient concentration of Na^+ ions compared with that in the extracellular fluid.

To investigate the generation of a spike, Hodgkin and Huxley developed an original method called the "method of fixing potential." Electrodes are introduced into the cell and are used for feedback, thus allowing determination of the membrane potential at any given level. Thus, it is possible to simulate various phases of the action potential as if freezing them. This system allows researchers to measure ionic currents. By varying the ionic composition of the medium (i.e., the composition of intracellular fluid is changed through the microcapillaries), one can identify the ions that cause the current. Investigators have discovered that even the passive transport of ions (according to the gradient of concentrations and the direction of the electric field) is subject to specific regularities. For their explanation, Hodgkin and Huxley hypothesized the existence of specific ion channels in the membrane that transport the ions. With the positive deviation of the membrane potential from the potential in a resting state (value of the potential decreases in absolute value), channels for Na^+ are activated letting through these ions into the cell. Na^+ ions move in the direction of the electric field and to the area with a lower concentration. The channels act as selective (polling) filters letting through the Na^+ ions and become impermeable to other ions. Mechanisms leading to the opening and closing of channels are called "gates." At depolarization (i.e., a decrease of the absolute value of the membrane potential), the so-called "m-gates" for the sodium channels are activated and opened. The more depolarized the membrane, the greater the number of m-gates opened, which in turn increases depolarization.

With respect to long-term depolarization, sodium channels are inactivated. The so-called "h-gates" block the channels through a time order of 1 ms after the opening of m-gates. In 1980, Sigvors and Neher registered the currents through single sodium channels. It turned out that they are subject to the "all or nothing" law in that as instantaneously as they open, in approximately 1 ms they also almost instantaneously close. In fact, the time during which the ion channel is open is a random variable (Nicholls et al. 2003). Ion channels are protein molecules embedded in the lipid layer of the membrane.

In the area of the peak of the action potential, the majority of sodium channels are inactivated, i.e., the h-gates are closed. The problem lies in understanding how the membrane potential returns to its initial level, i.e., the resting state. Experimentally, Hodgkin and Huxley found that with some delay relative to the peak of the sodium current, there arises a potassium current. Potassium ions begin to leave the cell due to the activated potassium channels. The hypothesis attributes this to the opening of the "so-called n-gates." Potassium ions move into the area with lower concentration (passive transport). At the same time, in the area of overshoot (i.e., when the membrane potential is positive), motion is carried out in accordance with the direction of the electric field. The potassium current takes away the positive charge from the cell (more precisely, from the inside surface of the membrane), and the membrane is again polarized. It is believed that in contrast to sodium channels, potassium channels do not have inactivation gates. Activation n-gates close themselves when the membrane returns to the level of the resting potential (i.e., is again polarized).

It is important to note that the number of ions that pass through the membrane with a single action potential practically does not change the intracellular ion concentrations. Only tiny nerve cells may be an exception. By some estimates, with one action potential content of Na^+ inside the squid giant axon is changed only 1/100,000 from the initial state. If you block the activity of sodium pumps by some poison, the axon will still be able to generate several thousand pulses. Active transport is not directly involved in the generation of spikes. However, it maintains the concentration gradients at the proper level, and their presence leads to the emergence of action potentials.

Having described the mechanism of spike generation, we will now look at the system of equations describing this process as proposed by Hodgkin and Huxley (1939), see also Khodorov (1975).

1.4 System of Equations of Hodgkin–Huxley

Hodgkin and Huxley studied the squid's giant axon in a seawater environment, where it has the resting state of membrane potential, the value of which is denoted by E_r. This is calculated theoretically (see Nicholls et al. 2003) based on the formula denoted in (1.2.4). Thus, it is possible investigate the work of an active sodium pump.

The model is based on the electrochemical equation as denoted in (1.2.3). The membrane potential will be measured from its level E_r at rest, i.e., write $u = V + E_r$ or $V = u - E_r$. Similarly, we introduce the notation for the measure from E_r the equilibrium electrochemical potentials $V_{Na} = E_{Na} - E_r$ and $V_K = E_K - E_r$. It is clear that differences remain: $u - E_{Na} = V - V_{Na}$ and $u - E_K = V - V_K$. Except for ions of Na^+ and K^+, other ions (in particular, ions of Cl^-) also play a role in the creation of the membrane potential. $Na^+ K^+ Cl^-$. We denote their electrochemical potential by way of E_1, respectively, $V_1 = E_1 - E_r$. For the balance of the membrane currents, we add into Eq. (1.2.3) the term $i_1 = \bar{g}_1(V - V_1)$. This current is called "leakage current." For the squid giant axon in the work of Hodgkin and Huxley, the following values were experimentally found: $V_{Na} = 112$ mV, $V_K = -12$ mV, $V_1 = 10$ mV.

Electrochemical Eq. (1.2.3) is the essence of the ordinary differential equation of the first order. Even if its coefficients are considered as functions of the membrane potential V, it is easy to prove that all of its solutions tend to states of equilibrium at $t \to \infty$. It is impossible to explain within a single electrochemical equation the generation of the action potential.

Hence, it follows that conductivities $\bar{g}_{Na}$ and $\bar{g}_K$ should be considered as functions not only of the membrane potential V but also of time. The following construction of Hodgkin and Huxley is largely phenomenological, i.e., it is without sound theoretical basis but is based on a number of assumptions and guesses. Let us consider sodium conductivity. To describe its dependence on time and membrane potential, Hodgkin and Huxley introduced two new variables, $m(t)$ and $h(t)$, which characterize the processes of activation and inactivation of sodium channels, respectively. They wrote $\bar{g}_{Na} = g_{Na}m^3h$ where the coefficient g_{Na} was named the "maximum sodium conductivity." In the investigators' interpretation of the model, variable $m(t)$ is the probability of occurrence in the neighbourhood (i.e., in the mouth) of the sodium channel of subunit activating it, and $1 - h(t)$ is the probability of occurrence of inactivating subunit. The emergence of three activating subunits [the probability is equal to $m^3(t)$] and the absence of an inactivating subunit [the probability is equal to $h(t)$] is necessary to activate the channel. At present, searches for subunits of such type embedded in the membrane continue. However, according to Nicholls et al. (2003), the gates themselves have not been found, although a series of interesting hypotheses has been put forth. On the basis of experimental data and by the method of selection, Hodgkin and Huxley wrote out ordinary linear differential equations of the first order for variables $m(t)$ and $h(t)$. In this case, the variable $h(t)$ is not included in the equation for $m(t)$, and $m(t)$ is not included in the equation for $h(t)$. The coefficients of the equations depend only on the membrane potential V (a convincing justification for such dependence is missing).

To describe the changes in potassium conductivity $\bar{g}_K$, Hodgkin and Huxley introduced the function $n(t)$—the variable activation of potassium channels. We recall that according to the hypothesis of Hodgkin and Huxley, inactivation gates for potassium channels are missing. The investigators empirically selected linear ordinary differential equations for $n(t)$ and put $\bar{g}_K = \bar{g}_K n^4$. The coefficients of the equation for $n(t)$ depend only on the membrane potential V. Coefficient g_K is called the "maximum potassium conductivity." In the interpretation by Hodgkin and Huxley, the value of $n(t)$ is the probability of occurrence at the mouth of the channel of the activating subunit. To activate the channel, the four activating subunits must meet.

Relatively to conductivity $\bar{g}_1$ for leakage, Hodgkin and Huxley suggested it to be permanent: $\bar{g}_1 = g_1 = \text{const}$. As a result, a system of four ordinary differential equations has been proposed to describe the generation of action potentials.

The first of these is the electrochemical equation for the balance of the membrane currents: $i_c + i_{Na} + i_K + i_1 = 0$, where we are reminded that

$$i_c = c\frac{dV}{dt},$$

$$i_{Na} = g_{Na}m^3 h(V - V_{Na}),$$

$$i_K = g_K n^4(V - V_K),$$

$$i_1 = g_1(V - V_1).$$

The second and third equations describe, respectively, the variables $m(t)$ and $h(t)$ of activation and inactivation of sodium channels. Finally, the fourth equation reflects the dynamics of $n(t)$, a variable of the activation of potassium channels. The system of Hodgkin–Huxley equations has the following form:

$$c\frac{dV}{dt} = g_{Na}m^3 h(V_{Na} - V) + g_K n^4(V_K - V) + g_1(V_1 - V), \qquad (1.4.1)$$

$$\frac{dm}{dt} = \frac{m_\infty(V) - m}{\tau_m(V)}, \qquad (1.4.2)$$

$$\frac{dh}{dt} = \frac{h_\infty(V) - h}{\tau_h(V)}, \qquad (1.4.3)$$

$$\frac{dn}{dt} = \frac{n_\infty(V) - n}{\tau_n(V)}, \qquad (1.4.4)$$

where $V_{Na} = 112$, $V_K = -12$, $V_1 = 10$. In (1.4.2) through (1.4.4), the functions $m_\infty(V), h_\infty(V), n_\infty(V), \tau_m(V), \tau_h(V), \tau_n(V)$ are positive and continuous. Functions $m_\infty(V), h_\infty(V), n_\infty(V)$ belong to the class of the so-called "sigmoid functions," i.e., they are monotonic and have the following properties: $m_\infty(V) \to 1, h_\infty(V) \to 0$, $n_\infty(V) \to 1$ at $V \to \infty$ and $m_\infty(V) \to 0, h_\infty(V) \to 1, n_\infty(V) \to 0$ at $V \to -\infty$. It is easy to show that in (1.4.2) through (1.4.4), variables $m(t), h(t), n(t) \in (0,1)$, and therefore they have a probabilistic sense.

Hodgkin and Huxley indicated a specific form of entering functions having a complex structure into system (1.4.2) through (1.4.4). In reality, the system of equations allows only numeric investigation. We restrict ourselves to some comments on a qualitative level, thus illustrating the experience of numeric investigation. When entering functions into system (1.4.1) through (1.4.4), they are chosen such that a number of properties are implemented. The system has stable equilibrium state (V_*, m_*, h_*, n_*) where $V_* = 0$, m_*, h_*, n_* are positive. Let $V(t_0) > 0, m(t_o) \approx m_*, h(t_o) \approx h_*, n(t_o) \approx n_*$. If the value of $V(t_0)$ is not too great, by virtue of the stability of the equilibrium state over time we have $V(t) \to 0$.

There exists such a threshold value $V_p > 0$ that solutions of the system behave differently at $V(t_0) > V_p$. Initially $V(t)$ decreases. However, the value $m(t)$ begins to increase. The term $g_{Na}m^3 h(V_{Na} - V)$ in Eq. (1.4.1) is positive because $V_{Na} > 0$, and $V(t)$ is close to zero. At the same time, the potassium current, i.e., the term $g_K n^4 (V_K - V) < 0$, is small in absolute value. As a result, the derivative $\dot{V}_t$ becomes positive. The membrane potential begins to increase (the moment of the start of spike generation). Along with it, the value of $m(t)$ increases, which leads to the next increase of $\dot{V}_t$ and the acceleration of growth $V(t)$. However, a little later the value of $h(t)$ begins to decrease. This slows the growth of $V(t)$ all the more so that the difference $V_{Na} - V$ decreases. However, this delay increases the value of $n(t)$. The corresponding term $g_K n_4(V_K - V)$ in Eq. (1.4.1) is negative because $V_K < 0$ and $V > 0$. There exists a moment of time t_m where $\dot{V}(t_m) = 0$. This is the peak of the action potential. The value of the membrane potential $V(t_m)$ is relatively close to V_{Na}. After passing the peak, the value of the derivative $\dot{V}(t)$ is determined by the value of term $g_K n^4(V_K - V)$, and it becomes negative. The rate of decrease of the membrane potential on the descending part of the spike is less in absolute value than the rate of increase on the ascending part. The descending part is longer than the ascending part. Then, the value of $V(t)$ becomes negative, and after that $V(t) \to 0$. The values $m(t), h(t), n(t)$ also tend to their equilibrium values. The process of generating the spike is finished.

During the entire cycle of spike generation, the value $g_K n^4$ of potassium conductivity is delayed with respect to both sodium conductivity $g_{Na}m^4 n$ and the membrane potential $V(t)$ (see Khodorov 1975; Nicholls et al. 2003) itself. Potassium channels are called "delayed rectifier channels" (Nicholls et al. 2003). In the system of Hodgkin Huxley equations, the idea of delay is realized.

The equations describing the membrane of the squid's giant axon were modified for the node of Ranvier membrane (Khodorov 1975). Later, the equations were declaratively transferred to all nerve cells.

As already mentioned, the system of Hodgkin–Huxley equations is complex and not really amenable to analytical studies. In this regard, a number of works have been aimed at its simplification. There are papers where number of equations is reduced to three. In system (1.4.1) through (1.4.4), one can discard Eq. (1.4.4) for $n(t)$ and put in (1.4.1) $g_K = 0$. Return of the membrane potential to the neighborhood of zero (membrane repolarization) will ensure the leakage currents $g_1(V - V_1)$ (the last term in Eq. 1.4.1). This is a biologically reasonable course. According to data in the core nodes of Ranvier, membrane potassium channels are practically absent (see Nicholls et al. 2003). Thus, the number of equations can be reduced to three.

Furthermore, in Eqs. (1.4.1) through (1.4.4), (let Eq. (1.4.4) be retained), one can enter functions $\tau_m(V), \tau_h(V), \tau_n(V)$ by constants. Finally, the sigmoid functions $m_\infty(V), h_\infty(V), n_\infty(V)$ can be replaced by logistic ones. In particular, in

Khodorov 1975, it is assumed that the dependence of h_∞ on V can be described by the following empirical equation

$$h_\infty(V) = \frac{1}{1 + \exp[(V - V_h/7)]},$$

where for the squid giant axon the constant V_h is within the range of 1.5–3.5 mV. The scope of applicability of the model is an important problem. In this case, the system of Hodgkin–Huxley equations is used to describe the transition process, which is called the "generation of spike." The investigators themselves, realizing it, made no attempt to identify data, obtained by the model after the transition process, with the experiment. They successfully used the model to explain the propagation of nerve pulses along the fibers. This model is certainly a phenomenal achievement that embodies the physiological theory of the origin and conduction of excitation.

The extrapolation of the Hodgkin–Huxley model to all neural formations is, in our opinion, not quite justified. The pulse, as the response to external stimuli only, is inherent in neurons as detectors. However, many neurons, to say nothing of pacemaker neurons, have autorythmicity. Some researchers (see, for example, Khodorov 1975) assert that the coefficients in the system of equations of Hodgkin–Huxley can be chosen so that periodic regimes will be observed. This is correct because the system is sufficiently rich. However, such a choice of coefficients significantly complicates the analysis of the system.

1.5 Phenomenological Derivation of Equation with Delay for the Neuron

In our studies, we use our own model (Mayorov and Myshkin 1990a, b, 1993), which follows from the idea of delay and therefore has the same biological significance as the Hodgkin–Huxley model. The model is simple and explains, in particular, the autorythmicity of neurons. Phenomenological equations are often written without any explanation. Then we study their consequences and compare them with experimental results. The equation considered below (when it is written out) seems to be simple enough. However, the course of the arguments leading to the equation requires explanation. Note that the model takes into account only the sodium and potassium currents.

We use the highest level of polarization of the membrane (hyperpolarization state) as the reference point. Denote the positive deviation of the membrane potential from this level by u. Neglect leakage currents. We write the balance equation of the currents flowing through the membrane in the form:

$$c\dot{u} = I_{Na} + I_K. \tag{1.5.1}$$

Represent the sodium I_{Na} and potassium I_K currents as follows:

$$I_{Na} = \chi_{Na}u, \quad I_K = \chi_K u. \tag{1.5.2}$$

Here, the coefficients χ_{Na} and χ_K are functions of u. The presentation for I_K is no surprise because the value of the membrane potential in a state of hyperpolarization is close to the equilibrium value for potassium ions. For sodium I_{Na} current, according to generally accepted theories, should be written as follows $-I_{Na} = \chi_{Na}^*(u_{Na} - u)$—where u_{Na} is the equilibrium value of the potential for sodium ions (it is both positive and large). However, even for passive membranes, coefficient χ_{Na}^* is a nonlinear function of u. Moreover, the analysis (data given in Khodorov 1975) shows that under conditions of strong polarization of the membrane (u is small), the coefficient χ_{Na}^* is quite small. Therefore, it is conceivable that $\chi_{Na}^* = \chi_{Na}^{**}u$. The result is representation (1.5.2).

Another difficulty arises. Below we will conduct the analysis of ion flows, which shows what is necessary to require $\chi_{Na}(0) < 0$. Thus, in the absence of potassium current, the state of hyperpolarization must be stable. Experiments on fixation of the membrane potential with the simultaneous blocking of potassium channels do not confirm this fact.

However, spike generation is a dynamic process consisting of several stages. It seems that one cannot freeze the state of polarization. At one time, Hodgkin and Huxley (see Hodgkin 1965; Khodorov 1969) doubted that the results of experiments with the fixation potential could be transferred to the behavior of membranes under normal conditions.

We will call coefficients χ_{Na} and χ_K in (1.5.2) the sodium and potassium conductivities, respectively. According to this, assume the hypothesis that potassium conductivity is delayed relative to the current value of the membrane potential for the time h, i.e., $\chi_K = \chi_K(u(t - h))$. Turn to the analysis of the dependence $\chi_{Na}(u)$. In a state of strong polarization ($u \ll 1$) on the inner surface of the membrane, there is an excess of sodium ions, which are pumped out of the cell. Because sodium ions are positively charged, we conclude that this process decreases the membrane potential, i.e., $\chi_{Na}(u) < 0$ at $u \ll 1$. Conversely, the spike is associated with a current of positive sodium ions into the cell, which leads to an increase in the value u. Thus, with a value of u greater than a certain level, conductivity is $\chi_{Na}(u) > 0$. Then follows the electrochemical equilibrium condition $u_{Na} \gg 1$ for sodium ions: $\chi_{Na}u_{Na} = 0$. The result is a form of dependence (Kashchenko and Mayorov 1992) of conductivity $\chi_{Na}(u)$, which is depicted in Fig. 1.1a.

The nature of the dependence of potassium conductivity on the membrane potential is somewhat different. In a state of strong polarization, the flow of potassium ions is directed into the cell, which promotes the increase of the membrane potential, and therefore $\chi_K(u) > 0$ at $u \ll 1$. After passing the peak of the action potential, the flow of potassium ions changes direction. Thus, there exists such a level that at a value of $u(t - h)$, the conductivity mentioned above is $\chi_K(u(t - h)) < 0$. These arguments have led us (see Kashchenko and Mayorov 1992) to the form of dependence $\chi_K(u)$, which is depicted in Fig. 1.1b. It has also

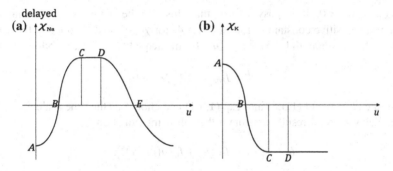

Fig. 1.1 View of sodium (**a**) and potassium (**b**) conductivities

become clear that the value of the delay h of potassium conductivity is close to the duration of the increasing phase of the action potential.

In a state of the strong polarization, the membrane potential slowly increases, so

$$\chi_{Na}(0) + \chi_K(0) > 0. \tag{1.5.3}$$

It is difficult to predict in greater detail the form of curves $\chi_{Na}(u)$ and $\chi_K(u)$. At the same time, this is not necessary because the methods presented later in the text allow us to investigate equations for a broad class of nonlinearities, and the results have a reasonable biological interpretation.

Pay attention to the areas of relative stabilization of the conductivities $\chi_{Na}(u)$ and $\chi_K(u)$ (areas of CD graphs). Conductivities change signs at already relatively small values of u due to the low level of potential, from which begins spike generation.

Simultaneously, the areas of relative stability must be large enough. Furthermore, analysis of the biological data shows that the peak of the action potential is less than the equilibrium potential for sodium ions. We make the assumption that during the operation of the neuron, only part AD of the curve $\chi_{Na}(u)$ is involved. In this regard, we ignore the ascending section of DE and write:

$$\chi_{Na}(u) = a - f^*_{Na}(u), \tag{1.5.4}$$

$$\chi_K(u) = f^*_K(u) - b, \tag{1.5.5}$$

where $a > 0$ and $b > 0$ are constants, and f^*_{Na}, f^*_K are sufficiently smooth functions having zero limits as $u \to \infty$. At the same time, $f^*_{Na}(0) > 0$ and $f^*_K(0) > 0$, which follows from the form of curves of conductivities. In (1.5.4) and (1.5.5), according to [95, 96, 97], $b > a$. This is due to the fact that after reaching the maximum value functions, $f^*_{Na}(u)$ and $f^*_K(u(t - h))$ are small, and according to experimental data the membrane potential u exponentially decreases. Here we make a remark: According to Hodgkin and Huxley, the directions of the flow of sodium and potassium do not change in spike generation (the work of the potassium–sodium pump is not taken into account). In our discussion, it would be necessary to require $\chi_{Na}(u) > 0$ at

$u < u_{\text{Na}}$, $\chi_K(u) < 0$. It is easy to see that this can be achieved by adding the appropriate positive constant to formula (1.5.4) for $\chi_{\text{Na}}(u)$ and subtracting the same constant from formula (1.5.5) for $\chi_K(u)$. From inequality (1.5.3), it follows that

$$a - b - f_{\text{Na}}^*(0) + f_K^*(0) > 0. \tag{1.5.6}$$

Substitute (1.5.4) and (1.5.5) into Eq. (1.5.1) and choose as the time scale the value of the delay h. As a result, we obtain the following equation:

$$\dot{u} = \lambda[-1 - f_{\text{Na}}(u) + f_K(u(t-1))]u. \tag{1.5.7}$$

Here $\lambda = h(b-a)/c > 0$ and $f_{\text{Na}}(u) = f_{\text{Na}}^*(u)/(b-a)$, $f_K(u) = f_K^*(u)/(b-a)$. The important parameter in (1.5.7) plays the role

$$\alpha = f_K(0) - f_{\text{Na}}(0) - 1 > 0. \tag{1.5.8}$$

Its positivity follows from (1.5.6) and ensures that the highly polarized membrane tends to depolarization. By the foregoing, we believe that $f_{\text{Na}}(0) > 0, f_K(0) > 0$, and also $f_{\text{Na}}(u) \to 0, f_K(u) \to 0$ as $u \to \infty$, are quicker than $O(u^{-1})$.

Let us turn to the preliminary discussion of the research problem of the solutions of Eq. (1.5.7). As we already noted, the ability of many neurons (in particular, pacemakers) to periodically generate spikes plays an important role. In this context, the problem of periodic solutions of Eq. (1.5.7) is essential.

Let us first consider the problem of the positive states of equilibrium of Eq. (1.5.7). Of the listed constraints on the coefficients of the equation, it immediately follows that there surely must be such. Let u_* be one of the equilibrium states. Having linearized the equation on it and having written out the characteristic quasi-polynomial by simple reasoning, it is easy to see that the state of equilibrium u_* is unstable for all $\lambda > 0$ if $f_K'(u_*) - f_{\text{Na}}'(u_*) > 0$. On the contrary, the state of equilibrium u_* is stable for all $\lambda > 0$ if $f_{\text{Na}}'(u_*) > 0$ and $|f_K'(u_*)| < f_{\text{Na}}'(u_*)$. In all other difficult cases for Eq. (1.5.7), there exists such value $\lambda_* > 0$ that at $\lambda < \lambda_*$ the state of equilibrium u_* is stable, and at $\lambda > \lambda_*$ it is unstable. Loss of stability occurs in an oscillatory manner, and from the equilibrium state there arises a stable periodic solution. Its parameters can be calculated by the standard methods of bifurcation theory. Immediately after its "birth," this solution is close to sinusoidal. Oscillations of this form are not typical for normally functioning neurons. The dynamics of the membrane potential have the nature of relaxation. In such a way, periodic solutions, bifurcating from the equilibrium state, are unlikely to have profound biological significance.

Because the electrical processes in the neuron occur explosively, it is natural to assume that in Eq. (1.5.7) the value of the parameter λ is both positive and large. To study the dynamics of solutions of Eq. (1.4.1) for such values λ, we apply a special method for determining the large parameter [56, 57]. The essence of this method is as follows: In the infinite-dimensional phase space of Eq. (1.5.7), on the basis of biological considerations, a sufficiently narrow set of S initial functions is selected. After this, we construct asymptotics at $\lambda \to \infty$ of all solutions (1.5.7) with the initial conditions from the set of S. The central point is the introduction of a special type of operator Π, i.e., the successor operator. With the help of this operator, we can show that $\Pi S \subset S$ and thus formulate conclusions about the attractors of Eq. (1.5.7) belonging to S.

Pay attention to the fact that the modes with characteristic high peak (spike), constructed later by the method of large parameter, have a profound biological significance and meet the substance of the case. It should be noted, however, that the dynamics of the solutions of Eq. (1.5.7) are much richer. In particular, it has modes with no spike structure. Biologically they mean drift, i.e., no pulse evolution of the membrane potential, which is indeed sometimes observed in experiments. Modes are called "plateau potentials," and their role is not completely clear.

1.6 Asymptotic Analysis of the Equation of the Neuron

We formulate some biologically consistent, meaning a priori, considerations, thus allowing us to choose the set of initial conditions S. First, as already mentioned, the value of the variable u is nonnegative. Second, after a spike during a time period not less than the value of the delay (taken as a unit) of potassium conductivity, the membrane is hyperpolarized, i.e., the value u is very small. Third, during the output of the neuron from the state of hyperpolarization, the value of u increases exponentially with the exponent of the order of λ, which follows from the form of Eq. (1.5.7). Fix the positive number γ, the smaller of the positive roots of equation

$$f_{Na}(v) = f_K(0) - 1 \qquad (1.6.1)$$

if there are any; otherwise the value of $\gamma > 0$ is arbitrary. In the set S, we refer to all of the continuous functions $\varphi(s)$, defined for $s \in [-1,0]$, which have the following properties: $0 < \varphi(s) \leq \gamma \exp(\lambda \alpha s/2)$ and $\varphi(0) = \gamma$ where positive number α is defined by formula (1.5.8). Denote the solution of Eq. (1.5.7) with the initial condition $u(s) = \varphi(s)$ for $s \in [-1,0]$ and $\varphi(s) \in S$ by $u(t, \varphi)$.

We introduce the sequence of nonnegative numbers $v_0, v_1, v_2, \ldots, v_m$, where $v_0 = 0$. It is trivial, i.e., the number of its first and last element is $m = 0$ if Eq. (1.6.1) has no positive roots. Otherwise, we put v_1 equal to the smallest positive root of this equation. Next, the elements $v_2, v_3, \ldots$ are defined recursively. Consider the following sequence of differential equations:

$$\dot{v} = \lambda \left[-1 - f_{Na}(v) + f_K(v_{j-1}) \right] v, \quad (j = 2, 3, \ldots). \qquad (1.6.2)$$

If solution (1.6.2) with initial condition $v(0) = v_{j-1}$ indefinitely increases, then we complete the construction of the set of numbers assuming that the number of the last element is $m = j - 1$. If this solution tends to the equilibrium state, we take it for v_j.

Construction also terminates if $v_j = 0$. In this case, of course, the number of the last element is $m = j$. We assume that the set of numbers v_j is finite.

We should distinguish three cases (see Kashchenko and Mayorov 1992, 1993):

1. $m = 0$, i.e. sequence v_j consists of one member v_0,
2. $m > 0$ and $v_m > 0$,
3. $m > 0$, but $v_m = 0$.

In the first two, there exists the solution of Eq. (1.5.7) of pulse type with an exponentially large peak at λ. The third case corresponds to no pulse evolution of the membrane potential.

Consider the first two cases. Along with the positive number α, given by formula (1.5.8), we introduce positive numbers

$$\alpha_1 = f_K(v_m) - 1, \quad \alpha_2 = f_{Na}(0) + 1, \tag{1.6.3}$$

$$T_1 = 1 + \alpha_1, \quad T_2 = T_1 + 1 + \alpha_2/\alpha. \tag{1.6.4}$$

Note that with the help of coefficients of α_1 and α_2, we will identify indicators of the exponential change of $u(t, \varphi)$ at different time stations, and the value of T_1 is close to (when $\lambda \to \infty$) the time duration of the spike of the function.

Denote by $t_1(\varphi), t_2(\varphi), \ldots$ the consecutive positive roots of equation

$$u(t, \varphi) = \gamma, \tag{1.6.5}$$

belonging to the interval $(m + 1, \infty)$, by δ, which is an arbitrary and sufficiently small but independent of the λ constant.

Expression of the form $o(1)$ is understood below uniformly with respect to $\varphi(s) \in S$ and relatively indicated in each case values of time t.

Theorem 1.6.1 *Let $m = 0$. Then for the indicated below values of t hold asymptotic as $\lambda \to \infty$ equalities:*

$$u(t, \varphi) = \exp \lambda \alpha_1(t + o(1)), \quad t \in [\delta, 1 - \delta], \tag{1.6.6}$$

$$u(t, \varphi) = \exp \lambda(\alpha_1 - (t - 1) + o(1)), \quad t \in [1 + \delta, T_1 - \delta], \tag{1.6.7}$$

$$u(t, \varphi) = \exp[-\lambda \alpha_2(t - T_1 + o(1))], \quad t \in [1 + \delta, T_1 + 1 - \delta], \tag{1.6.8}$$

$$u(t, \varphi) = \exp \lambda(\alpha(t - T_1 - 1) - \alpha_2 + o(1)), \quad t \in [T_1 + 1 + \delta, t_2(\varphi)], \tag{1.6.9}$$

and values $t_1(\varphi)$ and $t_2(\varphi)$ are simple roots of Eq. (1.6.5) and satisfy the relations:

$$t_1(\varphi) = T_1 + o(1), \quad t_2(\varphi) = T_2 + o(1). \tag{1.6.10}$$

This theorem confirms the ideas expressed by Mayorov and Myshkin (1990a), who believed that the specific form of functions $f_{Na}(u)$ and $f_K(u)$ under certain assumptions is not significant. The resulting form of the solutions is shown in

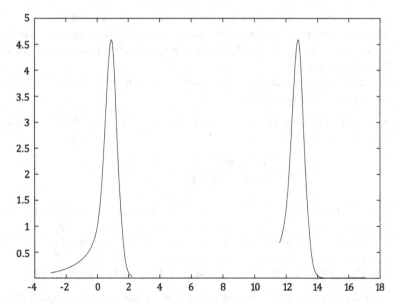

Fig. 1.2 View of the solution of Eq. (1.5.7) at $m = 0$ (Theorem 1.6.1)

Fig. 1.2 and is consistent with physiological data (see Cole 1968; Katz 1968; Ochs 1969; Khodorov 1975). Therefore, to the values $\lambda, f_{Na}(0), f_K(0)$ of the coefficients of Eq. (1.5.7), one can give natural biological interpretation, thus allowing us to locate their value.

Constant λ by formula (1.6.7) characterizes the rate (in logarithmic scale) of the restoration of the membrane potential, i.e., the indicator of the exponent on the descending part of the spike. Furthermore, $\lambda\alpha_1$ according to (1.6.6), $\lambda\alpha_1$ is the maximum speed of the process of depolarization, i.e., the indicator of the exponent approximates the ascending part of the spike. The time duration of this section is taken as a unit and coincides with the value of the delay in Eq. (1.5.7).

Note that according to Ochs (1969), the membrane slowly depolarizes after hyperpolarization. In this regard, the positive value of α, given by formula (1.5.8), is small. The meaning of the constant $\lambda\alpha$ is an indicator of the exponent that approximates the part of the change in membrane potential after hyperpolarization up to the next spike.

Thus arises a simple and natural way of determining approximate values $\lambda, f_{Na}(0), f_K(0)$ of the coefficients of Eq. (1.5.7). According to experimental data, approximating by exponents the ascending and descending fronts of the spike, as well as part of the slow change of the membrane potential, we find indicators of these exponents. They are, respectively, the values $\lambda(f_K(0) - 1, \lambda)$ and $f_K(0) - f_{Na}(0) - 1$. It is clear that now how values $\lambda, f_{Na}(0)$ and $f_K(0)$ are calculated.

Note the results of the paper (see Mayorov and Myshkin 1990a), in which Eq. (1.5.7) was analyzed for the specific type of functions $f_{Na}(u)$ and $f_K(u)$, which are contained in Theorem 1.6.1. In the mentioned paper, the researchers did not put forward a hypotheses about the behavior of functions $f_{Na}(u)$ and $f_K(u)$ to avoid difficulties in determining the parameters. However, doing so is useful to clarify the behavior of the solutions of Eq. (1.5.7), which may lead to new meaningful biological interpretations.

In this connection, we consider the second case when the previously entered numbers were $m > 0$ and $v_m > 0$.

Theorem 1.6.2 *Let* $m > 0$ *and* $v_m > 0$. *Then as* $\lambda \to \infty$ *hold asymptotic equalities:*

$$t_1(\varphi) = m + T_1 + o(1),$$
$$t_2(\varphi) = m + T_2 + o(1),$$

$$
\begin{aligned}
u(t, \varphi) &= v_j + 0(1), & t &\in [j - 1 + \delta, j - \delta], j = 1, \ldots, m, \\
u(t, \varphi) &= \exp \lambda \alpha_1(t + o(1)), & t &\in [m + \delta, m + 1 - \delta], \\
u(t, \varphi) &= \exp \lambda(\alpha_1 - t + m + 1 + o(1)), & t &\in [m + 1 + \delta, m + T_1 - \delta], \\
u(t, \varphi) &= \exp[-\lambda \alpha_2(t - T_1 - m + o(1))], & t &\in [m + T_1 + \delta, m + T_1 + 1 - \delta], \\
u(t, \varphi) &= \exp \lambda(\alpha(t - T_1 - m - 1) - \alpha_2 + o(1)), & t &\in [m + T_1 + 1 + \delta, t_2(\varphi)].
\end{aligned}
$$

As described in Theorem 1.6.2, the evolution of the membrane potential (see Fig. 1.3) is indeed sometimes seen in experiments. After some absolute increase of the potential, its drift is observed: It is quite abrupt, but it has a small range of values and unsystematic changes that sooner or later end with spike. From an experimental point of view, it is unclear whether small jumps of values of the potential are explained by obstacles or if they are fundamental.

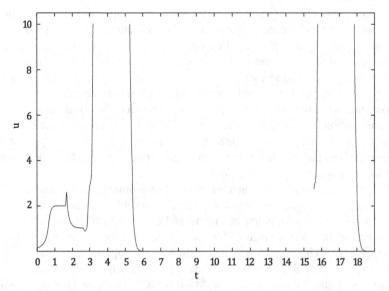

Fig. 1.3 View of the solution of Eq. (1.5.7) at $m = 2$ and $v_m > 0$ (Theorem 1.6.2). There are discontinuous changes of the membrane potential. The peak is truncated, and transient processes are observed near the ledges

One of the possible hypotheses here is as follows. A neuron is a distributed structure. It's as if it were the nervous system in miniature. Waves of excitation and inhibition caused by various factors, including the architecture of the cell itself as well as its links, can propagate on the membrane. Just these waves alone cause an abrupt change of the local membrane potential.

In this connection, we can make an assumption that sequence $v_1, \ldots, v_m$ is a unique code stored by the neuron. It is repeated cyclically, and the spike signals the end of the cycle. Hypothetically, the role of the spike (except as a signal of the end of the code) can also be explained. Propagating by branching of the axon, it has an influence on other neurons. As a result of stimulation, the value of the potential changes abruptly, possibly leading to a sequence of random walks, which is different from the natural course. Thus, the noncore code of the neuron, which has come under the influence of the potential, is activated. The fact is that we built sequence $v_1, \ldots, v_m$ for the solutions of Eq. (1.5.7) with initial conditions of class S. For other initial conditions, the solution of (1.5.7) on a finite interval of time will have a similar structure, but it may possibly have another sequence of random walks. The process of changing the code can be interpreted as scanning the information field. The hypothesis advanced may be too bold and have no place in biological neurons, but it must be kept in mind when considering the synthetic analogues.

Consider the third case: $m > 0$ and $v_m = 0$. Let $t_2(\varphi)$ be the smaller root of Eq. (1.6.5) on the interval $[m, \infty)$, and, unlike the previous one, we write

$$\alpha_2 = f_{Na}(0) + 1 - f_K(v_m - 1),$$

which at a general position can be considered positive.

Theorem 1.6.3 *Let* $m > 0$ *and* $v_m = 0$. *Then as* $\lambda \to \infty$ *hold asymptotic equalities:*

$$t_2(\varphi) = m + \alpha_2/\alpha + o(1),$$
$$u(t, \varphi) = v_j + o(1), t \in [j - 1 + \delta, j - \delta], \quad j = 1, \ldots, m - 1,$$
$$u(t, \varphi) = \exp \lambda[-\lambda \alpha_2(t - m - 1 + o(1))], \quad t \in [m - 1 + \delta, m - \delta],$$
$$u(t, \varphi) = \exp \lambda(\alpha(t - m) - \alpha_2 + o(1)), \quad t \in [m + \delta, t_2(\varphi)],$$

where α is defined by formula (1.5.8).

This theorem describes the evolution of the no-spike membrane potential (Fig. 1.4), which is characteristic of neurons as detectors (see [136]) because they are not capable of generating pulses without external action.

Let the conditions of Theorem 1.6.3 be fulfilled. Assume that in Eq. (1.5.7) the functions $f_K(u)$ and $f_{Na}(u)$ are nonnegative and that external stimuli lead to a sufficiently rapid increase of the membrane potential. Then, from the form of Eq. (1.5.7) and condition $f_{Na}(u) \to 0$ at $u \to \infty$, the following is implied. If the action is sufficiently intense and holds the value of $u(t, \varphi)$ above a certain threshold value, then follows a splash of function $u(t, \varphi)$ with an exponentially large

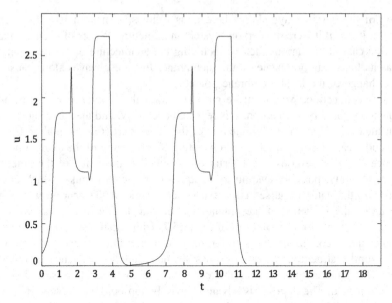

Fig. 1.4 View of the solution of Eq. (1.5.7) at $m = 4$ and $v_m = 0$ (Theorem 1.6.3). There is no high-amplitude pulse (*spike*). Transient processes are observed in the neighbourhood of ledges

on-lambda peak. At the same time, the necessary force of perturbations at different stages of the evolution of $u(t, \varphi)$ is different. Thus, in this case we can assume Eq. (1.5.7) as the model of the neuron as detector. For nerve cells, the force of action causing the spike really depends on the stage of the evolution of the membrane potential.

Note also that the consideration of the situation in Theorem 1.6.3 applies to everything stated previously with respect to the hypothetical possibility of storing information in the set of numbers $v_1, \ldots, v_m$.

We turn to the discussion of Theorems 1.6.1 through 1.6.3. Their proofs are quite simple but lengthy. They are based on the use of standard asymptotic analysis of ordinary differential equations, into which (1.5.7) is integrated by step method. We outline calculations corresponding to the conditions of Theorem 1.6.2.

At $t \in [0, 1 - \delta]$, where δ is arbitrary and small and $\varphi(s) \in S$, Eq. (1.5.7) takes the form:

$$\dot{u} = \lambda[-1 - f_{Na}(u) + f_K(0) + o(1)]u.$$

Because $u(0, \varphi) = \gamma < v_1$, and v_1 is a stable equilibrium state of equation $\dot{u} = \lambda[-1 - f_{Na}(u) + f_K(0)]u$, the domains of attraction, to which belong all its solutions with the initial conditions $0 < u(0) < v_1$, then as $\lambda \to \infty$ uniformly with respect to $t \in [\delta, 1 - \delta]$ holds equality at $u(t, \varphi) = v_1 + o(1)$. Asymptotics $u(t, \varphi)$ at $t \in [j - 1 + \delta, j - \delta]$ $(j = 2, \ldots, m)$ are constructed similarly. We obtain $u(t, \varphi) = v_j + o(1)$.

At $t \in [m+\delta, m+1-\delta]$ for the determination of $u(t, \varphi)$, we arrive at equation:

$$\dot{u} = \lambda[-1 - f_K(v_m) + o(1)]u,$$

from which follows $u(t, \varphi) = \exp \lambda \alpha_1(t - m + o(1))$, where $\alpha_1 = f_K(v_m) - 1$ (see formula 1.6.3). In turn, at $t \in [m+1+\delta, m+T_1-\delta]$, the equation for $u(t, \varphi)$ has the form

$$\dot{u} = \lambda[-1 + o(1)]u.$$

As a result, we obtain $u(t, \varphi) = \exp -\lambda(t - m - 1 + o(1))$. Finally, the relevant equations on the intervals $t \in [m+T_1+\delta, m+1+T_1-\delta]$ and $t \in [m+1+T_1+\delta, t_2(\varphi)]$ are as follows:

$$\dot{u} = \lambda[-1 - f_{Na}(0) + o(1)]u,$$

$$\dot{u} = \lambda[-1 - f_{Na}(0) + f_K(0) + o(1)]u.$$

From the latter equations, it follows that $u(t, \varphi) = \exp[-\lambda\alpha_2(t - T_1 - m + o(1))]$, at $t \in [m+T_1+\delta, m+T_1+1-\delta]$ and $u(t, \varphi) = \exp \lambda(\alpha(t - T_1 - m - 1) - \alpha_2 + o(1))$, and at $t \in [m+T_1+1+\delta, t_2(\varphi)]$. Here, $\alpha_2 = 1 + f_{Na}(0)$ and $\alpha = f_K(0) - f_{Na}(0) - 1$ [formulas (1.6.3) and (1.5.8)].

As given in Theorems 1.6.1 through 1.6.3, the formulas for $u(t, \varphi)$ do not cover time intervals of the order of δ (δ is arbitrarily small). To construct uniform approximation on the entire interval $0, t_2(\varphi)$, it is sufficient to use a well-developed method of boundary functions [20]. However, the corresponding expressions are lengthy. The moment is purely technical in nature because the theorems give a complete picture of the behavior of solutions $u(t, \varphi)$ of Eq. (1.5.7).

Consider the corollaries of Theorems 1.6.1 through 1.6.3. We introduce the successor operator Π by the rule $\Pi(\varphi(s)) = u(t_2(\varphi) + s, \varphi)(s \in [-1, 0], \varphi(s) \in S)$. From the above-written asymptotic, it follows that $\Pi(\varphi(s)) \in S$, that is, $\Pi(S) \subset S$. Hence, in view of the known results about the existence of fixed points, we find that there is an element $\varphi_0(s) \in S$, for which $\Pi(\varphi_0(s)) = \varphi_0(s)$. Thus, the function $u_0(t) = u(t, \varphi_0)$ is a periodic solution of Eq. (1.5.7) with period $t_2(\varphi)$, and Theorems 1.6.1 through 1.6.3 disclose the asymptotics of this solution in different cases.

It should be noted that solutions (1.5.7) with initial conditions from ΠS form the attractor, to which also belongs the periodic solution. The attractor is quite narrow because all the elements of $\varphi(s) \in \Pi S$ have as $\lambda \to \infty$ general asymptotics: $\varphi(s) = \exp(\lambda \alpha s + o(1))$. Simultaneously, the attractor has a broad domain of attraction. The problem of the stability of periodic solution from the biological point of view is of no interest because of the identity of the principal terms of the asymptotics of solutions forming the attractor.

Asymptotics of solutions belonging to the attractor are stitched together from several exponentials and constants. In this connection, we will call Eq. (1.5.7) the "model of exponential approximation."

Apparently the studied solutions do not exhaust all the established modes of Eq. (1.5.7). For example, for the arbitrarily large time interval (for sufficiently large λ), this equation has solutions that are close (in the integral sense) to functions defined by the trajectories of a one-dimensional map:

$$f_{Na}(u) + 1 = f_K(u(t-1)).$$

Note that the dynamics of such mappings can be complex. Solutions of this type in various problems have been studied (Sharkovskii et al. 1986; Ivanov 1987). The complexity of their interpretation for the neural systems is the lack of high splash (spike).

1.7   Model of a Pure Potassium Current Delay

By the term "pure potassium current delay," we mean the delay of the entire potassium current and not only of the coefficient, which characterizes the conductivity. We again consider the balanced equation for currents:

$$c\dot{u} = i_1 + i_N(u) + i_K(u(t-h)). \tag{1.7.1}$$

Here, c is the capacity; $i_1 = -g_1(u - u_1)$ is the value of the leakage current where u_1 is electrochemical potential for the ions involved in leakage current; and g_1 is the conductivity. Further, as stated previously, $i_{Na}(u) = (a - f_{Na}(u))u$, where $a > 0$, $f_{Na}(0) > 0$. Finally, in contrast to stated previously, $i_K(u) = f_K(u) - b$, where $b > 0$, $f_K(0) > b$. In the corresponding term, h is the delay time of the potassium current.

As before, assume that $f_{Na}(u) \to 0, f_K(u) \to 0$ as $u \to \infty$. Suppose that at the current time the values of $u(t)$ and $u(t-h)$ are large. Then, by virtue of the assumption made concerning the functions $f_{Na}(u)$ and $f_K(u)$, we obtain the following approximate equation for the membrane potential:

$$c\dot{u} = (a - g_1)u + g_1u_1 - b.$$

As already mentioned, in this situation the membrane potential exponentially approaches its value in a state of strong polarization, which is taken by us as the origin. In this regard, $a - g_1 < 0$, and we write $g_1u_1 - b = 0$.

Choosing as the time scale the value of the delay of the potassium current, we can rewrite Eq. (1.7.1) as:

$$\dot{u} = -\gamma u - \lambda[f_{Na}(u)u - f_K(u(t-1))], \tag{1.7.2}$$

where $\gamma = (g_1 - a)h/c, \lambda = h/c$.

A special form $(f_{Na}(u) \equiv 0, f_K(u) = R\exp(-u^2))$ of the latter equation was first proposed in [83] to describe the dynamics of the membrane potential averaged over a set of neurons. The investigators believe that their equation simultaneously describes the change in the so-called "slow cortical potentials." The equation was developed by Lebedev at the qualitative level of the theory of storing and processing information in the brain in the form of stable combinations of wave packets (Lebedev 1990; Lebedev et al. 1991). The given ideology hypothetically made possible explanations of some phenomena of the human psyche, e.g., the spread of response time, variability in the amount of memory, etc.

As stated previously, we assume in (1.7.2) that the parameter λ is large. It turns out that the speed of functions $f_{Na}(u)$ and $f_K(u)$ tend to zero as $u \to \infty$ is important. For example, the power-law decay $f_{Na}(u) = O(u^{-n}), f_K(u) = O(u^{-n})$ is not sufficient for use of the asymptotic method. Suitable and sufficiently natural are the following conditions:

$$f_{Na,K}(u) \leq d \exp(-\beta u^n) \quad (d > 0, \beta > 0, \eta > 0).$$

To illustrate the distinctive properties of the solutions of Eqs. (1.7.2) and (1.5.7), we will restrict ourselves to the simpler class of functions $f_{Na}(u)$ and $f_K(u)$ selected by the requirement of finiteness:

$$f_{Na}(u) = \begin{cases} F_{Na}(u) > 0, & |u| \leq A, \\ 0, & |u| > A. \end{cases}$$

$$f_K(u) = \begin{cases} F_K(u) > 0, & |u| \leq A, \\ 0, & |u| > A. \end{cases}$$

Denote by S the set of continuous functions $\varphi(s) \in C[-1,0]$, for which

$$\varphi(s) = A[1 + \psi(s)] \exp(-\gamma s),$$

where $\psi(0) = 0, |\psi(s)| \leq \lambda^{-1}$. The solution (1.7.2), with initial conditions from S, are denoted by $u(t, \varphi)$, and the consecutive positive roots of the equation $u(t, \varphi) = A$ are denoted by $t_1(\varphi), t_2(\varphi), \ldots$ Let us assume that equation

$$F_K(0) - F_{Na}(v)v = 0 \tag{1.7.3}$$

has no roots on the interval $(0, A)$.

Theorem 1.6.4 *Hold asymptotic as $\lambda \to \infty$ equalities:*

$$t_1(\varphi) = 1 + o(1), t_2(\varphi) = [(\ln \lambda)/\gamma](1 + o(1)), \tag{1.7.4}$$

$$u(t, \varphi) = o(1), \quad t \in [\delta, 1], \tag{1.7.5}$$

$$u(t, \varphi) = \lambda \frac{F_K(0)}{\gamma} (1 - \exp[-\gamma(t - 1)] + o(1)), \quad t \in [1 + \delta, 2], \tag{1.7.6}$$

$$u(t, \varphi) = \lambda \frac{F_K(0)}{\gamma} (1 - \exp[-\gamma] + o(1)) \exp[-\gamma(t - 2)], \quad t \in [2 + \delta, t_2(\varphi)].$$

$$(1.7.7)$$

Introduce the successor operator Π by the rule:

$$\Pi(\varphi(s)) = u(t_2(\varphi) + s, \varphi) \quad s \in [-1, 0), \quad \varphi(s) \in S.$$

We check that $\Pi S \in S$, and hence it is possible to draw a conclusion about the existence in S of a fixed point $\varphi_0(s)$ of this operator. Function $u_0(t) = u(t, \varphi_0(s))$ is a periodic solution of Eq. (1.7.2). The period of solution coincides with the number $t_2(\varphi_0)$, the asymptotics of which for $\lambda \to \infty$ are given by (1.7.4). Note that a similar result is given in [55]. In addition, from [55], it follows that in S is the only fixed point and that the solution $u_0(t)$ is stable (at sufficiently large λ). The curves of neuronal activity, such as those described by (1.7.5) through (1.7.7), are sometimes observed in experiments.

In the case when Eq. (1.7.3) has its roots on the interval $(0, A)$, there is a sequence of random walks of the membrane potential. The corresponding solutions of Eq. (1.7.2) can have both a spike and no-spike structure.

Let us say few more words about another way of introducing the large parameter λ. Consider the case when λ on the right-hand side of (1.7.2) is multiplied only by $f_K(u(t - 1))$. The changes here are insignificant. Formula (1.7.5) takes the form $u(t, \varphi) = v(t) + o(1)(t \in [\delta, 1])$, where $v(t)$ is the solution of equation $\dot{v} = -[\gamma + F_{Na}(v)]v$ with initial condition $v(0) = A$. The analogues of formulas (1.7.6) and (1.7.7) are as follows:

$$u(t, \varphi) = \lambda \left[\int_1^t \exp[-\gamma(t - s)] F_K(v(s - 1)) ds + o(1) \right], \quad t \in [1 + \delta, 2],$$

$$u(t, \varphi) = \lambda \left[\int_0^1 \exp[-\gamma(1 - s)] F_K(v(s)) ds + o(1) \right] \exp[-\gamma(t - 2)]$$

$$t \in [2 + \delta, t_2(\varphi)].$$

The essential difference of the periodic solutions of Eqs. (1.5.7) and (1.7.2) is that the period of the latter has no finite limit as $\lambda \to \infty$. In this case, the duration of the spike is $O(\ln(\lambda))$, which is determined mainly by the duration of its descending phase. In reality, the ascending phase is really much shorter than the descending phase. Nevertheless, there are difficulties with biological interpretation of the mathematical results. In particular, it is unclear how much larger the parameter λ can be in reality. Therefore, we accept Eq. (1.5.7) as the model of the neuron.

References

Bernstein, N. A. (1990). *Physiology of movements and activity.* Moscow: Nauka.

Blum F., Leiserson A., & Hofstefer L. 1988. *Brain, mind and behaviour.* Moscow: Mir.

Borisyuk, G. N., Borisyuk, R. M., Kazanovich, Y. B., & Ivanitsky, G. R. (2002). Models of neural dynamics in brain information processing—Results of "decade". *UFMN, 172*(10), 1189–1214.

Burns, B. N. (1968). *Uncertainty in the nervous system.* Moscow: Mir.

Cole, K. S. (1968). Nerve impulse (theory and experiment). *Theoretical and Mathematical Biology,* 154–193 (Mir).

Deutsch, S. (1970). *Models of the nervous system.* Moscow: Mir.

Eckert, R., Randall, D., & Augustine, G. (1991). *Animal physiology* (Vol. 1). Moscow: Mir.

Green, H., Stout, W., & Taylor, D. (1993). *Biology* (Vol. 2). Moscow: Mir.

Guselnikov, V. I., & Suping, A. J. (1968). *Rhythmic activity of the brain.* Moscow: Moscow State University Press.

Hebb, D. O. (1949). *The organization of behaviour neuropsychological theory* (335 p.). New York: Wiley.

Hodgkin A. L. 1965. *Nerve impulse.* Moscow: Mir.

Hodgkin, A. L., & Huxley, A. F. (1939). Action potentials recorded from inside a nerve fiber. *Nature, 144,* 710–711.

Hopfield, J. J. (1982). Neural networks and physical systems with emergent collective computational abilities. *Proceedings of the National Academy of Sciences, 79*(8), 2554–2558.

Hopfield, J. J. (1984). Neurons with gradual response have collective computational properties like those of two-state Neurons. *Ibid, 81,* 3088–3092.

Hubel, D. (1990). *Eye, brain, vision.* Moscow: Mir.

Ivanov, A.D. (1987). *Attracting cycles of period two to display the interval and asymptotically stable periodic solutions of singularly perturbed differential-difference equations.* Kiev: IM AN USSR, Preprint 85.64.

Izhikevich, E. M., & Malinetskii, G. G. (1992). On possible role of chaos in neural systems. *Reports of the Academy of Sciences of Russia, 326*(4), 626–632.

Izhikevich, E. M., & Malinetskii, G. G. (1993). *A neural network model with chaotic behaviour.* Moscow: Russian Academy of Sciences, Institute of Applied Mathematics named after Keldysh, Preprint, 17.

Kashchenko, S. A., & Mayorov, V. V. (1992). *Investigation of difference-differential equations, modelling the dynamics of neuron.* Moscow: Russian Academy of Sciences, Institute of Applied Mathematics named after Keldysh, Preprint, 8.

Kashchenko, S. A., & Mayorov, V. V. (1993). On a differential-difference equation, modelling neuron impulse activity. *Mathematical modelling, 5*(12), 13–25.

Katz, B. (1968). *Nerve, muscle, synapse.* Moscow: Mir.

Khodorov, B. I. (1969). *The problem of excitability.* Leningrad: Medicine.

Khodorov, B. I. (1975). *General physiology of excitable membranes.* Moscow: Nauka.

Lebedev, A. N. (1990). Human memory, its mechanisms and limits. *Research of memory* (pp. 104–118). Moscow: Nauka.

Lebedev, A. N., & Lutsky, V. A. (1972). EEG rhythms are the result of the interaction of oscillating processes. *Biophysics, 17*(3), 556–558.

Lebedev, A. N., Mayorov, V. V., & Myshkin, I. Y. (1991). The wave model of memory. In A. V. Holden & V. I. Kryukov (Eds.), *Neurocomputers and attention. neurobiology, synchronisation and Chaos* (Vol. 1, pp. 53–59). Manchester: Manchester University Press.

Livanov, M. N. (1972). *Spatial organization of the processes of the brain.* Moscow: Nauka.

Mayorov, V. V., & Myshkin, I. Y. (1990a). On one model of functioning of the neural network. *Modelling of Dynamic Populations,* 70–78 (Nizhny Novgorod).

Mayorov, V. V., & Myshkin, I. Y. (1990b). Mathematical modelling of neurons of the networks based on equations with delay. *Mathematical Modelling, 2*(11), 64–76.

Mayorov, V. V., & Myshkin, I. Y. (1993). M-machine is a model of short-term memory. *Neurocomputer as the basis of thinking computers* (pp. 137–146). Moscow: Nauka.

Mayorov, V. V., & Shabarshina, G. V. (1997). Report on networks of W-neurons. *Modelling and Analysis of Information Systems, 4*, 37–50. (Yaroslavl).

Mayorov, V. V., & Shabarshina, G. V. (1999). Simplest modes of burst wave activity in the networks of W-neurons. *Modelling and Analysis of Information Systems, 6*(1), 36–39. (Yaroslavl).

Mayorov, V. V., & Shabarshina, G. V. (2001). Networks of W-neurons in the associative memory problem. *Journal of Computational Mathematics and Mathematical Physics, 41*(8), 1289–1299.

Mayorov, V. V., Shabarshina, G. V., & Anisimova, I. M. (2005). Neuronet solution to the problem of planning of optimal paths for point robots. In *Proceedings of the VII All-Russian Conference "Neuroinformatics- 2005"*, Part I (pp. 197–202). Moscow: Moscow Engineering Physics Institute.

McCulloch, W. A., & Pitts, W. (1943). A logical calculus of ideas immanent in nervous activity. *The Bulletin of Mathematical Biophysics, 5*, 115–133.

Minsky, M., & Papert, S. (1971). *Perceptrons*. Moscow: Mir.

Murray, A. F. (1989, December). Arithmetic in VLSI neural networks. *IEEE Micro*, 64–74.

Nicholls, J. G., Martin, A. R., Wallace, B. J., & Fuchs, P. A. (2003). *From neuron to brain*. Moscow: Editorial URSS.

Ochs, S. (1969). *Fundamentals of Neurology*. Moscow: Mir.

Plonsi, R., & Barr, R. (1992). *Bioelectricity*. Moscow: Mir.

Rosenblatt, F. (1965). *Principles of neurodynamics*. Moscow: Mir.

Sharkovskii, A. I., Maistrenko, Y. L., & Romanenko, E. Y. (1986). *Difference equations and their applications*. Kiev: Naukova Dumka.

Singer, W., Tretter, F., & Cynader, M. (1975). Organization of cat striate cortex: A correlation of receptive-field properties with afferent and efferent connections. *Journal of Neurophysiology, 10*(3), 311–330.

Tasaki, I. (1971). *Nervous excitement*. Moscow: Mir.

Timofeev, E. A. (1997). Modelling of neuron, transmitting information by density of the flow of pulses. *Automation and Remote Control, 3*, 190–199.

Wiener, N., & Rosenbluth, A. (1961). Conduction of impulses in the heart muscle. *Cybernetic Collection, 3*, 7–56 (Moscow: IL).

Chapter 2
Model of the Interaction of Neurons

The complexity of the pattern of electrical activity of neural ensembles is due to the interaction of their elements. The interaction is associated with the presence of functional contacts between neurons. Since the time of Charles Sherrington, zones of contacts have been called "synapses." Depending on the manner of conducting signals, synapses are divided into two types: electrical and chemical. The synapse of the former type is the zone of direct electrical interaction of the neurons. In this area of the membrane, neurons are in contact with each other, and currents are associated with the gradient of the potentials.

The functional structure of chemical synapses is another one. The branches of the axon are the only centrifugal appendages of the body of the neuron, and they end in the immediate vicinity of the surface of the bodies or dendrites of other neurons. In a synapse, under the action of an incoming nerve impulse, chemical intermediaries called "mediators" are generated. They provide stimulating or inhibitory effects on ion channels of neuron-receiver membranes. Thus, mediators either zoom in or out at the start of the spike of the neuron-receiver. Processes taking place in the electrical synapses are similar to those observed in experiments on the artificial stimulation of neurons by means of electrodes.

2.1 Response to Electrical Action

The artificial stimulation of neurons is carried out with microelectrodes, which are usually placed near the outer surface of the cell. There are variants of experiments with intracellular introduction. In any case, into the balanced equation of currents (1.5.1) flowing through the membrane, we should add the term $I_V(t)$, to characterize the external influence:

$$c\dot{u} = I_{Na}(t) + I_K(t) + I_V(t).$$

Here, as presented previously, u indicates the current value of the membrane potential. The specific form of $I_V(t)$ depends both on the experimental scheme as well as its mathematical model. We assume that $I_V(t) = g^*(v(t) - u(t))$. The value

© Springer International Publishing Switzerland 2015
S. Kashchenko, *Models of Wave Memory*,
Lecture Notes in Morphogenesis, DOI 10.1007/978-3-319-19866-8_2

$v(t)$ can be interpreted as the effective value of the potential, which attempts to impose external action on the membrane. The constant $g^* > 0$ is the corresponding coefficient of conductivity. Because we have agreed to count the membrane potential from the minimum possible value (large and negative in absolute value), then the imposed value is $v(t) \geq 0$. Suppose that for sodium $I_{Na}(t)$ and potassium $I_K(t)$, all of the assumptions of Sect. 1.5 are fulfilled. Taking as time scale the value of the delay of the potassium conductivity (denoted previously as h), and having performed the transformation as described in Sect. 1.5, we change the balanced equation to the form:

$$\dot{u} = \lambda[-1 - f_{Na}(u) + f_K(u(t-1))]u + g(v(t) - u). \tag{2.1.1}$$

Here, $g = g^* h/c$ is the normalized coefficient of conductivity. The meaning of all other parameters and functions is the same as for Eq. (1.5.7). Below are given all the notations of indicators of exponents and constants approximating the solutions of the equation.

We will begin the study of Eq. (2.1.1) with the case when $v(t) = v_0$ for $t \in [0, T]$ and $v(t) = 0$ at $t \notin [0, T]$, where v_0 is independent of λ constant, and T is a sufficiently large time of action. Let us assume that in Eq. (2.1.1) $f_{Na}(u) > 0, f_K(u) > 0$, and also monotonically and quite rapidly,

$$f_{Na}(u) \to 0, \quad f_K(u) \to 0 \text{ at } u \to \infty. \tag{2.1.2}$$

As the initial condition for the solution of Eq. (2.1.1), we choose the function $\varphi(s) \in S \subset C[-1, 0]$. Recall that class S includes continuous functions, for which $\varphi(0) = \gamma$ and $\varphi(s) \leq \gamma \exp(\lambda \alpha s/2)$ for $s \in [-1, 0]$. Here the number α is given by formula (1.5.8), i.e., $\alpha = f_K(0) - f_{Na}(0) - 1$. In this case, algebraic Equation (1.6.1) has no positive roots. We can write $\gamma = 1$. Conventionally, we associate the start and the end of the spike with moments of time when the solution of Eq. (2.1.1) or (1.5.7) crosses the unit value with positive and negative speed, respectively. Denote by $u_v(t, \varphi)$ the solution of Eq. (2.1.1) with the initial condition $u_v(s, \varphi) = \varphi(s) \in S$. As before, δ is an arbitrarily small number.

Equation (2.1.1) is integrated by steps with an asymptotic as $\lambda \to \infty$ simplification of formulas. For the time interval $t \in [\delta, 1 - \delta]$, Eq. (2.1.1) takes the form:

$$\dot{u} = \lambda(\alpha_1 + o(1))u + gv_o, \tag{2.1.3}$$

where $\alpha_1 = f_K(0) - 1$. In the asymptotic integration of Eq. (2.1.3), the term gv_0 can be neglected. We obtain for the ascending part of spike the approximate formula from Theorem 1.6.1:

$$u_v(t, \varphi) = \exp \lambda \alpha_1(t + o(1)), \quad t \in [\delta, 1 - \delta]. \tag{2.1.4}$$

By virtue of (2.1.2), in the time interval $t \in [1+\delta, T_1 - \delta](T_1 = \alpha_1 + 1)$, Eq. (2.1.1) goes over into

$$\dot{u} = \lambda(-1 + o(1))u + gv_0,$$

from which, in accordance with Theorem 1.6.1,

$$u_v(t, \varphi) = \exp \lambda(\alpha_1 - (t-1) + o(1)), \quad t \in [1+\delta, T_1 - \delta]. \tag{2.1.5}$$

In the next time interval $t \in [T_1 - \delta, T_1 + 1 + \delta]$ for $u_v(t, \varphi)$, we obtain equation:

$$\dot{u} = \lambda[-1 - f_{Na}(u) + o(1)]u + g(v_0 - u).$$

The assumptions made about the properties of functions $f_{Na}(u)$, $f_K(u)$, guarantee the existence of a unique exponentially stable equilibrium state $u^* > 0$ for the latter equation. Its asymptotic approximation is at $\lambda \to \infty : gv_0/\lambda\alpha_2$ where $\alpha_2 = f_{Na}(0) + 1$. As a result, we obtain:

$$u_v(t, \varphi) = gv_0/\lambda\alpha_2 + O(1/\lambda), \quad t \in [T_1 + \delta, T_1 + 1 - \delta]. \tag{2.1.6}$$

This asymptotic representation significantly differs from the appropriate formula (1.6.8) of Theorem 1.6.1. In the time interval $t \in [T_1 + 1 + \delta, T_1 + 2 - \delta]$, Eq. (2.1.1) again passes into (2.1.3). We obtain asymptotic formula (2.1.4) with a time shift on $T_1 + 1$. The sequence of operations can be continued as long as $t < T$.

The solution $u_v(t, \varphi)$ of Eq. (2.1.1) has properties other than the solution $u(t, \varphi)$ of Eq. (1.5.7). Asymptotic formulas (2.1.4)–(2.1.6) for $u_v(t, \varphi)$ are cyclically repeated through a period of time $T_v = \alpha_1 + 2$ [in contrast to $T_2 = \alpha_1 + \alpha_2/\alpha + 2$ for (1.5.7)]. The duration of spike, as before for (1.5.7), is close to T_1, but the time interval between the end of one pulse and the beginning of the next one is nearly equal to the lag length (for (1.5.7) it is close to $\alpha_2/\alpha + 1$).

The described type of neural activity wherein the pulses follow frequently is called "bursting." It has been observed in biological experiments. The neuron responds with salvo bursts (Khodorov 1975, p. 229) on the constant electric depolarizing action, thus decreasing the absolute value of the membrane potential.

If in Eq. (2.1.1) we consider the value of the external action v_0 to be small and to agree with the parameter $\lambda(v_0 = \exp(-\lambda\sigma)$, where $0 < \sigma < \alpha_2)$, then the time interval between the end of one spike and the start of the next spike will be asymptotically close to $\sigma/\alpha + 1$. We assume that the action time T is sufficiently long.

A similar construction can be applied to neurons with a resting state of the membrane potential (i.e., neurons as detectors). Here the external action should be considered as large and consistent with the parameter $\lambda : v_0 = \lambda\omega$. We describe the results on a qualitative level. One can distinguish the class of functions $f_{Na}(u)$ and $f_K(u)$ such that Eq. (2.1.1) will possess the following property: For $g = 0$, it has a stable positive equilibrium state u_0. In this case, however, if in the initial condition $u(0)$ is greater than some constant independent of λ, then the corresponding solution will have a peak of exponentially large amplitude at λ. Furthermore, suppose that for $t < 0$ $u_v(t, \varphi) \equiv u_0$. There exists such a threshold value ω_p that at $\omega < \omega_p$ and $t \in [0, T]$, Eq. (2.1.8) also has a stable positive equilibrium state. If $\omega > \omega_p$, then we observe a spike at $t > 0$ for the solution $u_v(t, \varphi)$. At a greater value of the parameter ω, the spike is cyclically repeated. However, its amplitude, being exponentially large at λ, is different than that of the spike starting at time zero. The time interval between the end of one spike and the start of the next one is close to 1.

The described behavior of the solutions of Eq. (2.1.1) is consistent with the results of biological experiments. The weak action on the neuron whose membrane potential is at rest does not lead to the generation of a spike. Such an action is described as "sub-threshold." If the force of depolarizing current flowing through the membrane exceeds the threshold, the neuron responds with a pulse. Increased action leads to a burst response (see Khodorov 1975; Eckert et al. 1991).

A similar pattern of the behavior of the solution $u_v(t, \varphi)$ of (2.1.1) can be observed in the case when the functions $f_{Na}(u)$ and $f_K(u)$ satisfy the conditions of Theorem 1.6.3, which describes a no-spike evolution of the membrane potential. The difference lies in the fact that at the sub-threshold values of the external action, the solution $u_v(t, \varphi)$ does not tend to the equilibrium state but abruptly wanders in the system of ledges.

Consider the general case of action on the neuron of arbitrary signal v_t. We assume that in Eq. (2.1.1), the functions $f_{Na}(u)$ and $f_K(u)$ are such that the conditions of Theorem 1.6.1 or 1.6.2 are fulfilled, i.e., the modes of the pulse structure are realized without external action. The results of comparison of the solution $u_v(t, \varphi)$ of Eq. (1.5.7) with $u_v(t, \varphi)$ gives:

Lemma 2.1.1 *At every fixed function $v(t)$ uniformly with respect to $\varphi(s) \in S$ for all $t \in [0, m + T_1]$ holds asymptotic equality as $\lambda \to \infty$: $u(t, \varphi)/u_v(t, \varphi) = 1 + o(1)$, and at $t \in [m + T_1 + \delta, m + T_1 + 1 - \delta]$ respectively, $u(t, \varphi)/u_v(t, \varphi) = o(1)$ and besides, $u_v(t, \varphi) = o(1/\lambda)$.*

Thus, the time interval $t \in [0, m + T_1 + 1]$ is an area of low susceptibility to external influence. One can use formulas from Theorems 1.6.1 and 1.6.2 as the asymptotic representation for the solution $u(t, \varphi)$ at $t \in [0, m + T_1]$, Pay attention to important details. In the time interval $t \in [m + T_1 + \delta, m + T_1 + 1 - \delta]$, the solution $u(t, \varphi)$ of (1.5.7) is exponentially small, and the solution $u_v(t, \varphi)$ of Eq. (2.1.1) has the order of smallness $o(1/\lambda)$. This is consistent with biological data. After a spike, the external depolarizing action counteracts hyperpolarization of the membrane [54, 151]. The fact is that $u_v(t, \varphi) = o(1/\lambda)$ at $t \in [m + T_1 + \delta, m + T_1 + 1 - \delta]$ has one more corollary. Let $t_{2v}(\varphi)$ be the second positive root of equation $u_v(t, \varphi) = \gamma$. We have valid asymptotic as $\lambda \to \infty$ equality $t_{2v}(\varphi) = m + T_1 + 1 + o(1)$, which differs substantially from the asymptotic representation of the corresponding moment of time $t_2(\varphi)$ for the solution $u(t, \varphi)$ [by virtue of Theorems 1.6.1 and 1.6.2 $t_2(\varphi) = m + T_1 + 1 + \alpha_2/\alpha + o(1)$]. Note that $u_v(t_{2v}(\varphi) + s, \varphi) \notin S$, but nevertheless we can continue construction of the asymptotics $u_v(t, \varphi)$ for $t > t_{2v}(\varphi)$.

Thus, the response of the neuron to external action for the time interval $t \in [m + T_1 + 1, t_2(\varphi)]$ has a completely different character. To illustrate this more clearly, assume that the continuously differentiable nonnegative function v_t is as follows: $v_t = 0$ at $0 \leq t \leq t_0$, where $m + T_1 + 1 < t_0 < t_2(\varphi)$ and $\ddot{v}t_0 > 0$.

Lemma 2.1.2 *Holds for $\lambda \to \infty$ equality*

$$t_{2v}(\varphi) = t_0 + 3(\alpha\lambda)^{-1}(\ln\lambda)(1 + o(1)). \tag{2.1.7}$$

For proof, it suffices to note that for $t > t_0$ Eq. (2.1.1) has the form:

$$\dot{u} = \lambda(\alpha + o(1))u + v(t).$$

Its solution is approximately

$$u_v(t, \varphi) = \exp[\lambda\alpha(t - t_0)]u_v(t_0, \varphi) + \exp[\lambda\alpha(t - t_0)] \int\limits_0^{t - t_0} \exp[-\lambda\alpha\tau]v(\tau + t_0)d\tau,$$

where $u_v(t_0, \varphi)$ is exponentially small. Neglecting the first term and using the Laplace approximation for the integral, we obtain:

$$u_v(t, \varphi) \approx \exp[\lambda\alpha(t - t_0)]\ddot{v}(t_0)/(\lambda\alpha)^3.$$

Hence and from equation $u_v(t, \varphi) = \gamma$ follows formula (2.1.7).

Thus, according to (2.1.7), even short-term and relatively low external action on the area $(m + T_1 + 1, t_2(\varphi))$ can dramatically change the behaviour of the solutions (compared with $u(t, \varphi)$): The mechanism relocating for $u_v(t, \varphi)$ to the interval $[t_0, t_0 + m + T_1]$ almost instantly "starts" the values of $u(t, \varphi)$ from the interval $[0, m + T_1]$.

The above-mentioned mechanism leads to an important conclusion: Periodic with period $T_0 \in (m + T_1 + 1, t_2(\varphi))$, external action imposes its period on the solutions of (2.1.1). In this case, the leading terms of all but one of the main characteristics of the asymptotic expansions of $u(t, \varphi)$ and $u_v(t, \varphi)$ are the same. The only difference lies in the time of the slow exponential growth of these functions after passing through a minimum. For $u(t, \varphi)$, this time is as close as $\lambda \to \infty$ is to $t_2(\varphi) - (m + T_1 + 1)$. In turn, for $u_v(t, \varphi)$ this is determined by external action $v(t)$. Let, for example, $v(t)$ periodic pulse be a function with the period $T_0 \in (m + T_1 + 1, t_2(\varphi))$ given for $t \in [0, T_0]$ by the relations:

$$v(t) = \begin{cases} \psi_0(t) > 0, & t \in (0, t^0) \\ 0, & t \in (t^0, T_0) \end{cases}, \tag{2.1.8}$$

where $0 < t^0 < m + T_1$. Then the mentioned period of time is close to $T_0 - (m + T_1)$.

The described phenomenon of imposing frequency of impulsation with the help of an external stimulus is well known in physiology (Livanov 1972). For the visual cortex, it is actively being studied, for example, by Singer et al. (1975, 1976) who —by the method of probing electrodes—in particular found that periodic flashes of light impose their frequency on biopotentials of the cortex.

Because according to current views the columns of the visual cortex are formed by the neuron-detectors (Hubel and Wiesel 1977), it is necessary to make the following remark. As already mentioned, for neurons as detectors, the spike is a response to an external and sufficiently strong action. Within the framework of our model, this is possible if for Eq. (1.5.7) the conditions of Theorem 1.6.3 are met or if Eq. (1.5.7) has a stable equilibrium state. In such a situation, strong $(v(t) = \lambda\omega(t))$ external periodic action, generally speaking, will impose its frequency on the solution of Eq. (2.1.1). The response to action can be complex, e.g., generating structures with ledges. Such multimode responses have been observed in biological experiments.

2.2   Model of Electrical Synapse

In physiological studies on the giant axons of annelids and crabs, it is has been shown that there are synapses with electrical transfer of excitation (see Pappas and Waxman 1973; Eckert et al. 1991; Green et al. 1993). The most important criterion for the identification of neural connection, such as the electrical synapse, is the presence of close contact between membranes of the neurons. The presence of contact, generally speaking, does not mean that the neurons must interact. However, in a number of experiments, previously associated cells were disjoined. As a result, the correlation between the dynamics of their membrane potentials disappeared.

Two alternative hypotheses about the nature of the electric interaction are conceivable: Either the interaction is related to electrical cross-talk, or it is conditioned by conduction currents. Preference is given to the latter assumption (Nicholls et al. 2003). In experiments with dyes, it was found that in the contact zone, the intracellular substance of one neuron is able to diffuse into another neuron. Thus, the channels have been found to be relating to the contents of the cells with which it is in contact. At the same time, it was found that intercellular fluid is present in the contact gap. According to Plonsi and Barr (1992), two coupled neurons can be thought of simply as two tanks connected by a system of parallel tubes. Tanks and tubes themselves are washed by intercellular fluid. The transport of ions is possible through the walls of the tanks. To effectively take into account the electric interaction, we can add terms depending on the difference of the membrane potentials in balanced equations for the membrane currents.

We consider two identical neurons in contact with each other. Denote the membrane potentials by u_1, u_2. Balanced equations for membrane currents have the form:

$$c\dot{u}_1 = I_{Na}(u_1) + I_K(u_1) + I_1,$$
$$c\dot{u}_2 = I_{Na}(u_2) + I_K(u_2) + I_2.$$

Here, $c\dot{u}$ is the current through the capacitance; $I_{Na}(u)$ and $I_K(u)$ are sodium and potassium currents, respectively; and I_1 and I_2 are currents caused by the difference of the membrane potentials. It would seem that $I_1 = -I_2$; however, as biological data show, in many cases there is no symmetry in the interaction [124, 172, 33].

One can put forward a number of hypotheses about the nature of the dependence of I_1 and I_2 on the membrane potentials. The simplest of them are the ohmic and the semiconductor hypotheses. According to the former, the currents are proportional to the potential difference, i.e., they obey Ohm's law: $I_1 = g_1^*(u_2 - u_1), I_2 = g_2^*(u_1 - u_2)$, where g_1^* and g_2^* are conductivities. According to the latter, the currents occur only if the potential difference is positive, whereas they have a tendency to saturation with the growth of potential difference. For this reason, we can assume that $I_1 = g_1^*\chi((u_2 - u_1)/u_p), I_2 = g_2^*\chi((u_1 - u_2)/u_p)$.   Here $\chi(u) = 0$ at $u < 0, \chi(u) = u$ at $0 \le u \le 1$ and $\chi(u) = 1$ at $u > 1$. Value u_p is the potential difference from which saturation begins. Semiconducting properties have, for example, electrical synapses in the giant axons of crayfish [124]. However, in general, it is believed that electrical synapses obey Ohm's law.

Balanced equations for membrane currents can be transformed as was performed in Sect. 1.5. As a result, we obtain for ohmic synapses the following system of equations:

$$\dot{u}_1 = \lambda[-1 - f_{Na}(u_1) + f_K(u_1(t-1))]u_1 + g_1(u_2 - u_1), \qquad (2.2.1)$$

$$\dot{u}_2 = \lambda[-1 - f_{Na}(u_2) + f_K(u_2(t-1))]u_2 + g_2(u_1 - u_2). \qquad (2.2.2)$$

In turn, for the semiconductor synapses, this system has the following form:

$$\dot{u}_1 = \lambda[-1 - f_{Na}(u_1) + f_K(u_1(t-1))]u_1 + g_1\chi((u_2 - u_1)/u_p), \qquad (2.2.3)$$

$$\dot{u}_2 = \lambda[-1 - f_{Na}(u_2) + f_K(u_2(t-1))]u_2 + g_2\chi((u_1 - u_2)/u_p). \qquad (2.2.4)$$

In these equations $g_1 = g_1^* h/c$ and $g_2 = g_2^* h/c$, the meaning of h and all other notation was explained earlier in Sect. 1.5.

We consider some peculiarities of the electric interaction between the neurons as described by the system of Eqs. (2.2.1) and (2.2.2). In this section, we assume that positive functions $f_{Na}(u)$ and $f_K(u)$ monotonically and quite rapidly tend to zero at $\lambda \to \infty$. For isolated neurons, the case corresponds to the conditions of Theorem 1. 6.1. In determining class S of the initial functions, we can write the number $\gamma = 1$. We associate the start and end of a spike with an intersection by the membrane potential of a unit value with both positive and negative speed. We introduce the class P_r of continuous functions $\psi(s)$ for $s \in [-1, 0]$, for which $0 < \psi(s) \le \exp(-r\lambda)(r > 0)$. These functions are uniformly exponentially small at

$\lambda \to \infty$. Let Eqs. (2.2.1) and (2.2.2) have the initial conditions $u_1 = \varphi(s) \in S$ and $u_2 = \psi(s) \in P_r$. Thus, the first spike of the neuron starts at zero moment. Assume that g_1 and g_2 are independent of λ constants. Then, at $\lambda \to \infty$, the spike of the first neuron begins at the time moment $\xi = O((\ln \lambda)/\lambda)$, i.e., almost instantly. For each of the functions, $u_1(t)$ and $u_2(t)$ are valid asymptotic formulas (1.6.6) and (1.6.9) of Theorem 1.6.1. A new delay of the start of the spike of the second neuron, relative to the first one, is the value $O((\ln \lambda)/\lambda)$.

A completely analogous picture is observed at the electric interaction between neurons, as described by the system of Eqs. (2.2.3) and (2.2.4), if we assume that the value $u_p > 0$ is independent of λ.

Let us discuss another aspect of the electrical interaction model. A neuron is a distributed system. The membrane potential has some effective characteristics. The contact zone can be relatively small. In this case, one can and should consider the coefficients of the conductivities to be small and consistent with the parameter λ. Now let us turn to the system of (2.2.1) and (2.2.2).

Thus, model representations about electric synapses explain their role in the synchronization of neurons. This is consistent with biological facts. Neurophysiologists have considered that "electric conducting is more convenient in cases when it is necessary to synchronize the electrical activity of some nerve cells or to encompass excitation of several cells" (cited in [172], p. 175).

Let $g_1 = d_1 \exp(-\lambda\sigma), g_2 = d_2 \exp(-\lambda\sigma)(\sigma > 0)$ and assume that $u_1 = \varphi(s) \in S$ and $u_2 = \psi(s) \in P_r$. We consider that constant r, allocating the class P_r, satisfies the inequality $r > \sigma$. Denote the start of the second spike of a neuron by ξ. We restrict ourselves to the case when $0 < \sigma < \alpha_1$, where $\alpha_1 = f_K(0) - 1$ [given by formula (1.6.3)], and $\alpha < \alpha_1$ [α is given by formula (1.5.8)]. Because $\lambda \to \infty$ is a valid asymptotic representation, $\xi = \sigma/\alpha_1 + o(1)$. If under the same assumptions we consider the model of electrical synapse, as described by the system of Eqs. (2.2.3) and (2.2.4), then at $\lambda \to \infty$, the moment of the start of a spike of the second neuron is $\xi = \sigma/\alpha + o(1)$.

In such a way, a spike of the second neuron occurs across time, which renders it practically independent of the state in which it was before the action of the first spike of the neuron. According to model representations, the electrical synapse has been proven to be very reliable in signal transmission. The same argument, based on experimental data, is given by J. Eccles [170].

We give an example of numerical study of the system of Eqs. (2.2.1) and (2.2.2). Write $f_{Na}(u) = R_1 \exp(-u^2), f_K(u) = R_2 \exp(-u^2)$, thus concretizing the entrance of the system functions. With such choice as theirs for the equation of a single neuron, the conditions of Theorem 1.6.1 are fulfilled if, of course, $\alpha = R_2 - R_1 - 1 > 0$. Take $R_2 = 2.2, R_1 = 1, \lambda = 3$. Even with this relatively small value of λ, the solutions of the equation of the individual neuron have a pulse structure as shown in Fig. 1.2. We note that the value of the period of the solution is in good agreement with the statement of Theorem 1.6.1. Numerical calculation gives the value of a period of approximately 12, and by Theorem 1.6.1 we obtain 13.2.

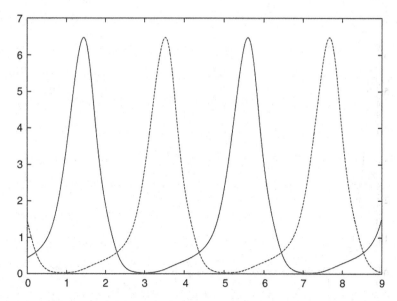

Fig. 2.1 Symmetric mode of system (2.2.1) and (2.2.2) for $G = 0.03, \lambda = 3, R_1 = 1, R_2 = 2.2$. The period of functions $u_1(t), u_2(t)$ is considerably less than that at $G = 0$.

Write (2.2.1) $g_1 = g_2 = G$. At fixed values of the parameters R_1, R_2, λ in (2.2.1) and (2.2.2), we vary the coupling coefficient G. For its small values (G of the order 0.01–0.1), the homogeneous solution ($u_1(t) \equiv u_2(t)$) of systems (2.2.1) and (2.2.2) is unstable, or perhaps the domain of its attraction is small. Solutions with nonoverlapping time pulses are stable formations. Graphs $u_1(t)$ and $u_2(t)$ for $G = 0.03$ are shown in Fig. 2.1. It is easy to see that the period of these functions is much less than that at $G = 0$. The corresponding phase trajectory $(u_1(t), u_2(t))$ on the plane (u_1, u_2) is shown in Fig. 2.2. It is located at the first quadrant, and it is symmetric with respect to its bisector. Later, the existence of such modes will be shown by asymptotic methods.

With an increase of the coefficient G from the symmetric modes, two non-symmetric modes arise. The phase trajectory of one of them at $G = 0.1$ is shown in Fig. 2.3. In this case, the symmetric mode itself loses stability and possibly disappears. In this situation, the pulse of one neuron precedes for a time the spike of another neuron, i.e., there is a causal relationship between discharges. The period of the nonsymmetric modes is also much smaller than the period of the solution of the equation of the isolated neuron. This is due to two factors that have biological meaning. Let the spike of the first neuron anticipate the spike of the second neuron, which is induced and therefore begins earlier. Simultaneously, the spike of the second neuron falls on the descending part of the spike of the first neuron and increases the minimum of its membrane potential (i.e., it prevents the membrane's strong polarization). As a result, the next spike of the first neuron occurs earlier.

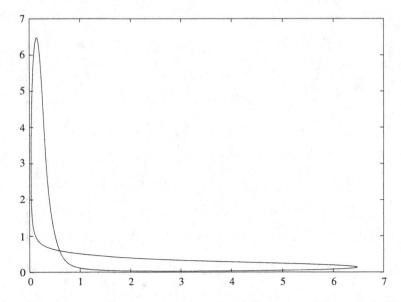

Fig. 2.2 Phase trajectory of the symmetric mode of the system (2.2.1) and (2.2.2) for $G = 0.03, R_1 = 1, R_2 = 2$

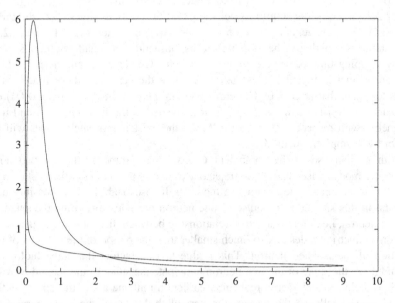

Fig. 2.3 Phase trajectory of the nonsymmetric mode of the system (2.2.1) and (2.2.2) at $G = 0.1 (\lambda = 3, f_{Na}(u) = R_1 \exp(-u^2), f_K(u) = R_2 \exp(-u^2)$, where $R_1 = 1, R_2 = 2.2)$

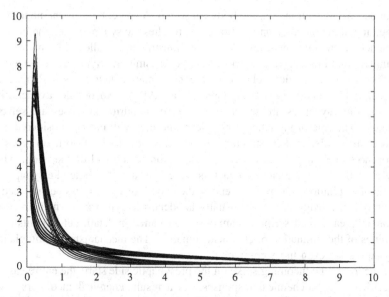

Fig. 2.4 Phase portrait on the plane u_1, u_2 of multi-winding attractor of system (2.2.1) and (2.2.2) at $G = 0.2(\lambda = 3, f_{Na}(u) = R_1 \exp(-u^2), f_K(u) = R_2 \exp(-u^2)$, where $R_1 = 1, R_2 = 2.2)$

A further increase of G leads to the fact that the causal mode impulses undergo a series of transformations and give rise to a multi-winding attractor, the phase portrait of which is shown for the value $G = 0.2$ in Fig. 2.4. Visually, this attractor can be described as follows. Let the spike of the second neuron immediately follow the spike of the first neuron. At the next cycle of pulses, the time between the maxima peaks will be longer, but after a series of cycles the pulse of the second neuron anticipates the discharge of the first neuron, and thus the cause and effect are exchanged. Then the process goes in the reverse direction. Note that at large values of the parameter $\lambda(\lambda \approx 5)$, a multi-winding attractor can not be detected. Apparently, its existence is the effect of a finite parameter.

The described attractor at $G = 0.2$ coexists with a stable homogeneous mode. The increase of the coefficient G (for example, $G = 0.3$) leads to the attractor, thus becoming part of the transition process to a stable homogeneous mode.

2.3   Model of Chemical Synapse

A classic presentation of the mechanisms of interaction of neurons is associated with processes taking place in the chemical synapses. In contrast to electrical synapses, chemical synapses have a unidirectional effect. As already noted, the neuron is a distributed formation. A spike, relatively speaking, is generated in the central part of the neuron, i.e., the body (in soma). A spike, born by the neuron and propagating along the axon and its branches, reaches the synapses, i.e., the intervals of contacts with other neurons. A neuron-transmitter is called the "presynaptic transmitter" and the receiver the "postsynaptic transmitter." Synapses are classified according to their location relative to the postsynaptic neuron (see Pappas and Waxman 1973; Eckert et al. 1991; Green et al. 1993). Axosomatic, axodendritic, and axoaxonic synapses are selected. The dendrodendritic synapses are selected separately. The role of the latter is not clear, and they will not be considered here. Synapses also differ in their structure (Kositsyn 1976, 1993): Two main types are "button hole like" and "spinulose." The extension of the end of the axon, which forms a relatively large area of contact, is typical for a button hole like synapse. In the case of a spinulose synapse, the end of the axon looks as if it has been placed on the postsynaptic ledge (these are usually axodendritic synapses), which increases the contact area. This description cannot be considered in detail, but it points to the difficulties of the simulation of chemical signaling. The mechanisms taking place in the synapses are even more complex.

The action of the incoming spike on the presynaptic side starts the entire cascade of electrical and biochemical responses. As a result, chemical mediators called "neurotransmitters," which are enclosed in the vesicles, are liberated (see Pappas and Waxman 1973; Eckert et al. 1991; Green et al. 1993). The most widespread neurotransmitter is acetylcholine. The amount of neurotransmitter, liberated in the presynaptic side in response to a spike, rapidly increases, relatively stabilizes, and then is rapidly destroyed. Vesicles of neurotransmitters cross the synaptic cleft and act on the membrane of a postsynaptic neuron, which starts a new cascade of biochemical responds activating (or inactivating) ion channels. This results in local changes of the membrane potential of a postsynaptic neuron. These deviations influence the dynamics of the membrane potential in general.

Experiments (and their interpretation) are usually performed on biological neurons, the membrane potential of which is at rest u_0. Arising as a result of neurotransmitter action, currents shift the postsynaptical membrane potential to the new level, the so-called "reversal potential" u_{rev} (see Eckert et al. 1991). In experiments, the equilibrium potential u_0 can be controlled. It has long been clear that regardless of the sign $u_0 - u_{rev}$, the shift of the membrane potential occurs in the direction of u_{rev}, which is completely determined by the mechanism of the neurotransmitter's action.

To avoid misunderstanding, we note once more that we count the potential value of u from the minimum possible value, which is negative and large in absolute value. Thus, at $u > 0$, u indicates a decrease in the polarization of the membrane, and an increase indicates depolarization. If the value is $u_{rev} < u_0$, then the neurotransmitter's action polarizes the membrane. If $u_{rev} > u_0$, membrane depolarization, which can cause a spike, takes place. In accordance with this, the neurotransmitter's action is classified as inhibitory or excitatory. It is easy to see that the same mechanism of the neurotransmitter (one and the same u_{rev}) can lead to the inhibition of both neurons as well as the generation of spikes depending on the sign of the difference $u_0 - u_{rev}$. This has been observed in protozoa. For some neurons, the neurotransmitter acetylcholineis excitatory, and for others it is inhibitory (see Eckert et al. 1991). One of the nearest tasks is to explain the described phenomena in the context of the model. Before doing so, it is also necessary to cite a series of facts reflecting the peculiarities of neurotransmitter action. It is believed that during a spike, a neurotransmitter has almost no influence on the dynamics of the membrane potential of the postsynaptic neuron. The view of several researchers (Katz 1968; Blum et al. 1988; Eckert et al. 1991; Plonsi and Barr 1992) on the role of the neurotransmitter after a spike is different. Some believe that the presence of the neurotransmitter at a certain time interval after the spike does not influence the dynamics of the potential. Others say that as a result of the action of the neurotransmitter, a new spike of the neuron can begin. The time interval in which the neuron is not sensitive to the action of neurotransmitter is called the "zone of refractoriness." By unanimous opinion, it is considered that at the time of spike generation, the neuron enters a zone of refractoriness. It is possible that the answer to the question of refractoriness of a neuron after a spike depends on the type of neuron involved.

Note that the functioning of the mechanism of neurotransmitter action assumes the presence of delays. However, from biological data (Eccles 1966), these delays are usually not large. It is not difficult to take into account the synaptic delays in the model.

We also note that in the opinion of several researchers (Pappas and Waxman 1973; Eckert et al. 1991), small presynaptic fibers can excite large postsynaptic cells at chemical transmission of the signal. Regarding this connection, even in a model of the simplest situation, the action of one neuron on another through a single synapse is of interest.

Let us turn to the model of synaptic action. Consider two neurons, one of which acts on the second through a single synapse. Let the membrane potential of the postsynaptic neuron be denoted by u and that of the presynaptic one by ω. For simplicity, we assume that the evolution of the membrane potential ω of presynaptic neuron is given as $\lambda \to \infty$ by the asymptotic formulas of Theorem 1.6.1. We neglect the propagation delay of the nerve impulse along the axon. For the neurotransmitter, we could write one more equation, but this would complicate the model. We proceed differently. As already mentioned, the amount of neurotransmitter released from the presynaptic neuron spike on some torus-time interval is relatively stable. We assume that on this interval with the duration T_ν the amount of neurotransmitter is constant. This is the usual method of replacing the variable value with some effective value. Represent the amount of the neurotransmitter $V(t)$ in the functional form $V(t) = \ell(\omega(t+s))$, where $s \in [-T_\nu, 0]$, $\ell(*)$ is a continuous nonlinear functional defined on the space of continuous functions $C[-T_\nu, 0]$. We describe the functional $\ell(*)$, corresponding to the essence of the problem. Choose the constant ω_0, i.e., the level of potential of the presynaptic neuron, the exceedance of which begins the release of the neurotransmitter. Associate the start and end of a spike of the presynaptic neuron with moments of time when the potential $\omega(t)$ also crosses the value is ω_0, respectively, with positive and negative speed. According to Theorem 1.6.1, the duration of the spike is close as $\lambda \to \infty$ to T_1. From the viewpoint of asymptotic theory, in this case the number ω_0 is arbitrary, and we can write $\omega_0 = 1$.

Let t_v be the time of the start of a spike of the presynaptic neuron and the effective time of the presence of the neurotransmitter be $T_v > T_1$. Let on the interval $t \in [t_v, t_v + T_*]$, where $T_* > T_v$ is not observed, be a new spike of the presynaptic neuron. The functional $\ell(\omega)$ is written out from simple considerations: For the time moment $t \in [t_v, t_v + T_*]$, its value is equal to V_0 (effective constant) if on the interval $[t - T_v, t]$ there are points t_*, where $\omega(t_*) > 1$ but in the absence of such points the value of functional is $\ell(\omega) = 0$. The specific form of the functional may be different. Let us consider, for example, the following method of presentation. As in Sect. 2.2 of this chapter, introduce the function $\chi(\omega)$ with the following properties: $\chi(\omega) = 0$ as $\omega < 0$, $\chi(\omega) = \omega$ at $0 < \omega < 1$ and $\chi(\omega) = 1$ as $\omega \geq 1$. Denote $\chi_\varepsilon(\omega) = \chi(\omega/\varepsilon)$ by $\chi_\varepsilon(\omega)$, where positive ε is sufficiently small. Note that pointwise $\varepsilon \to 0$, function $\chi_\varepsilon(\omega) \to \theta(\omega)$, where $\theta(\omega)$ is the function of Heaviside. Now we write:

$$l(\omega) = V_0\chi_\varepsilon(\chi_\varepsilon(\omega(t) - 1) + \chi_\varepsilon(\omega(t - (T_1 - \varepsilon_1)) - 1) \\ + \cdots + \chi_\varepsilon(\omega(t - k(T_1 - \varepsilon_1)) - 1) - 0.5), \tag{2.3.1}$$

where $0 < \varepsilon_1 < T_1$. For $T_v > T_1$, integer k and real ε_1 can be ordered so that $\ell(\omega) = V_0$ at $t \in [t_v + \delta, t_v + T_v - \delta]$ and $\ell(\omega) = 0$, if $t_v + T_v < t < T_*$. Here, δ is an arbitrarily small number. It is sufficient to require $(k+1)T_1 + k\varepsilon_1 = T_v$.

If the effective time of action of the neurotransmitter $T_v < T_1$ (T_1 is asymptotic for the duration of spike), then we can take:

$$\ell(\omega) = V_0\chi_\varepsilon(\omega(t) - 1)\chi_\varepsilon(1 - \omega(t - T_v)). \tag{2.3.2}$$

In model representations (2.2.1) and (2.2.2), the stabilization of the amount of neurotransmitter released in response to the spike of the presynaptic neuron occurs in time $O(\varepsilon/\lambda)$, i.e., it occurs quickly. One can pass to the limit as $\varepsilon \to 0$. As a result, we obtain the expression for the functional $\ell(\omega)$ by way of the Heaviside function $\Theta(*)$:

$$\ell(\omega) = V_0\Theta(\Theta(\omega(t) - 1) + \Theta(\omega(t - (T_1 - \varepsilon_1)) - 1) \\ + \cdots + \Theta(\omega(t - k(T_1 - \varepsilon_1)) - 1) - 0.5), \tag{2.3.3}$$

$$\ell(\omega) = V_0\Theta((\omega(t) - 1)\Theta(1 - \omega(t - T_v)), \tag{2.3.4}$$

respectively, for the cases $T_v > T_1$ and $T_v < T_1$. Below we assume that $V_0 = 1$ and that the presentation

$$V(t) = \ell(\omega) \tag{2.3.5}$$

considers the presence of the neurotransmitter as the indicator of the presence of a mediator. In case of necessity, we introduce normalization factors.

Consider the postsynaptic neuron in the time interval of the effective presence of the neurotransmitter $(V = 1)$. In the balanced equation, we should add the following term taking into account the currents initialized by neurotransmitter:

$$c\dot{u} = I_{Na} + I_K + I_v, \qquad (2.3.6)$$

where $c\dot{u}$ is the capacitive current, I_{Na} is the sodium current, and I_K is the potassium current. Finally, I_v is the current flowing through ion channels activated by the neurotransmitter. In full analogy with Sect. 1.5, we write $I_v = \chi^*(u, V)u$. Because the neuron during a spike is refractory to the effects of neurotransmitter, then $\chi^*(u, V) \to 0$ as $u \to \infty$. If there is no neurotransmitter, current I_v is not observed, i.e., $\chi^*(u, 0) = 0$. Following the steps described in Sect. 1.5, change Eq. (2.3.6) of the current balance to the following form:

$$\dot{u} = \lambda[-1 - f_{Na}(u) + f_K(u(t - 1)) + \chi_v(u, V)]u, \qquad (2.3.7)$$

where $\chi_v(u, V) = \chi^*(u, V)(h/c(b - a))$. The meaning of all other parameters and functions, as well as their expected properties, are explained in the analysis of Eq. (1.5.7).

Explain within the framework of Eq. (2.3.7) the appearance of the above-mentioned reversal of potential under the neurotransmitter's action. Physiological experiments [172] to study the neurotransmitter's action can be interpreted as follows. Special agents inhibit the potentially dependent (i.e., controlled by the membrane potential of the neuron) sodium and potassium channels, which ensure the functioning of the neuron under normal conditions. At the same time, the neurotransmitter-dependent ion channels are not influenced. In the presence of the neurotransmitter, the membrane potential achieves some stable level u_{rev}. In the situation reflecting the experimental conditions, Eq. (2.3.7) takes the form:

$$\dot{u} = \lambda \chi_v(u, V)u$$

and it must have stable equilibrium state $u_{rev} > 0$. Then in some neighbourhood of the equilibrium state, $\chi_v = g_v(u_{rev} - u)$, where g_v is the coefficient of conductivity.

In the absence of a neurotransmitter $(V = 0)$, let Eq. (2.3.7) have a stable equilibrium state u_0, i.e., the membrane potential is at rest. In the presence of the neurotransmitter, Eq. (2.3.7) takes the form:

$$\dot{u} = \lambda[-1 - f_{Na}(u) + f_K(u(t - 1)) + g_v(u_{rev} - u)]u.$$

If $u_{rev} < u_0$, then $1 - f_{Na}(u_0) + f_K(u_0) + g_v(u_{rev} - u_0) < 0$. As a result, the equilibrium state of this equation shifts toward u_{rev}. As noted, this phenomenon has been observed in biological experiments. If $u_{rev} > u_0$, then the equilibrium state increases shifting to u_{rev} or disappearing, thus leading to the generation of a spike. This is also consistent with biological data. Thus, the reversal potential is explained in terms of Eq. (2.3.7).

Discuss other aspects of Eq. (2.3.7) of the model neuron, which is under neurotransmitter action. It is necessary to clarify the properties of function $\chi_v(u, V)$. We
state here a biological fact: If the membrane is strongly polarized (i.e., u is small),
then the neurotransmitter has almost no effect on the neuron. In this connection, we
write:

$$\chi_v(u, V) = GV(t)f_v(u)\chi_\varepsilon((u - u_*)/u_*), \qquad (2.3.8)$$

where $\chi_\varepsilon(*)$ is the function defined before formula (2.3.1), and u_* is the critical
value of the membrane potential such that at $u < u_*$ the neurotransmitter has no
effect on the neuron. We consider that the threshold value is exponentially small for
$\lambda : u_* = \exp(-\lambda p)$. Here, $0 < p < \alpha_2$, where the number $\alpha_2 = f_{Na}(0) + 1$ (formula
(1.6.3)). Assume that in (2.3.8), the positive function $f_v(u) \to 0$ occurs as $u \to \infty$.
One can consider that $f_v(0) = 1$. Coefficient G characterizes the efficiency of the
synapse. Call it the "synaptic weight." Below we find out that at $G > 0$, we naturally consider the synapse to be excitatory; if $G < 0$, we consider the synapse is
called inhibitory. Note that in (2.3.8), instead of function $\chi_\varepsilon((u - u_*)/u_*)$, in some
cases we can take the Heaviside function: $\Theta(u - u_*)$.

Assume that in Eq. (2.3.7), functions $f_{Na}(u)$ and $f_K(u)$ are positive and quickly
become $f_{Na}(u) \to 0$ and $f_K(u) \to 0$ as $u \to \infty$. For Eq. (2.3.7), in the absence of
action $(V = 0)$, the conditions of Theorem 1.6.1 are fulfilled. In the determination
of class S of the initial functions, number $\gamma > 0$ is arbitrary. Associate the start and
end of a spike with moments of time when the membrane potential crosses the
value γ with both positive and negative speed, respectively.

Denote by $u_v(t, \varphi)$ the solution of Eq. (2.3.7) with the initial condition
$u_v(s, \varphi) = \varphi(s) \in S$. Let t_v be the beginning of the release of the neurotransmitter,
i.e., function $V(t) = 1$ at $t \in [t_v, t_v + T_v]$ and $V(t) = 0$ at $t \notin [t_v, t_v + T_v]$.

Suppose that $t_v \le 0$ and $t_v + T_v > T_1 + A(A > 0)$, i.e., the transmitter appeared
later than the spike began and decays later (let significantly $A \ge 1$), which should
end the spike. Then because $\lambda \to \infty$ uniformly with respect to $\varphi \in S$, the specified
time intervals are valid asymptotic formulas:

$$u(t, \varphi) = \exp \lambda\alpha_1(t + o(1)), \quad t \in [\delta, 1 - \delta],$$

$$u(t, \varphi) = \exp \lambda(\alpha_1 - (t - 1) + o(1)), \quad t \in [1 + \delta, T_1 - \delta],$$

coinciding with formulas (1.6.6) and (1.6.7) from Theorem 1.6.1. As expected, in
the the interval of spike generation, the neurotransmitter practically does not act on
the neuron.

For $T_1 - \delta < t < t_v + T_v$ the situation strongly varies. Let $G > 0$. The asymptotics
for solution $u_v(t, \varphi)$ are defined mainly from following equation:

$$\dot{u} = \lambda[-1 - f_{Na}(u) + Gf_v(u)\chi_\varepsilon((u - u_*)/u_*)]u. \qquad (2.3.9)$$

It may happen that this equation has no positive equilibrium states or that there may be no such states. It is easy to see that the latter case occurs, for example, if $G > \alpha_2 = 1 + f_{Na}(0)$.

First, let us consider the case when there are no positive equilibrium states. Additionally, let $\alpha_2 > G > \alpha_2 - p$. Then, on that part of the interval $t \in [T_1 + \delta, t_v + T_v]$, where $u_v(t, \varphi) < \gamma$, the following formula is valid:

$$u_v(t, \varphi) = u(t, \varphi) \exp(\lambda G(t - T_1 + 0(1))).$$

Here, $u(t, \varphi)$ is the solution of Eq. (1.5.7) for the neuron without external action (the asymptotics of this solution reveal Theorem 1.6.1). Function $u(t, \varphi)$ on the considered interval is exponentially large in comparison with $u(t, \varphi)$. The action should be called "excitatory." The picture is consistent with biological data: In some cases, excitatory synapses after a spike prevent hyperpolarization of the membrane, i.e., they have a residual depolarizing action. Note also that the solution $u(t, \varphi)$ was cross-linked from several exponentials. This corresponds to the ideology of the exponential model of approximation.

Now let $0 < G < \alpha_2 - p$. Introduce the following numbers:

$$T_p^1 = T_1 + p/(\alpha_2 - G), \quad T_p^2 = T_1 + 1 + \alpha_2(1 - p/(\alpha_2 - G))/\alpha.$$

If the time duration of the presence of the neurotransmitter is sufficiently long, then as $\lambda \to \infty$, the asymptotic formulas hold true:

$$u_v(t, \varphi) = \exp[-\lambda(\alpha_2 - G)(t - T_1 + o(1))], \quad t \in \left[T_1 + \delta, T_p^1 - \delta\right],$$

$$u_v(t, \varphi) = \exp\left[-\lambda\left(p + \alpha_2\left(t - T_p^1\right) + o(1)\right)\right], \quad t \in \left[T_p^1 + \delta, T_1 + 1 - \delta\right],$$

$$u_v(t, \varphi) = \exp\lambda\left[\begin{array}{c} \alpha(t - T_1 - 1) - \\ -p - \alpha_2\left(T_1 + 1 - T_p^1\right) + o(1)) \end{array}\right], \quad t \in \left[t_1 + 1 + \delta, T_p^2 - \delta\right].$$

Furthermore, at $T_p^2 + \delta < t < t_v + T_v$, as $u_v(t, \varphi) < \gamma$, holds representation

$$u_v(t, \varphi) = \exp\lambda\left[(\alpha + G)\left(t - T_p^2\right) - p + o(1)\right].$$

As before, the neurotransmitter stimulates action: The values of function $u_v(t, \varphi)$ are exponentially greater as $\lambda \to \infty$ in comparison with $u(t, \varphi)$. However, the ratio $u_v(t, \varphi)/u(t, \varphi)$ has become smaller. This is natural because the synaptic weight of G is less. Note that at moments of time $T_p^1 + o(1)$ and $T_p^2 + o(1)$, solution $u_v(t, \varphi)$ crosses the threshold value $u_* = \exp(-\lambda p)$, respectively, with both negative and positive speed. In the interval of time $t \in [T_p^1 + o(1), \in [T_p^2 + o(1)]$, the neurotransmitter does not influence the membrane potential as shown by the above-mentioned formulas. They do not include the synaptic coefficient G for this interval. We note the obvious fact that the excitatory action approximates the start of the next spike of the neuron.

Consider now the second case when Eq. (2.3.9) has positive stable equilibrium states. Let u_v be the largest of these. Let now assume that the number γ, which singles out the zone of spike, is larger than u_v.

In the interval of time $t \in [T_1 + \delta, T_1 + 1 - \delta]$, the solution $u_v(t, \varphi)$ of Eq. (2.3.7) is close to the equilibrium state u_v of Eq. (2.3.7). In turn, for $t \in [T_1 + 1 + \delta, T_1 + 2 - \delta]$, the solution $u_v(t, \varphi)$ asymptotically differs little from the solution of the following equation

$$\dot{u} = \lambda[-1 - f_{\mathrm{Na}}(u) + f_{\mathrm{K}}(u_v) + Gf_v(u)]u \tag{2.3.10}$$

with initial condition $u + (T_1 + 1) = u_v$. We assume that the solution $u(t)$ grows without limit when $t \to \infty$. In addition, let number

$$\alpha_v = f_{\mathrm{K}}(u_v) - 1 > 0.$$

Then Eq. (2.3.10) asymptotically integrates as $\lambda \to \infty$ for $t \in [T_1 + 1 + \delta, T_1 + 2 - \delta]$. Such step-by-step actions can easily be continued further. Let us formulate the results of the calculations as follows.

Lemma 2.3.1 *Under these conditions, uniformly with respect to t from the specified periods, and $\lambda \to \infty$ hold the formulas:*

$u_v(t, \varphi) = u_v + o(1), \quad t \in [T_1 + \delta, T_1 + 1 - \delta],$

$u_v(t, \varphi) = u_v \exp[\lambda \alpha_v(t - T_1 - 1 + o(1))], \quad t \in [T_1 + 1 + \delta, T_1 + 2 - \delta],$

$u_v(t, \varphi) = u_v \exp[-\lambda(t - T_1 - 2 - \alpha_v + o(1))], \quad t \in [T_1 + 2 + \delta, T_1 + 2 + \alpha_v - \delta],$

$u_v(t, \varphi) = u_v + o(1), \quad t \in [T_1 + 2 + \alpha_v + \delta, T_1 + 3 + \alpha_v - \delta].$

This lemma proves the existence of solutions of a specific type for Eq. (2.3.7), which are the result of sufficiently long and intensive external actions. For this type, it is characteristic that the time interval between the end of one spike and the start of the next spike is close to one. Thus, the pulses follow often and form the burst. Therefore, this type of activity has been named "above bursting." In the burst, the amplitude of the spikes, beginning with the second spike, is asymptotically close to $\exp(\lambda \alpha_v)$. In general, it differs from the amplitude of the first spike, which is close to $\exp(\lambda \alpha_1)$. The duration of the spikes starting from the second spike asymptotically differs little from the number $T_1^v = \alpha_v + 1$, which also differs from the asymptotic duration $T_1 = \alpha_1 + 1$ of the first spike in the burst. For nerve cells, the bursting type of response to stimulation often takes place (Khodorov 1975).

The existence of the bursting type of activity is associated with the length of the refractory period. According to some biologists (Khodorov 1975), a neuron has a short interval of increased excitability after a spike, which leads to the emergence of the burst response. Other investigators (Green et al. 1993) consider that there is no such period of increased excitability. A spike plus a certain time interval after it form the zone of refractoriness. It is possible that we are talking about different types of neurons. The model of the synapse, in which the refractory period continues after the spike, will be considered here and used in subsequent chapters.

The results formulated previously allow us to understand the model the role of the exciting action. In its absence of an exciting action $(G = 0)$, a neuron rarely spontaneously generates spikes. Relatively weak excitatory action only decreases interspike intervals. Strong excitatory action gives rise to burst activity. All of this follows from the above-presented analysis of the model adopted by us, but simultaneously it is fully consistent with biological data.

Thus, the considered model of the neuron has the properties of the neuron as detector, which is the result of functioning, the expression of which is burst activity. Detection proceeds according to the force and the duration of the external action. At the same time, excitation must enroll on quite definite stage of the internal processes of the neuron for it to start functioning. The neuron should capture the spike zone. If a neurotransmitter is released in a moment of time t_v when $u(t_v) < u_*$, then initially the neuron would not feel the presence of the neurotransmitter. Thus, the neuron follows the time of the appearance of the signal.

A burst response to a sufficiently strong excitatory action is observed if the solution $u(t)$ of Eq. (2.3.10) with initial condition $u(T_1 + 1) = u_v$ grows unboundedly with increasing t. If this solution is bounded, then as $\lambda \to \infty$, it tends very rapidly to the equilibrium state of Eq. (2.3.10). The solution $u_v(t, \varphi)$ of Eq. (2.3.7) for $t \in [T_1 + 1 + \delta, T_1 + 2 - \delta]$ will be close to the same equilibrium state. In such a way, we will observe the process of the walking of the membrane potential according to the system of ledges. On each ledge, the membrane potential will be delayed during the time interval close to one.

In the case when the functions $f_{Na}(u)$ and $f_K(u)$ in Eq. (2.3.7) satisfy the conditions of Theorem 1.6.2, the result of action of the exciting neurotransmitter is, in many respects, analogous to that already described. The interval of spike generation is the zone of refractoriness. All of the above formulas are saved on the interval of time after the spike, and these are valid conclusions to draw from them. Lemma 2.3. 1, which describes burst neuron activity observed at sufficiently large value of synaptic weight G, is held. Differences arise during the interval of time in which, in the absence of mediator action, the membrane potential walks the system of ledges. The presence of the neurotransmitter changes the system of steps and can cause a premature spike. Here, the specific type of functions $f_{Na}(u)$, $f_K(u)$ and $f_v(u)$. are important. The sequence of steps is defined recurrently in the same way as in Theorem 1.6.2, of course taking into account that changes that have appeared on the right side of equation.

Note that simulation of the exciting action of a neurotransmitter occurs when in Eq. (2.3.7) instead of $\chi_\varepsilon((u - u_*)/u_*$, we use the Heaviside function $\Theta(u - u_*)$.

Let us turn to the simulation results of the action of an inhibitory neurotransmitter on the neuron. Let in Eq. (2.3.7) the positive functions $f_{Na}(u)$ and $f_K(u)$ sufficiently and rapidly tend to zero as $u \to \infty$. Number $\alpha = f_K(0) - f_{Na}(0) - 1 > 0$. As before, we assume that $f_v(u) > 0, f_v(0) = 1$ and $f_v(u) \to 0$ as $u \to \infty$; however, unlike in the previous equation, the synaptic weight is $G < 0$. Let the beginning of the release of the neurotransmitter be $t_v < 0$, and let the time of its effective action T_v be sufficiently long. Recall that in Eq. (2.3.7) $V(t) = 1$ at $t \in [t_v, t_v + T_v]$,; otherwise, this function takes the value zero.

In the problem of action of the inhibitory neurotransmitter, we consider a number of different cases. Let, as above, P_r be the class of continuous on the interval $s \in [-1, 0]$ of functions $\psi(s)$, for which $\psi(s) \leq \exp(-\lambda r)$, where $r > 0$, i.e., these functions are uniformly exponentially small as $\lambda \to \infty$. Denote by $u_v(t, \psi)$ the solution of Eq. (2.3.7) with the initial condition $u_v(s, \psi) = \psi(s) \in P_r$.

Lemma 2.3.2 *Let $G < 0$ and $|G| > \alpha$. Then as $\lambda \to \infty$ for all $t \in [0, t_v + T_v]$, we have the inclusion $u_v(t + s, \psi) \in P_r$. If the time of the effective action of neurotransmitter $T_v \to \infty$, then $u_v(t, \psi) \to u_*(1 + \varepsilon \alpha/|G|)$ as $t \to \infty$.*

From a biological point of view, the meaning of Lemma 2.3.2 is clear. If the membrane is sufficiently strongly polarized, the presence of a highly active inhibitory neurotransmitter does not allow the membrane to leave this state. Moreover, if the action of the inhibitory neurotransmitter is continuous, then an equilibrium state of the membrane potential is established. This is caused by a compromise between the neural processes leading to depolarization as well as by external inhibition. Effective braking action, according to Eccles (1971), can be quite continuous.

If $|G| < \alpha$, then sooner or later, even in the presence of the inhibitory neurotransmitter, the membrane of the neuron will be released from the state of strong polarization. This means that with time, the value is $u_v(t, \psi) = O(1)$. Let in Eq. (2.3.7) the functions $f_{Na}(u)$ and $f_K(u)$ be positive and monotonically tend to zero as $u \to \infty$. For the neuron in the absence of action, the case is covered by Theorem 1.6.1. Introduce number $\alpha_* = \alpha - |G| > 0$ and arbitrary number $\gamma > 0$. Associate the spike with the interval of time where $u_v(t, \psi) > \gamma$. Denote by S_* the set of continuous on the interval $s \in [-1, 0]$ functions $\varphi_*(s)$, for which $\varphi_*(0) = \gamma$ and $\psi(s) \leq \gamma \exp(\lambda \alpha_* s/2)$. Consider the solution $u_v(t, \varphi_*)$ of Eq. (2.3.7) with the initial condition $u_v(s, \varphi_*) = \varphi_*(s) \in S$ (i.e., the spike starts at zero time). Let its end be in a moment of time $t_1 \varphi_*$, and $t_2 \varphi_*$ be the beginning of the next spike. Denote by $t_p^1(\varphi_*)$ and $t_p^2(\varphi_*)$ the moments of time when the solution $u_v(t, \varphi_*)$ crosses the value $u_* = \exp(-\lambda p)(0 < p < \alpha_2$ with both negative and positive speeds. Consider the time of the effective action of the neurotransmitter T_v to be sufficiently continuous. Write:

$$T_p^1 = T_1 + p/(\alpha_2 + |G|),$$

$$T_p^2 = T_1 + 1 + \alpha_2(1 - p/(\alpha_2 + |G|))/\alpha,$$
$$T_{2v} = T_p^2 + p/(\alpha - |G|).$$

Lemma 2.3.3 *As $\lambda \to \infty$ hold asymptotic equalities:*

$$t_1(\varphi_*) = T_1 + o(1), \quad t_2(\varphi_*) = T_{2v} + o(1), \tag{2.3.11}$$

$$t_p^1(\varphi_*) = T_p^1 + o(1), \quad t_p^2(\varphi_*) = T_p^2 + o(1), \tag{2.3.12}$$

$$u_v(t, \varphi_*) = \exp \lambda \alpha_1 (t + o(1)), \quad t \in [\delta, 1 - \delta], \tag{2.3.13}$$

$$u_v(t, \varphi_*) = \exp \lambda(\alpha_1 - (t - 1) + o(1)), \quad t \in [1 + \delta, T_1 - \delta], \tag{2.3.14}$$

$$u_v(t, \varphi_*) = \exp[-\lambda(\alpha_2 + |G|)(t - T_1 + o(1))], \quad t \in [T_1 + \delta, T_p^1 - \delta], \tag{2.3.15}$$

$$u_v(t, \varphi_*) = \exp\left[-\lambda\left(\alpha_2(t - T_p^1) + p + o(1)\right)\right], \quad t \in \left[T_p^1 + \delta, T_1 + 1 - \delta\right], \tag{2.3.16}$$

$$u_v(t, \varphi_*) = \exp \lambda \left(\begin{array}{c} \alpha(t - T_1 - 1) - \\ -\alpha_2\left(T_1 + 1 - T_p^1\right) - p + o(1) \end{array} \right), \quad t \in \left[T_1 + 1 + \delta, T_p^2 - \delta\right], \tag{2.3.17}$$

$$u_v(t, \varphi_*) = \exp \lambda\left[(\alpha - |G|)\left(t - T_p^2\right) - p + o(1))\right], \quad t \in [T_p^2 + \delta, T_{2v} - \delta]. \tag{2.3.18}$$

From formulas (2.3.13) and (2.3.14) in Lemma 2.3.3, it follows that inhibitory action on the part of a spike does not influence the membrane potential. Immediately after the spike, the presence of the the inhibitory neurotransmitter, by virtue of formula (2.3.15), accelerates the process of polarization (i.e., it increases the absolute value of the negative indicator of the exponent). Time interval $t \in \left[t_p^1(\varphi_*), t_p^2(\varphi_*)\right]$, according to formulas (2.3.16) and (2.3.17), is the interval of resistance to the inhibitory action. In view of formula (2.3.18), the process of polarization is sloweron the interval $t \in [T_p^2 + \delta, T_{2v} - \delta]$, i.e., it decreases the indicator of the exponent approximating the solution. As a result of the inhibitory action, the new spike of the neuron begins later than in the absence of a neurotransmitter. At intervals of susceptibility, the result of the neurotransmitter's action agrees with the ideology of the model of exponential approximation.

Let us now turn briefly to individual aspects of the influence of the inhibitory action when the functions $f_{Na}(u)$ and $f_K(u)$ in Eq. (2.3.7) satisfy the conditions of Theorem 1.6.2. Let in Eq. (2.3.7) $f_v(v) > 0, f_v(0) = 1, G < 0$, and the number $\alpha_* = \alpha - |G| < 0$, i.e., the synaptic weight of the action, not be too large. We assume that the beginning of the neurotransmitter's release is $t_v < 0$ and that the effective time of action is sufficiently large.

Under the conditions of the assumptions made about the functions $f_{Na}(u), f_K(u), f_v(u)$ and the synaptic coefficient G, equation

$$\dot{v} = \lambda[-1 - f_{Na}(v) + f_K(0) + Gf_v(v)]v$$

has a positive equilibrium state v_1^* and a domain of attraction that is always adjacent to zero. Here $v^* < v_1$, where v_1 is the first term of the sequence in Theorem 1.6.2. Recall that the solution of Eq. (1.5.7) for the neuron without external action on the time interval $[\delta, 1 - \delta]$ is close to v_1.

In determining the set S_* of initial conditions for the solutions $u_v(t, \varphi_*)$ of Eq. (2.3.7), choose the number $\gamma < v_1^*$. Then, on the interval of time $t \in [\delta, 1 - \delta]$ as $\lambda \to \infty$, the solution is $u_v(t, \varphi_*) = v_1^* + o(1) < u(t, \varphi)$. Thus, the neurotransmitter exhibits inhibitory properties. From step v_1^* begins the process of the walking of the membrane potential on the system ledges, which is enforced by external action. The process of the walk can be completed or not completed by the spike. This depends on the specific type of functions in Eq. (2.3.7). The algorithm of asymptotic construction of the solution $u_v(t, \varphi_*)$ is completely analogous to that used in the construction of the approximation of the solution of Eq. (1.5.7) in the conditions of Theorem 1.6.2.

References

Blum, F., Leiserson, A., & Hofstefer, L. (1988). *Brain, mind and behaviour*. Moscow: Mir.

Eccles, J. (1966). *The physiology of synapses*. Moscow: Mir.

Eccles, J. (1971). *Inhibitory pathways in the central nervous system*. Moscow: Mir.

Eckert R., Randall D., & Augustine G. (1991). *Animal Physiology* (Vol. 1). Moscow: Mir.

Green, H., Stout, W., & Taylor, D. (1993). *Biology* (Vol. 2). Moscow: Mir.

Hubel D. H., & Wiesel T. N. 1977. Functional architecture of macaque monkey cortex (Vol. 198, pp. 1–59). In *Proceedings of the Royal Society*, London.

Katz, B. (1968). *Nerve, muscle, synapse*. Moscow: Mir.

Khodorov, B. I. (1975). *General physiology of excitable membranes*. Moscow: Nauka.

Kositsyn, N. S. (1976). *The microstructure of the dendrites and axodendritic connections in the central nervous system*. Moscow: Nauka.

Kositsyn, N. S. (1993). Features of the structural organization of nerve cells and interneuronal connections, providing information processing in the central nervous system. *Neurocomputer as the basis of thinking computers* (pp. 10–22). Moscow: Nauka.

Livanov, M. N. (1972). *Spatial organization of the processes of the brain*. Moscow: Nauka.

Nicholls, J. G., Martin, A. R., Wallace, B. J., & Fuchs, P. A. (2003). *From neuron to brain*. Moscow: Editorial URSS.

Pappas, G., & Waxman, S. (1973). Ultrastructure of synapses. *Physiology and pharmacology of synaptic transmission* (pp. 7–30). Leningrad: Nauka.

Plonsi, R., & Barr, R. (1992). *Bioelectricity*. Moscow: Mir.

Singer, W., Tretter, F., & Cynader, M. (1975). Organization of cat striate cortex: a correlation of receptive-field properties with afferent and efferent connections. *Journal of Neurophysiology, 10*(3), 311–330.

Singer, W., Tretter, F., & Cynader, M. (1976). The effect of reticular stimulation on spontaneous and evoked activity on the cat visual cortex. *Brain Research, 102*(1), 71–90.

Chapter 3
Model of Wave Propagation in the Ring Neural Structures with Chemical Synapses

The presence in the brain of closed neural structures has long been observed (Lorento de Nó 1938; Eccles 1966; Rosenblatt 1965), and this points to the importance of the role that closed neural structures may play in ring neural formation in mechanisms of memory. For example, models of neural networks, the architecture of which includes ring elements, are considered in the work of Frolov and Shulgina (1983).

This chapter examines the problems of conducting cyclic excitation through closed structures. Neurons are described by the equations obtained in Chap. 1. They interact with each other by way of chemical synapses, the model representations of which are given in Chap. 2. In this chapter, on the basis of asymptotic analysis of the model, we obtain the values of the synaptic coefficients, which guarantee the existence of waves of activity with predetermined structure in neural formations. We show that oscillatory modes of various types coexist in the system. On the basis of this analysis, we can give interpretations of the known biological phenomena associated with the propagation of excitation, the structure of memory traces, and the problem of attention as well as hypothesize about the quantization of the frequencies of the brain.

3.1 Model of Population of Neurons Connected by Chemical Synapses

Consider a population consisting of N identical neurons interconnected by chemical synapses. Denote the membrane potential of the ith neuron by u_i. Without taking into account interaction, the values u_i obey Eq. (1.5.7). The neuron with the number j acts on the ith neuron by a single chemical synapse. The result of this action is described by Eq. (2.3.7). As we noted in Sect. 2.3, according to biological data (Schade and Ford 1976), in some cases the neuron is refractory to the neurotransmitter's action both during the spike and for some time interval immediately after the spike (i.e., there is no increased excitability). The refractory period (i.e., the period of resistance) is opened by the spike and ends when the membrane potential

© Springer International Publishing Switzerland 2015
S. Kashchenko, *Models of Wave Memory*,
Lecture Notes in Morphogenesis, DOI 10.1007/978-3-319-19866-8_3

crosses with positive speed some threshold value u_*. Below we take into account the phenomenon of refractoriness by modifying the term responsible for the interaction. We assume that the threshold level u_* is identical for all neurons.

Based on model (2.3.7) of synaptic action, the system of equations for the population, consisting of neurons, has the following form:

$$\dot{u}_i = \lambda \left[-1 - f_{Na}(u_i) + f_K(u_i(t-1)) + \sum_{j=1}^{n} \chi_{ij}(u_i, V_j) \right] u_i, \qquad (3.1.1)$$

where $i = 1, \ldots, N$. We suppose that in system (3.1.1), positive functions $f_{Na}(u)$ and $f_K(u)$ monotonically and sufficiently rapidly tend to zero as $u \to \infty$. Assume that speed parameter $\lambda \gg 1$ is large and that the number given by formula (1.5.8) is $\alpha = f_K(0) - f_{Na}(0) - 1 > 0$. Then, for the isolated neuron, the conclusions of Theorem 1.6.1 are valid. We arbitrarily associate the start and end of spike with moments of time when the membrane potential crosses the unit value with positive and negative speed, respectively. According to Theorem 1.6.1, the duration of the spike as $\lambda \to \infty$ is close to the number $T_1 = 1 + \alpha_1$ where $\alpha_1 = f_K(0) - 1$.

In system (3.1.1), V_j is indicator of the presence of the neurotransmitter, which is released as a result of the spike of the jth presynaptic neuron. The value $V_j(t)$ is given as a functional $V_j(t) = \ell(u_j(t+s))$, where $s \in [-T_v, 0]$, $\ell(*)$ is a continuous nonlinear functional defined by formula (2.3.1) or (2.3.3). Here, T_v is the effective time of the neurotransmitter's action. Recall that according to Formula (2.1.3) or (2.3.3), which defines the functional, its value is $\ell(u_j(t+s)) = 1$ if at least at one point $t_* \in [t - T_v, t]$ holds the inequality $u_j(t_*) > 1$. In other words, the interval of time $[t - T_v, t]$ overlaps with the interval of the spike of the jth neuron. Otherwise, $\ell(u_j(t+s)) = 0$. Below, for simplicity, we assume that the effective time of action of the neurotransmitter coincides with the duration of the spike: $T_v = T_1$. Formula (2.3.1) for the functional $\ell(*)$ takes a very simple form $V_j(t) = \chi_\varepsilon(u_j - 1)$, or, taking the limit as $\varepsilon \to 0$, we obtain

$$V_j(t) = \Theta(u_j - 1), \qquad (3.1.2)$$

where $\Theta(*)$ is the Heaviside function. Note that all constructions carried out below are valid, and, in the general case, that $T_v > T_1$.

We agreed earlier to consider the case when the refractory period, which is opened by the spike of the neuron, captures some interval immediately following the spike. It is necessary to make changes to form (2.3.8) of the function $\chi_{ij}(u_i, V_j)$, reflecting the neurotransmitter's influence on the dynamics of the membrane potential. This can be done in different ways. Modernize formula (2.3.8) for $\chi_{ij}(u_i, V_j)$ in the following way:

$$\chi_{ij}(u_i, V_j) = \alpha g_{ij} V_j(t) H(u_i), \qquad (3.1.3)$$

where the functional

$$H(u_i) = f_v(u_i)\chi_\varepsilon(1 - u_i(t - h_s))\chi_\varepsilon((u_i - u_*)/u_*) \qquad (3.1.4)$$

guarantees the existence of a refractory period. In (3.1.3), $V_j(t)$ is the indicator of the presence of the neurotransmitter given by formula (3.1.2). Furthermore, αg_{ij} is the normalization factor where $\alpha = f_K(0) - f_{Na}(0) - 1$, and we call the coefficient g_{ij} the "synaptic weight." This is defined by morphology, i.e., the location and size of the synapse and the power of release of the neurotransmitter vesicles. We explain the meaning of each factor. Here the function $f_v(u_i)$ is positive, i.e., $f_v(0) = 1$. Assume that monotonically and quite rapidly, $f_v(u_i) \to 0$ as $u_i \to \infty$. The presence of this factor provides a weak susceptibility of the neuron to the neurotransmitter's action during generation of the spike. Note that we can replace function $f_v(u_i)$ in (3.1.4) by $\chi_\varepsilon(1 - u_i)$. Factor $\chi_\varepsilon(1 - u_i(t - h_s))$, where $0 < h_s < T_1$, guarantees the resistance of the neuron to action during time close to h_s after completion of the spike. We can interpret the value h_s as the recovery time of synaptic contact. Here it is appropriate to note that the spike for the neuron is a kind of stress, and it takes time for the neuron to return to its original state. During the time h_s the membrane potential u_i can be less than the threshold level u_*, and the neuron becomes no susceptive to the action of neurotransmitter. It turns out that it suffices to choose $h_s = 1$. The latter factor $\chi_\varepsilon((u_i - u_*)/u_*)$ in (3.1.4) reflects, as described in 2.3, the model representations about the synaptic contact: Because the value of the membrane potential u_i is less than the threshold level u_*, the neuron is not susceptive to the synaptic action. Assume that the threshold value u_* is consistent with the parameter λ and is small: $u_* = \exp(-\lambda p)$, where $0 < p < \alpha_2$. Here we have $\alpha_2 = f_{Na}(0) + 1$. Recall that by Theorem 1.6.1, the minimum value is $u_{min} = \exp[-\lambda(\alpha_2 + o(1))]$, i.e., $u_{min} < u_* < 1$.

In this chapter, we consider only excitatory synapses. In this case, as in Sect. 2.3, in formula (3.1.4) one can write a limit transition as $\varepsilon \to 0$. We preliminary carry out the above-mentioned replacement $f_v(u_i) = \chi_\varepsilon(1 - u_i)$. As a result, we obtain a convenient representation of the functional

$$H(u_i) = \Theta(1 - u_i)\Theta(1 - u_i(t - 1))\Theta(u_i - u_*) \qquad (3.1.5)$$

through the Heaviside functions $\Theta(*)$. Here it is taken into account that we can take $h_s = 1$.

In Sect. 2.3, which is devoted to models of the chemical synapse, we did not analyze the case when the neuron remains in a state of refractoriness after the spike and that there is no depolarizing effect on it. We consider the problem of external exciting action on the neuron channeled through a single synapse. Assume that after the spike the neuron is refractory. Let the effective time of action of neurotransmitter $T_v = T_1$ coincide with the asymptotic duration of the spike. General system

(3.1.1), which describes the interaction of the neurotransmitter, taking into account representations (3.1.2), (3.1.3) and (3.1.5), reduces to the system 3.1 to the following equation:

$$\ddot{u} = \lambda \left[\begin{array}{c} -1 - f_{\mathrm{Na}}(u) + f_{\mathrm{K}}(u(t-1)) + \\ + ag\Theta(1 - u(t))\Theta(1 - u(t-1))\Theta(u(t) - u_*)V(t - t_v) \end{array} \right] u, \qquad (3.1.6)$$

where the synaptic weight is $g > 0$. In (3.1.6), function $V(t) = 1$ at $t \in [0, T_1]$, $V(t) = 0$ at $t \notin [0, T_1]$, and $t_v \in [0, T_2]$ is the time of the neurotransmitter's appearance. The number T_2 according to Theorem 1.6.1 is an asymptotic approximation of the period of the solution of Eq. (1.5.7) of the isolated neuron. As the initial condition of the solution of Eq. (3.1.6), we take $u(s) = \varphi(s) \in S$ for $s \in [-1, 0]$, i.e., we confine the start of the spike to the zero moment of time. Denote this solution by $u_v(t, \varphi)$.

Let T_R be a moment of time ($T_R \in [0, T_2]$) when the function $u_v(t, \varphi)$ crosses the threshold value u_* with positive speed. On the interval of time $t \in [0, T_R]$, Eq. (3.1.6) turns into (1.5.7). By virtue of Theorem 1.6.1, the equation is valid with up to $o(1)$ asymptotic equality:

$$T_R = T_2 - p/\alpha. \qquad (3.1.7)$$

where time T_R is the duration of the refractory period.

Let $T_R < t_v < T_2$. On the interval of time $t \in [0, t_v]$ the solution $u_v(t, \varphi)$ of Eq. (3.1.6) coincides with the solution $u(t, \varphi)$ of the isolated neuron ($u_v(t, \varphi) = u(t, \varphi)$), the asymptotic behaviour of which as $\lambda \to \infty$ reveals Theorem 1.6.1. In this case, the interval $t \in [t_v, t_v + T_1]$ until $u_v(t, \varphi) < 1$ (this did not start generation of the spike) is valid as $\lambda \to \infty$ asymptotic formula $u_v(t, \varphi) = u(t, \varphi) \exp[\lambda(ag(t - t_v) + o(1)]$, which is consistent with the concept of the model of exponential approximation. From this formula, it follows that the presence of an excitatory neurotransmitter on the part of susceptibility approximates the moment of time t_S of the start of the next spike.

Introduce the value $Q(t_v, g, \varphi) = t_S - t_v$ i.e., the time between the beginning of action and the start of the induced spike. In neurophysiology, this value is called the "latent period" or the "period of hidden response." We calculate from Eq. (3.1.6) the asymptotic value $Q(t_v, g, \varphi)$. We immediately note an obvious fact. If $t_v < T_R - T_1$, i.e., and the continuous presence of the neurotransmitter falls in the refractory period of the neuron, then $Q(t_v, g, \varphi) = T_2 - t_v + o(1)$.

To find the value $Q(t_v, g, \varphi)$, we integrate Eq. (3.1.6) by the standard method of steps, of course taking into account the specifics of its right side. Introduce the value $\rho(g)$ by equality $\rho(g) = T_2 - T_R - T_1(1 + g)$. This, in a certain sense, characterizes the force, duration, and threshold value of the action. It turns out that we must distinguish two cases: $\rho(g) > 0$ and $\rho(g) < 0$.

In the first case ($\rho(g) > 0$) as $\lambda \to \infty$ for $Q(t_v, g, \varphi)$ the following asymptotic equalities are valid and accurate to $o(1)$:

$$Q(t_v, g, \varphi) = T_2 - t_v, \quad t_v \in [0, T_R - T_1], \tag{3.1.8}$$

$$Q(t_v, g, \varphi) = T_2 - t_v - g(t_v + T_1 - T_R) \quad t_v \in [T_R - T_1, T_R], \tag{3.1.9}$$

$$Q(t_v, g, \varphi) = T_2 - t_v - gT_1, \quad t_v \in [T_R, T_R + \rho(g)], \tag{3.1.10}$$

$$Q(t_v, g, \varphi) = (T_2 - t_v)/(1 + g), \quad t_v \in [T_R + \rho(g), T_2]. \tag{3.1.11}$$

Schematically, the dependence of the latent period for this case is shown in Fig. 3.1. The approximate curve (without the members $o(1)$) is a polygonal line of four links. The form of the curve is consistent with biological data [172].

In the second case $(\rho(g) < 0)$ formulas for $Q(t_v, g, \varphi)$ acquire accuracy to $o(1)$ form the following:

$$Q(t_v, g, \varphi) = T_2 - t_v, \quad t_v \in [0, T_R - T_1], \tag{3.1.12}$$

$$Q(t_v, g, \varphi) = T_2 - t_v - g(t_v + T_1 - T_R), \quad t_v \in [T_R - T_1, T_R + \rho(g)], \tag{3.1.13}$$

$$Q(t_v, g, \varphi) = (T_2 + gT_R)/(1 + g) - t_v, \quad t_v \in [T_R + \rho(g), T_R], \tag{3.1.14}$$

$$Q(t_v, g, \varphi) = (T_2 - t_v)/(1 + g), \quad t_v \in [T_R, T_2]. \tag{3.1.15}$$

The corresponding curve has a structure similar to that shown in Fig. 3.1 but with different points of gluing links of the polygonal line. We note an important fact. Asymptotic integration of Eq. (3.1.6) shows that for its solution it is valid to include $u(t_S + s) \in S$ at $s \in [-1, 0]$. Thus, the obtained formulas can be used in the

Fig. 3.1 Graph of the latent period of the neuron illustrating asymptotic formulas (3.1.8) through (3.1.11)

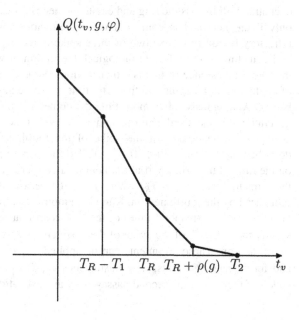

prediction of outcome of the action for $t_v > t_S$ with the natural replacement of t_v on $t_v - t_S$.

From the above-mentioned formulas (3.1.8) through (3.1.15), it follows that the presence of a neurotransmitter on the interval of susceptibility approximates the next spike of the neuron on a time proportional to the duration of action of neurotransmitter on this interval. The proportionality coefficient is equal to g.

Furthermore, from formulas (3.1.8) through (3.1.15), it follows that the required value of synaptic weight g in order for the neuron to spike occurs over time, does not exceed T_1 (the effective time of action of the mediator), and decreases with t_v. In other words, the effectiveness of action increases with distance from the start of the spike. This is consistent with facts (Eckert et al. 1991).

3.2 Model of the Ring Structure of Four Neurons

By its nature, the chemical synapse has a one-way action. Imagine a situation when the first neuron has its synaptic termination on the second neuron and simultaneously carries on its body the ends of the terminals of the second neuron. In this case, we say that these two neurons interact with each other.

Consider the ring structure of four neurons in which the first neuron interacts with the second and fourth neurons, the second neuron with the first and third neurons, the third neuron with the second and fourth neurons, and, finally, the fourth neuron with the first and third neurons. We need the situation to explain the idea of excitation propagation around the ring of neurons. The choice of four neurons is associated with the following fact: Generated by some neuron, a wave of excitation will bypass the ring and excite the neuron (i.e., will continue to move on) only if the neuron has come out of the refractory state. It is believed that the refractory period results in two or three spike durations (Schade and Ford 1976).

Let us turn to the informal biological description of wave excitation passage on the ring. The essence of this is rather obvious. Suppose that at zero time, the spike of the first neuron begins and that after time ξ_2, spike generation the second neuron begins. Analogously, after more time ξ_3, comes the moment of the start of spike generation by the third neuron, and after even time ξ_4, i.e., during time $t_4 = \xi_2 + \xi_3 + \xi_4$, generation of the spike of the fourth neuron begins. The process described on the interval $t \in [0, t_4]$ is called the "single passage of wave excitation on the ring." Furthermore, if at moment of time t_4 the first neuron has come out of the refractory state ($t_4 > T_R$, where T_R is the period of refractoriness), then it is influenced by the fourth neuron. Knowing empirically, for example, the value of the latent period Qt_4 (respond time) of the first neuron at moment of time t_4, one can specify the delay ξ_1' of the spike of the first neuron with respect to the spike of the fourth neuron. These arguments can be applied sequentially for the second, third, and fourth neurons. Thus, new mismatches ξ_2', ξ_3', ξ_4' of times of the start of the spikes of neurons at the second passage of wave excitation are identified. If $Q(t)$ the latent period of the neuron (the beginning of action follows after time t after the start of the spike), then in view of this:

$$\xi_1' = Q(\xi_2 + \xi_3 + \xi_4), \quad \xi_2' = Q(\xi_3 + \xi_4 + \xi_1'),$$

$$\xi_3' = Q(\xi_4 + \xi_1' + \xi_2'), \quad \xi_4' = Q(\xi_1' + \xi_2' + \xi_3').$$

This is the iterative process. The system is closed, and the natural hypothesis is that the process stabilizes. If the limits are the essence $\xi_1^0, \xi_2^0, \xi_3^0,$ and ξ_4^0, then these numbers are mismatches between the spikes of neurons, and the full period of the system is $T_s = \xi_1^0 + \xi_2^0 + \xi_3^0 + \xi_4^0$. At the same time, we must remember two facts. First, the excitation from any neuron, returning to it, must enter a period of susceptibility, i.e., $T_S - \xi_t^0 > T_R$ for all $i = 1, \ldots, 4$. Second, because all neurons have their own autorythmicity with the period T_2, then in $T_S < T_2$, all of the spikes are induced by this character.

In this case, the described biological ideology is the basis of the algorithm of the mathematical model (3.1.1), (3.1.2), and (3.1.3). The system of equations for airing of four neurons takes the form:

$$\dot{u}_1 = \lambda \left[\begin{array}{l} -1 - f_{Na}(u_1) + f_K(u_1(t-1)) + \\ + aH(u_1)\left(g_{1,4}\theta(u_4 - 1) + g_{1,2}\theta(u_2 - 1)\right) \end{array} \right] u_1, \qquad (3.2.1)$$

$$\dot{u}_2 = \lambda \left[\begin{array}{l} -1 - f_{Na}(u_2) + f_K(u_2(t-1)) + \\ + aH(u_2)\left(g_{2,1}\theta(u_1 - 1) + g_{1,3}\theta(u_3 - 1)\right) \end{array} \right] u_2, \qquad (3.2.2)$$

$$\dot{u}_3 = \lambda \left[\begin{array}{l} -1 - f_{Na}(u_3) + f_K(u_3(t-1)) + \\ + aH(u_3)\left(g_{3,2}\theta(u_2 - 1) + g_{3,4}\theta(u_4 - 1)\right) \end{array} \right] u_3, \qquad (3.2.3)$$

$$\dot{u}_4 = \lambda \left[\begin{array}{l} -1 - f_{Na}(u_4) + f_K(u_4(t-1)) + \\ + aH(u_4)\left(g_{4,3}\theta(u_3 - 1) + g_{4,1}\theta(u_1 - 1)\right) \end{array} \right] u_4, \qquad (3.2.4)$$

Here $g_{i,j}$ are synaptic weights, and for the functional $H(u)$ we take representation (3.1.5), i.e.,

$$H(u) = \theta(1 - u(t))\theta(1 - u(t-1))\theta(u(t) - u_*).$$

The threshold value $u_* = \exp(-\lambda p)$, where $0 < p < \alpha_2$.

Substantial study (Rabinovich and Trubetskov 1984; Anishchenko et al. 1986; Zaslavsky and Sagdeev 1988; Dmitriev and Kislov 1989; Chua and Yang 1988a, b; Osipov et al. 1992) has been devoted to the problem of nonlinear chains, lattices, and ring systems including those with delay. The distinctive feature of the approach used below is that for the investigation of systems of Eqs. (3.2.1) through (3.2.4), a special asymptotic method is applied based on papers (Kashchenko 1990a, b). With its help, we can prove the existence of complex (nonlocal) relaxation oscillations.

Below we refer to waves of excitation in which the spikes of neighbouring neurons overlap in time. In our opinion, this is the case when neurons respond directly to excitation.

Let us turn to the specification of initial conditions. For application of the asymptotic method, we require biological information about the behaviour of neurons on the time interval of the duration of the unit. When considering the individual neuron, this is the moment of start of the spike and the "prehistory" of the effect of the process on the unit interval of time. In the case of a system of several neurons, we also know how the membrane potential behaves on each of the neurons in the analogous interval before the start of the spike. Unfortunately, the moments of the start of the spikes are different. Thus, difficulties arise in specifying a biologically meaningful set of initial conditions (such as class S) as well as subsequent application of the method of steps for the construction of asymptotic solutions.

In this connection, we propose a special method allowing us to construct asymptotics of the solutions of system (3.2.1) through (3.2.4). The initial conditions for each of the neurons are specified in different, previously unknown moments of time [61]. These moments (i.e., the starts of of neuron spikes) will appear below as the most important parameter. Precisely their values precisely determine the behaviour of the solutions of system (3.2.1) through (3.2.4).

Choose four points in time $t_i (i = 1, \ldots, 4)$ as the starts of the spikes of neurons at an initial cycle of propagation of the excitation wave. Consider that

$$t_1 = 0,$$

$$t_{i-1} < t_i < t_{i-1} + T_1 \quad (i = 2, 3, 4), \tag{3.2.5}$$

$$T_R < t_4 < T_2, \tag{3.2.6}$$

where T_1, T_R, T_2 are asymptotic durations of the spike, the refractory period, and the period of oscillations of isolated neuron, respectively. The values T_1 and T_2 are defined in Theorem 1.6.1, and the number T_R is given by formula (3.1.7). As already mentioned, the refractory period accounts for two or three durations of the spike.

We consider: $2T_1 < T_R < T_2$. Hence and from (3.2.5), it follows that

$$t_i + T_1 < t_{i-1} + T_R \quad (i = 2, 3, 4), \tag{3.2.7}$$

Introduce the initial conditions for Eqs. (3.2.1) through (3.2.4) $u_i(t_i + s) = \varphi_i(s) \in S$, where $s \in [-1, 0]$, and S is the class of initial functions defined in Sect. 2.3. Below we propose an algorithm that allows us to uniquely construct the solution of the system of equations (3.2.1) through (3.2.4) according to these conditions for $t > t_4$. We denote this solution as $u_i(t, \varphi, \tau)$, where $\varphi = (\varphi_1, \ldots, \varphi_4), \tau = (\tau_1, \ldots, \tau_4)$ and $i = 1, \ldots, 4$. The mode of action requires that the function $u_4(t, \varphi, \tau)$ on the first step has been defined for all $t \in [T_R, t_4]$. On the interval $t \in [t_4 - 1, t_4]$ is defined $u_4(t_4 + s) = \varphi_4(s)$, where $s \in [-1, 0]$. If $T_R < t_4 - 1$, then we extend the definition $u_4(t, \varphi, \tau)$ arbitrarily in compliance with the condition $0 < u_4(t, \varphi, \tau) < 1$.

Drawing a parallel with previously given informal reasoning, we specified the initial cycle of the passage of the excitation wave. Condition (3.2.6) means that at the moment of the start of the spike of the fourth neuron, the first neuron has come out of the refractory state $(u_1(t_4, \varphi, \tau) > u_*)$, but has not yet started generation of its own spike.

Note that because $H(u) = 0$ at $u > 1$, then each of Eqs. (3.2.1) through (3.2.4) can be integrated (independently) on the interval $t \in [t_i, t_i^0]$, where t_i^0 is the moment of the end of the spike, i.e., the time when $u_i(t)$ crosses the single value with a negative rate. By virtue of Theorem 1.6., $t_i^0 - t_i = T_1 + o(1)$. Now we can comment on inequalities (3.2.5) through (3.2.6). The meaning of the first inequality lies in the fact that during the initial cycle, the wave of excitation propagates from the first to the second, the third, and finally, the fourth neuron. At the same time, the spikes of neighbouring neurons overlap in time. In turn, inequality (3.2.7) means that the spike of the ith neuron $(i = 2, 3, 4)$ ends before that of the i-1st neuron and manages to exit the state of refractoriness in which the function $H(u_i - 1) = 0$. Consequently, the ith neuron does not influence the previous one.

Let t_1', t_2', t_3', t_4' be the alleged moments of time of the start of the spikes at the next passage of the wave of excitation. We make the following a priori assumptions:

$$t_4 < t_1' < t_4 + T_1,$$

$$t_{i-1}' < t_i' < t_{i-1}' + T_1 \quad (i = 1, 2, 3), \qquad (3.2.8)$$

$$T_R < t_1' - t_{i+1} < T_2 \quad (i = 1, 2, 3),$$

$$T_R < t_4' - t_1' < T_2, \qquad (3.2.9)$$

the satisfiability of which will be mentioned later in the text. Inequality (3.2.8) guarantees the preservation of the direction of wave propagation and the overlapping of the spikes of neighbouring neurons in the second cycle. Inequality (3.2.9) says that at the start of the spike of the ith neuron, the $(i + 1)$th neuron has come out of the refractory state but has not started generating its own spike $(u_* < u_{i+1}(t_i') < 1)$. As before, from (3.2.8) it follows that

$$t_4' + T_1 < t_4 + T_R, \quad t_1' + T_1 < t_{i-1}' + T_R \quad (i = 1, 2, 3, 4),$$

i.e., the spike of the ith neuron falls on the zone of refractoriness of the preceding neuron $(H(u_{i-1}) = 0)$ during the second cycle of propagation of the wave.

As already noted, from Eq. (3.2.4) we find $u_4(t, \varphi, \tau)$ on the interval $t \in [t_4, t_4^0]$ of the spike's duration. In view of the described peculiarities of the action of the next neuron on the previous one, Eq. (3.2.1) for the first neuron is already considered in

Sect. 3.2 of the given model of synaptic action on the first neuron by the fourth one. The moment t_1' of the start of the spike is determined by the value of latent period $Q(t_4, g_{1,4}, \varphi_1)$. Furthermore, the value $u_1(t, \varphi, \tau)$ on the spike interval is found from Eq. (3.2.1) because on this interval $H(u_i) = 0$.

The arguments are repeated sequentially for the second, third, and fourth neurons (of Eq. (3.2.2) through (3.2.4), respectively). Thus, we build the functions $u_i(t, \varphi, \tau)$ $(i = 1, \ldots, 4)$ as the solutions of the corresponding equations on the intervals $t \in [t_4, t_i']$. Numbers themselves t_i' are expressed in through the values of latent periods:

$$t_1' - t_4 = Q(t_4, g_{1,4}, \varphi_1), \tag{3.2.10}$$

$$t_i' - t_{i-1}' = Q(t_{i-1}' - t_i, g_{i,i-1}, \varphi_1), \quad i = 2, 3, 4. \tag{3.2.11}$$

In Sect. 3.2 of this chapter, it was emphasized that the functions

$$u_i(t_i' + s, \varphi, \tau) = \varphi_i'(s) \in S \quad (i = 1, \ldots, 4, s \in [-1, 0]). \tag{3.2.12}$$

We can take one more step of the algorithm to denote the expected start time of the spikes through $t_1'', \ldots, t_4''$ and accepting a priori assumptions (3.2.8) and (3.2.9) with the appropriate change of variables. For the initial set of moments, we should take $\tau' = (t_1', \ldots, t_4')$, and as the initial functions we must use $\varphi' = (\varphi_1'(s), \ldots, \varphi_4'(s))$ where $\varphi_i'(s)$ are given by formulas (3.2.12). As a result, we will build functions $u_i(t, \varphi, \tau)$ $(i = 1, \ldots, 4)$ defined for $t \in [t_i', t_i'']$ and on the interval $t \in [t_4, t_1'']$ will be the solution of the system of equations (3.2.1) through (3.2.4).

The algorithm uniquely allows us to construct the solution of the system of equations (3.2.1) through (3.2.4) according to given initial conditions $\tau = (\tau_1, \ldots, \tau_4)$ and $\varphi = (\varphi_1, \ldots, \varphi_4)$ in any finite time interval. Mapping (3.2.10) and (3.2.11) gives the start times of the spikes, which are transformed.

By virtue of the a priori assumptions that the spikes of neighbouring neurons overlap in time and the spike of the ith neuron starts at the time when the $(i + 1)$th neuron has come out of the refractory state, the value $Q(*, *, *)$ must be calculated by formula (3.1.11) or (3.1.15). The latter have the same form, so for (3.2.10) and (3.2.11) we obtain as $\lambda \to \infty$:

$$t_1' - t_4 = (T_2 - t_4)/(1 + g_{1,4}) + o(1), \tag{3.2.13}$$

$$t_i' - t_{i-1}' = [T_2 - (t_{i-1}' - t_i)]/(1 + g_{i,i-1}) + o(1), \quad (i = 2, 3, 4). \tag{3.2.14}$$

Introduce the variables $\xi_i = t_i - t_{i-1}$ $(i = 2, 3, 4)$, $\xi_1' = t_1' - t_4$, and $\xi_i' = t_i' - t_{i-1}'$ $(i = 2, 3, 4)$. The values ξ_i are time intervals between the starts of the spikes of the ith and $(i - 1)$th neurons at the initial cycle of conducting wave excitation wave, and ξ_1' is the response time of the first neuron in response to the action of the fourth neuron. Finally, ξ_1' is a mismatch for time between the spikes at the repeated cycle of the passage of the wave excitation. For new variables, a priori assumptions (3.2.5), (3.2.6), (3.2.8), and (3.2.9) denote that

$$0 < \xi'_i < T_1 \quad (i = 2, 3, 4),$$
$$0 < \xi'_i < T_1 \quad (i = 1, \ldots, 4),$$

(3.2.15)

$$T_R < \xi_2 + \xi_3 + \xi_4 < T_2,$$
$$T_R < \xi_3 + \xi_4 + \xi'_1 < T_2,$$
$$T_R < \xi_4 + \xi'_1 + \xi'_2 < T_2,$$
$$T_R < \xi'_1 + \xi'_2 + \xi'_3 < T_2,$$
$$T_R < \xi'_2 + \xi'_3 + \xi'_4 < T$$

(3.2.16)

We rewrite formulas (3.2.13) and (3.2.14) for new variables:

$$\left(1 + g_{1,4}\right)\xi'_1 + \xi_2 + \xi_3 + \xi_4 = T_2 + o(1),$$

(3.2.17)

$$\xi'_1 + \left(1 + g_{2,1}\right)\xi'_2 + \xi_3 + \xi_4 = T_2 + o(1),$$

(3.2.18)

$$\xi'_1 + \xi'_2 + \left(1 + g_{3,2}\right)\xi'_3 + \xi_4 = T_2 + o(1),$$

(3.2.19)

$$\xi'_1 + \xi'_2 + \xi'_3 + \left(1 + g_{4,3}\right)\xi'_4 = T_2 + o(1).$$

(3.2.20)

Let us turn to new notations in (3.2.12):

$$\varphi'_1(s) = u_1\left(\xi_2 + \xi_3 + \xi_4 + \xi'_1 + s, \varphi, \tau\right) \in S,$$

(3.2.21)

$$\varphi'_2(s) = u_2\left(\xi_3 + \xi_4 + \xi'_1 + \xi'_2 + s, \varphi, \tau\right) \in S,$$

(3.2.22)

$$\varphi'_3(s) = u_3\left(\xi_4 + \xi'_1 + \xi'_2 + \xi'_3 + s, \varphi, \tau\right) \in S,$$

(3.2.23)

$$\varphi'_4(s) = u_4\left(\xi'_1 + \xi'_2 + \xi'_3 + \xi'_4 + s, \varphi, \tau\right) \in S,$$

(3.2.24)

where $s \in [-1, 0]$.

Using relationship (3.2.17), in formulas (3.2.18) through (3.2.24) we can eliminate the variable ξ'_1. As a result, the transformed expressions (3.2.18) through (3.2.20) yield mismatches between the starts of the spikes of neighbouring neurons at the second passage of the excitation wave (hence, the beginning of a new era of spikes), and formulas (3.2.21) though (3.2.24) are new initial functions for the system of equations (3.2.1) through (3.2.4). Thus, there will be a mapping for which we can study the problem of the existence of a fixed point. However, it turns out that technically it is easier not to perform the elimination of the variable ξ'_1 and consider the mapping, the finite component of which has the dimension by the unit as mentioned previously.

In formulas (3.2.17) through (3.2.20), we temporarily discard the terms $o(1)$. Linear mapping in a four-dimensional space arises. Assume that this space has a stable fixed point $\xi^0 = (\xi_1^0, \ldots, \xi_4^0)$ satisfying a priori conditions (3.2.15) and (3.2.16). Take the sphere $C(\xi^0)$ with the centre ξ^0, all points of which satisfy (3.2.15) and (3.2.16) (superscripts in inequalities are omitted). Formulas (3.2.17 (through (3.2.24) (together with the terms $o(1)$) can be regarded as the mapping Π, where $\xi = (\xi_1, \ldots, \xi_4)$, $\varphi = (\varphi_1, \ldots, \varphi_4)$. It is invariant with respect to $\xi \in C(\xi^0)$ and $\varphi \in S$ (the latter inclusion is understood coordinate-wise). In view of the known results [167] about the existence of the fixed point, we conclude that it exists in the mapping Π. This fixed point is the initial condition of the periodic solution of the system of equations (3.2.1) through (3.2.4), which is reconstructed on the length of the period with the help of the described algorithm. However, the solution itself belongs to a narrow attractor, which has a wide range of attraction. Each of the solutions incoming in the attractor has the same asymptotic behaviour as $\lambda \to \infty$ on any finite time interval. Therefore, a separate research of the stability of periodic solutions is of no interest from a biological point of view.

It remains to make sure that mapping (3.2.17) through (3.2.20), in which the terms $o(1)$, are discarded, had the desired stable fixed point. This is done by selecting the coefficients $g_{4,1}, g_{i,i-1}$ $(i = 2, 3, 4)$.

Let $\xi_1^0, \ldots, \xi_4^0$ be arbitrary positive numbers satisfying the following conditions. Introduce the value

$$T_S = \sum_{t=1}^{4} \xi_i^0.$$

Assume that

$$0 < \xi_i^0 < T_1 \quad (i = 1, \ldots, 4), \tag{3.2.25}$$

$$T_s < T_2, \quad T_s - \xi_i^0 > T_R \quad (i = 1, \ldots, 4). \tag{3.2.26}$$

Now we write

$$g_{1,4} = (T_2 - T_S)/\xi_1^0, \quad g_{i,i-1} = (T_2 - T_s)/\xi_i^0 \quad (i = 2, 3, 4) \tag{3.2.27}$$

It is easy to see that with this choice of weighting coefficients, mapping (3.2.17) through (3.2.20), in which the terms $o(1)$ are discarded, has a fixed point $\xi_1^0, \ldots, \xi_4^0$, thus satisfying the necessary requirements. The problem of sustainability is also easily solved. It suffices to note that the mapping represents the Seidel method of solutions of linear systems. We verify that the matrix of the system is positively defined and symmetric. In this case, the Seidel method converges [7].

Thus, for a ring of four neurons, described by the system of equations (3.2.1) through (3.2.4), we can specify the values of synaptic weights $g_{1,4}, g_{i,i-1}$,

$(i = 2, 3, 4)$ so that the system has an attractor of a given structure. Solutions forming it are the waves of excitation propagating in the direction of the numbers of neurons ascending (i.e., after the fourth neuron follows the first neuron). Time intervals between the spikes of neighbouring neurons up to $o(1)$ take preassigned values $\xi_1^0, \ldots, \xi_4^0$, which must satisfy inequalities (3.2.25) and (3.2.26).

The scheme of such a cycle of excitation is illustrated in Fig. 3.2. On it, one below the other, are arranged the graphs of neuronal activity from first to fourth neuron. Lines with arrows depict the moments of the beginning of action of the $(i - 1)$th neuron on the ith neuron. The cycle begins at zero moment of time with the spike of the first neuron. After a time ξ_2^0, the second neuron generates a pulse, under the influence of which at moment of time $\xi_2^0 + \xi_3^0$ begins the spike of the third neuron. The spike of the fourth neuron is induced by the third one with an additional delay ξ_4^0. Finally, the fourth neuron through time ξ_1^0, i.e., at moment of time $T_S = \xi_1^0 + \xi_2^0 + \xi_3^0 + \xi_4^0$, activates the first neuron. Then begins the second cycle of the wave excitation. Note that the beginning of action of the fourth neuron on the first one (as well as for other pairs) occurs after the end of refractory period, which is marked on the first graph by the point T_R.

So far we have not used the synaptic weights $g_{i,i+1}, i = (1, 2, 3), g_{4,1}$. Applying to them the method described previously, we can determine the attractor for system (3.2.1) through (3.2.4) consisting of the solutions, which represent waves of excitation, travelling in the opposite direction. This attractor exists simultaneously with the first one, and its parameters (time intervals between the spikes of neighbouring neurons) can be quite different.

Even as solved for the model case, the problem of the synthesis of the system that has attractors of the previously given structure makes sense by itself, but for studying neural structures it is of particular importance. According to the hypothesis of the wave nature of memory (Zabrodin and Lebedev 1977; Lebedev 1990; Lebedev et al. 1991), perceived data is recorded and stored in the brain in the form of stable combinations of different phases of coherent, persistent waves of neural activity. Each combination is a separate element of the neural code of memory and represents the cyclic sequence of waves. These are traces of memory, representations of which have been studied by Lebedev on a qualitative level. Structure, as proposed by us, can be regarded as the model of a neural ensemble that stores traces of memory.

We make one more important remark. System (3.2.1) through (3.2.4) has the homogeneous periodic solution $u_1 = u_2 = u_3 = u_4$. It belongs to the attractor, consisting of solutions, the components of which $u_i(t)(i = 1, \ldots, 4)$ have a common asymptotic behaviour given as $\lambda \to \infty$ by formulas from Theorem 1.6.1. The existence of such an attractor makes biological sense and is associated with the problem of attention (Kogan 1967, 1979; John 1982; Kryukov 1989, 1990). At present, structures of the brain should not be actualized for all traces of memory. The transition of ring structure into the homogeneous mode can be interpreted as one of the possible mechanisms of the inhibition of memory traces. This is

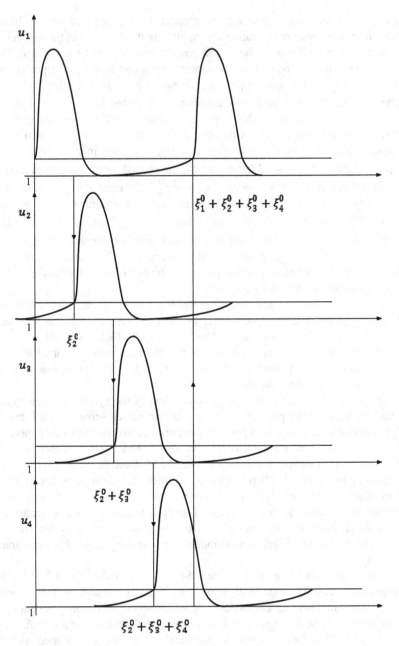

Fig. 3.2 Schema of propagation of the wave excitation along the ring of four structures

consistent with experimental data: In the quiescent state, neural structures of the brain tend to synchronization.

Note that the existence of a high-degree homogeneous attractor is the result of idealizations. We can take into account the synaptic delays replacing expression (3.1.2) for the indicator of the presence of a mediator $V_j(t) = \chi_\varepsilon\big(u_j(t - h_m) - 1\big)$ where $h_m > 0$ and is sufficiently small. Then for the existence of the attractor with a given mismatch of spike start times. Formula (2.3.27) for the values of synaptic weights should be changed slightly:

$$g_{1,4} = (T_2 - T_S)/(\xi_1^0 - h_m), g_{i,i-1} = (T_2 - T_S)/(\xi_i^0 - h_m) \quad i = 2, 3, 4.$$

In this case, instead of homogeneous attractors, "fluff attractors" will appear. In each of these functions, $u_i(t)$ are mismatched in phase by amount $O(h_m)$ and are described by the asymptotic formulas of Theorem 1.6.1 with the corresponding phase shift. At the same time there coexist several of such attractors.

In conclusion, we give one more reason why we chose the ring consisting of four neurons for the description of biological and mathematical ideas. Analysis of temporal characteristics shows that this is the minimum size of the ring, in which the umber neurons per cycle of a passage of the excitation wave have enough time to come out of the refractory state. This allows the wave to spread cyclically. If we assume that each neuron has a synaptic connection with only one subsequent neuron, but not with the preceding one, then we can confine ourselves to the ring of three neurons. At this point, we should drop the condition $T_R > 2T_1$. In this case, the mathematical construction is significantly simplified.

3.3 Model of the Ring Structure of N Neurons

Consider the ring of N identical neurons, in which the ith $(i = 1,\dots,N)$ neuron has synaptic connections with the $(i-1)$th and $(i+1)$th neurons. In this case, the $(N+1)$th neuron is identified with the first one, and predecessor of the first neuron is considered to be the Nth neuron. In this case, the system of equation (3.1.1) for the neuronal population, where synaptic interaction is modeled by formulas (3.1.2) and (3.1.3), takes the form:

$$\dot{u}_1 = \lambda\left[\begin{array}{c} -1 - f_{Na}(u_1) + f_K(u_1(t-1)) + \\ + \alpha H(u_1)\big(g_{1,N}\theta(u_N - 1) + g_{1,2}\theta(u_2 - 1)\big) \end{array}\right] u_1, \qquad (3.3.1)$$

$$\dot{u}_i = \lambda\left[\begin{array}{c} -1 - f_{Na}(u_i) + f_K(u_i(t-1)) + \\ + \alpha H(u_i)\big[g_{i,i-1}\theta(u_{i-1} - 1) + g_{i,i+1}\theta(u_{i+1} - 1)\big] \end{array}\right] u_i, \qquad (3.3.2)$$

$$\dot{u}_N = \lambda\left[\begin{array}{c} -1 - f_{Na}(u_N) + f_K(u_N(t-1)) + \\ + \alpha H(u_N)\big[g_{N,N-1}\theta(u_{N-1} - 1) + g_{N,1}\theta(u_1 - 1)\big] \end{array}\right] u_N. \qquad (3.3.3)$$

Here, as in the previous paragraph, $g_{i,j}$ are synaptic weights, and the functional $H(u)$ is given by expression (3.1.5):

$$H(u) = \theta(1 - u(t))\theta(1 - u(t - 1))\theta(u(t) - u_*).$$

The threshold value is $u_* = \exp(-\lambda p)$, where $0 < p < \alpha_2$.

All the constructions of the preceding paragraph are completely transferred to the general case. We will speak about the wave of excitation propagating in the direction of the increase of numbers of neurons. Denote by $\xi_1^0, \ldots, \xi_N^0$, as expected in the stationary mode, the asymptotic $\lambda \to \infty$ values of the time interval between the start of spikes of the Nth and first, ith, and $(i+1)$th neurons, respectively. Assume that the spikes overlap in time, i.e.,

$$0 < \xi_i^0 < T_1 \quad (i = 1, \ldots, N), \tag{3.3.4}$$

where T_1 is the asymptotic duration of spikes. The asymptotic period T_S of complete passage of the wave excitation, i.e., the time after which the wave, generated by the spike of the ith neuron, causes its next spike, is common for all neurons:

$$T_S = \sum_{i=1}^{n} \xi_i^0. \tag{3.3.5}$$

Note that the value T_S is asymptotically close to the period of oscillations of the ring neuron system as a whole. We assume that

$$T_S < T_2, \tag{3.3.6}$$

i.e., the period of the system is less than the period of a single neuron, the asymptotic value of which is T_2. As before, we assume that the wave generated by the spike of the ith neuron $(i = 1, \ldots, N)$, running on the ring, induces the spike of the $(i - 1)$th neuron already after the ith neuron has come out of the refractory state. It is guaranteed by the inequality:

$$T_S - \xi_i^0 > T_R, \quad (i = 1, \ldots, N), \tag{3.3.7}$$

where T_R is the value of the refractory period. Now, in complete analogy with (2.3.27), we assume in the system of equations (3.3.1) through (3.3.3):

$$g_{1,N} = \frac{T_2 - T_S}{\xi_1^0}, \quad g_{i,i-1} = \frac{T_2 - T_S}{\xi_i^0} \quad (i = 2, \ldots, N). \tag{3.3.8}$$

Let

$$T_R > 2T_1. \tag{3.3.9}$$

The coefficients $g_{i,i+1}$ $(i = 1, \ldots, N-1)$, $g_{N,1}$ are positive. Then the system of equations (3.3.1) through (3.3.3) has the attractor consisting of solutions for which the intervals between spikes of the $(i-1)$th and the ith neurons at $\lambda \to \infty$ are asymptotically close to numbers ξ_i^0. To check this statement, we construct mapping as described in the previous paragraph. Defined in this is a finite component. If its main, in terms of the asymptotic behaviour for λ part, will have the vector $(\xi_1^0, \ldots, \xi_N^0)$ as a stable fixed point, then this will imply the existence of an attractor (the corresponding arguments have already been put forth in Sect. 3.2 of this chapter).

We explain briefly the construction of the finite part. The time of the start of the spike of the first neuron is the zero moment of time. Let the spikes of the second and subsequent neurons have begun at moments of time $\xi_2, \xi_2 + \xi_3, \cdots, \xi_2 + \xi_3 + \cdots + \xi_N$, i.e., they follow each other through time $\xi_i (i = 2, \ldots, N)$. Assume that the numbers ξ_i and ξ_i^0 are sufficiently close. By virtue of this

$$0 < \xi_i < T_1 \quad \sum_{i=2}^{N} > T_R, \quad (i = 2, \ldots, N), \tag{3.3.10}$$

i.e., during the initial passage of the excitation wave, the spikes of neighbouring neurons overlap in time, and at the moment of the start of spike of the Nth neuron, the first neuron has come out of the refractory state.

Denote by ξ_1' the time between the starts of spikes of the Nth and first neurons. If

$$0 < \xi_1' < T_1, \tag{3.3.11}$$

i.e., the spikes overlap in time, and by virtue of the second inequality (3.2.10), value ξ_1' is calculated by the value of the latent period given by (3.2.11) (or (3.2.15)):

$$\xi_1' = \left(\left(T_2 - \sum_{j=2}^{N} \xi_j \right) \middle/ (1 + g_{1,N}) \right) + o(1). \tag{3.3.12}$$

The first equation of formula (3.3.8) can be rewritten in the form:

$$\xi_1^0 = \left(T_2 - \sum_{j=2}^{N} \xi_j^0 \right) \middle/ (1 + g_{1,N}).$$

Because ξ_i and $\xi_i^0 (i = 2, \ldots, N)$ were chosen close, then the numbers differ little, and consequently the a priori assumption (3.2.11) is satisfied. In addition, from (3.3.7) it follows that

$$\sum_{i=3}^{N} \xi_i + \xi_i' > T_R,$$

i.e., at the moment of start of spike of the first neuron, the second neuron is disinherited (no refractoriness). We also note the important role of inequality (3.3.9), which guarantees that the first neuron does not influence the Nth neuron at the second cycle of passage of the wave excitation.

Let $\xi_i'(i = 2, \ldots, N)$ be intervals of time between the starts of the spikes of the ith and i-first neurons at the second cycle of the wave of excitation. Repeating the arguments just described for the second, third, etc. neurons, we obtain the following recurrence relations:

$$\xi_i' = \left(\left(T_2 - \sum_{j=i+1}^{N} \xi_j - \sum_{j=1}^{i-1} \xi_j' \right) \middle/ \left(1 + g_{i,i-1} \right) \right) + o(1) \quad (i = 2, \ldots, N),$$

(3.3.13)

$$\xi_N' = \left(\left(T_2 - \sum_{j=1}^{N-1} \xi_j' \right) \middle/ \left(1 + g_{N,N-1} \right) \right) + o(1).$$

(3.3.14)

The main part of mapping (3.3.12) to (3.3.14) (after discarding the terms $o(1)$) can be rewritten in the equivalent form:

$$\left(1 + g_{1,N} \right) \xi_1' + \sum_{j=2}^{N} \xi_j = T_2,$$

$$\sum_{j=1}^{i-1} \xi_j' + \left(1 + g_{i,i-1} \right) \xi_i' + \sum_{j=i+1}^{N} \xi_j = T_2 (i = 2, \ldots, N - 1),$$

$$\sum_{j=1}^{N-1} \xi_j' + \left(1 + g_{N,N-1} \right) \xi_N' = T_2.$$

The latter mapping represents the method of the Seidel solution of a linear system with a positive definite symmetric matrix. It is known to be convergent. It is easy to see that the limit is the point $\xi_1^0, \ldots, \xi_N^0$.

Furthermore, as already noted, the reasoning, completely analogous to those presented in Sect. 2 of this chapter, convince us of the existence of the attractor, which consists of the solutions for which the pulse of the ith neuron begins later than the pulse of the $(i - 1)$th neuron for the time that is asymptotically close as $\lambda \to \infty$ to ξ_i^0. As a result, we arrive at the following statement.

Theorem 3.3.1 *Let in the system of equations* (3.3.1) *through* (3.3.3) *the synaptic weights* $g_{i,i-1}$ *defined by formula* (3.3.8), *in which the numbers* ξ_i^0 *satisfy the constraints* (3.3.4) *through* (3.3.7). *Then, for sufficiently large values of* λ, *the system of equations* (3.3.1) *through* (3.3.3) *has the attractor in which the spikes of neighbouring* $(i - 1)$th *and* ith *neurons start through time intervals* $\xi_i^0 + o(1)$, *and* *the form of the spikes is given by asymptotic formulas from Theorem* 1.6.1*(with an appropriate shift of time).*

Thus far we have not used the specific values of the synaptic coefficients $g_{i-1,i}(i = 2,\ldots,N), g_{N,1}$ characterizing the action of the next neuron on the previous one. Choosing them by the method described previously, we can generate in the system of equations the attractor consisting of the waves of excitation travelling in the direction of a decreasing number of neurons and having preassigned distribution of the sequence of spikes. Choose arbitrarily numbers

$$0 < \xi_i^{00} < T_1 \quad (i = 1,\ldots,N)$$

and write

$$T_S^{00} = \sum_{i=1}^{N} \xi_i^{00}.$$

We assume that $T_S^{00} < T_2$ and $T_S^{00} - \xi_i^{00} > T_R$ $(i = 1,\ldots,N)$. Define synaptic coefficients $g_{i-1,i}(i = 2,\ldots,N), g_{N,1}$ by formulas:

$$g_{i-1,i} = \frac{T_2 - T_S^{00}}{\xi_i^{00}} (i = 2,\ldots,N), g_{N,1} = \frac{T_2 - T_S^{00}}{\xi_1^{00}}.$$

Then the system of equations has the attractor consisting of solutions, for which as $\lambda \to \infty$, and the start of spike of the $(i-1)$th neuron is delayed with respect to the start of spike of the ith neuron by the value $\xi_i^{00} + o(1)$ for $i = 2,\ldots,N$. The Spike of the Nth neuron starts after the start of spike of the first neuron. If the number of neurons belonging to the ring structure is large, then the described waves do not exhaust the set of all the attractors of the system (3.3.1) through (3.3.3). In this case, there exist vibrational modes, which we call "multiple waves." We clarify this phenomenon in the particular case when in system (3.3.1) through (3.3.3), $g_{i-1,i} = 0(i = 2,\ldots,N), g_{N,1} = 0$ and $g_{i-1,i} = g(i = 2,\ldots,N), g_{N,1} = g$, i.e., in the ring each neuron acts only on the next after it, and all relevant synaptic coefficients are the same.

In such a situation, as already shown, the system of equations (3.3.1) through (3.3.3) has the attractor consisting of excitation waves travelling in the direction of ascending of numbers of neurons. Intervals of time between the starts of spikes of neighbouring neurons are asymptotically identical and close as $\lambda \to \infty$ to number

$$\xi^0 = \frac{T_2}{N+g}. \tag{3.3.15}$$

From (3.3.4) and (3.3.7) follows the condition of existence of the given attractor:

$$0 < \frac{T_2 - NT_1}{T_1} < g < \frac{T_2(N-1) - NT_R}{T_R}. \tag{3.3.16}$$

Another attractor, which is present along with the first attractor, has the following biological meaning under certain relations between the parameters. Suppose that at zero time started the first spike of the neuron, thus giving rise to the wave of excitation propagating in the direction of ascending of numbers of neurons. Suppose further that for some reason, at a moment of time when the first wave has not reached the Nth neuron, started a new spike of the first neuron. As a result, the secondary wave of excitation will travel after the first one on the ring structure. Such a mode of excitation we call the "double wave." From a biological point of, it is clear that the passage of a double wave has certain specifics because each of the waves propagates along already-organized environment.

Denote by $\xi_i (i = 2, \ldots, N)$ the intervals of time between the starts of spikes of neighbouring neurons for the first wave at the initial excitation wave. In turn, lets $\eta_i = (i, \ldots, N)$ be the delay of the secondary spike of the ith neuron in the same cycle. In other words, the values

$$0, \sum_{j=2}^{i} \xi_j \quad (i = 2, \ldots, N); \quad \eta_1, \sum_{i=2}^{i} \xi_j + \eta_i \quad (i = 2, \ldots, N)$$

are the starts of spikes for the first and second excitation waves in the cycle. Introduce numbers ξ_i' and $\eta_i' (i = 1, \ldots, N)$ having the same meaning for the next cycle of the excitation wave. Taking these as analogues to the previous a priori assumptions, we can obtain the following relations between the described variables:

$$\xi_1' = \left(\left(T_2 - \left(\sum_{j=2}^{N} \xi_j - \eta_1 \right) \right) \middle/ (1+g) \right) + o(1),$$

$$\eta_1' = \frac{T_2 + g(\eta_N - \xi_1')}{1+g} + o(1),$$

$$\xi_i' = \left(\left(T_2 - \left(\sum_{j=i+1}^{N} \xi_j + \sum_{j=1}^{i-1} \xi_j^1 - \eta_i \right) \right) \middle/ (1+g) \right) + o(1), \quad (i = 2, \ldots N-1),$$

$$\eta_i' = \frac{T_2 + g(\eta_{i-1}' - \xi_i')}{1+g} + o(1),$$

$$\xi_N' = \left(\left(T_2 - \left(\sum_{j=1}^{N-1} \xi_j - \eta_N \right) \right) \middle/ (1+g) \right) + o(1),$$

$$\eta_N' = \frac{T_2 + g(\eta_{N-1}' - \xi_N')}{1+g} + o(1), \quad (i = 2, \ldots, N).$$

The latter formulas, after discarding the terms $(o)1$, define the mapping in the finite space, which is convenient in terms of investigation to reduce to a system of difference equations:

$$\eta_i = \frac{T_2 + g(\eta_{i-1} - \xi_i)}{1 + g} \quad (i = 2, 3, \ldots), \tag{3.3.17}$$

$$\xi_i = \left(\left(T_2 - \left(\sum_{j=1}^{N-1} \xi_{i-j} - \eta_{i-N}\right)\right)\middle/ (1 + g)\right) \quad (i = N+1, N+2, \ldots). \tag{3.3.18}$$

It is easy to see the relationship between the variables $\xi_i'(i = 1, \ldots, N), \eta_i'$ in the finite-dimensional mapping and the values $\xi_i(i = N+1, \ldots, 2N), \eta_{N+1}$ in the system of difference Eqs. (3.3.17) and (3.3.18): $\xi_i' = \xi_{i+N}(i = 1, \ldots, N), \eta_1' = \eta_{N+1}$.
The system (3.3.17) and (3.3.18) has a stationary solution:

$$\xi_i = \xi^{00} \equiv \frac{2T_2}{N + 2g}, \quad \eta_i = \eta^{oo} \equiv \frac{T_2 N}{N + 2g}. \tag{3.3.19}$$

For its asymptotic stability, the roots of the characteristic polynomial of the homogeneous problem must be located on the complex plane inside of the unit circumference. The polynomial itself $P(\mu)$ can easily be written:

$$P(\mu) = (1 + g)^2 \mu^{N-1} + (1 - g^2) \mu^{N-2} + \sum_{j=0}^{N-3} \mu^j.$$

One can convert the polynomial $(\mu - 1)P(\mu)$ to the form:

$$(\mu - 1)P(\mu) = \mu^{N-1}[(1 + g)\mu - g]^2 - 1.$$

Assume that it has the root $|\mu| \geq 1$. Then

$$|(1 + g)\mu - g|^2 < 1.$$

The latter inequality highlights the complex plane of the circle with the center at $z_0 = g/(1 + g)$ and a radius $\frac{1}{(1 + g)^{0.5}}$. This circle lies entirely inside the unit circumference and touches it at the point $\mu = 1$. Consequently, the polynomial $\mu - 1P(\mu)$ has the number $\mu = 1$ as the single root not lying inside the unit circumference. It remains to note that $P(1) = N + 2g > 0$, and therefore all the roots of the polynomial $P(\mu)$ satisfy the necessary inequality $|\mu| < 1$. Thus, stationary solution (3.3.19) of the system of difference Eqs. (3.3.17) and (3.3.18) is asymptotically stable.

The above-mentioned a priori conditions, under which we constructed asymptotic behaviour of the attractor of system (3.3.1) 1 (3.3.3), mean the fact that the stationary solution ξ^{00}, η^{00} of the difference system (3.3.17) and (3.3.18) must satisfy the relations:

$$0 < \xi^{00} < T_1, \quad (N-1)\xi^{00} > \eta^{00} + T_R, \quad \eta^{00} > T_R.$$

The meaning of the first inequality is that the spikes of neighboring neurons for each of the waves overlap in time. In turn, the second inequality means that at the moment when the first wave comes to any of neurons, it has already come out of the refractory state after generation of the second spike. The last inequality means that the second spike of a neuron can not occur before the time T_R after generation of the first spike. From these inequalities, it follows that the synaptic coefficients g in the system of equations (3.3.1) through (3.3.3) satisfy inequality:

$$\frac{2T_2 - NT_1}{2T_1} < g < \frac{(N-2)T_2 - NT_R}{2T_R}. \tag{3.3.20}$$

For consistency of the last inequality, it is necessary that $2T_R < (N-2)T_1$. Hence follows the estimate for the number N of neurons in the ring structure: $N > 2T_R/T_1 + 2$. Considering that earlier was taken earlier ratio $T_R > 2T_1$, we conclude that the ring structure must contain more than six neurons.

So, if we fulfill conditions (3.3.20) on synaptic weights g, then in the considered case the system of equations (3.3.1) through (3.3.3) has the attractor consisting of two waves travelling side to side. The intervals between spikes of neighbouring neurons for each of them are close as $\lambda \to \infty$ to ξ^{00}. At the same time, the second wave is delayed with respect to the first wave for time asymptotically close to η^{00}.

Note that if the value g satisfies (3.3.20), then it also satisfies inequality (3.3.16), which guarantees the existence of the first attractor. Consequently, the second attractor coexists along with the first one. Then we can talk about triple, etc., waves of excitation. Thus, the length of the ring should increase.

We can give biological interpretation to the results. Regarding the experiments of Livanov (1972, 1989) with rhythmic light stimulation, a number of authors (Zabrodin and Lebedev 1977) have considered the evidence of quantization of natural frequencies of the brain. In our model, this is reflected in the coexistence of the attractors of different periods. Furthermore, the presence in the brain of wave structures with stable and quantized difference of phases is the underlying hypothesis of the wave theory of memory developed at a qualitative level by Lebedev (1990). This theory has been indirectly confirmed and well proven in psychology. We proposed a simple variant and a model of neural structure in which such waves exist.

Multiple waves exist in the ring neural structure and in the case when the synaptic weights differ. The process of finite-dimensional mapping for determining mismatches ξ_i and η_i is analogous that previously presented. We should only replace g for $g_{i,i-1}$ in corresponding equations.

3.4   Model of the Double Ring of N Neurons

Consider the ring structure consisting of N identical neurons, each of which acts on two neurons following it. Consider that the Nth neuron is followed by the first one. We call such network a "double-ring network." In this case, the initial system of equation (3.1.1) (taking into account formulas (3.1.2) and (3.1.3) for the synaptic interaction) takes the form:

$$\dot{u}_1 = \lambda \left[\begin{array}{c} -1 - f_{Na}(u_1) + f_K(u_1(t-1)) + \\ + \alpha H(u_1)\left(g_{1,N-1}\Theta(u_{N-1}-1) + g_{1,N}\Theta(u_N-1)\right) \end{array} \right] u_1, \qquad (3.4.1)$$

$$\dot{u}_2 = \lambda \left[\begin{array}{c} -1 - f_{Na}(u_2) + f_K(u_2(t-1)) + \\ + \alpha H(u_2)\left(g_{2,N}\Theta(u_N-1) + g_{2,1}\Theta(u_1-1)\right) \end{array} \right] u_2, \qquad (3.4.2)$$

$$\dot{u}_i = \lambda \left[\begin{array}{c} -1 - f_{Na}(u_i) + f_K(u_i(t-1)) + \\ + \alpha H(u_i)\left(g_{i,i-2}\Theta(u_{i-2}-1) + g_{i,i-1}\Theta(u_{i-1}-1)\right) \end{array} \right] u_i (i = 3, \ldots, N),$$

$$(3.4.3)$$

where, as before, the function $H(u)$ is given by expression (3.1.5), and $g_{i,j}$ are synaptic weights.

Let the number of neurons be odd: $N = 2m+1 (N \geq 5)$. The system (3.4.1) through (3.4.3) may have a sufficiently rich set of attractors. We describe two of them here. The first attractor corresponds to the case when the spikes of neurons occur in the natural ascending order. Propagation of the wave is specific because there exists anticipatory synaptic communication, i.e., in the process of excitation of the ith neuron are involved both $(i-1)$th and $(i-2)$th neurons. The phenomenon is widespread in the nervous system: It has been considered (Schade and Ford 1976; Eckert et al. 1991) that neurons carry out a summation of signals both over space and time.

The second attractor consists of the waves, the excitement of which is transmitted through a neuron, i.e., from the ith to the $(i+2)$th neuron. This is possible if during the spike of the ith neuron the $(i+1)$th neuron is in a refractory state and the $(i+2)$th is disinhibited.

We begin from the second case. By virtue of the odd number of neurons, the path of the excitation wave is closed. Starting with the spike of the first neuron, it sequentially calls the spikes of the odd-numbered neurons. After the spike of the Nth neuron, the spikes of four neurons are sequentially induced. Finally, the wave of excitation reaches the $(N-1)$th (even) neuron, which gives rise to spike of the first neuron. The case does not differ from the case analyzed in Sect. 3.3 of this chapter. It is only necessary to make sure that in a stationary mode at the time of spike of the ith neuron, the $(i+1)$th neuron is in a refractory state.

Let $\xi^0_{3,1}, \xi^0_{5,3}, \ldots, \xi^0_{N,N-2}, \xi^0_{2,N}, \xi^0_{4,2}, \ldots, \xi^0_{N-2,N-4}, \xi^0_{1,N-2}$ be the expected asymptotic values of time between the starts of spikes of the $(i-2)$th and ith neurons for the solutions belonging to attractor. We assume that

$$0 < \xi_{i,\mathrm{mod}(i-2,N)} < T_1 \quad (i = 1, \ldots, N), \tag{3.4.4}$$

where mod $(i, N) \in [1, N]$ is a positive remainder of the division of the first number on the second number, in particular, for integers k mod $(kN, N) = N$. Inequality (3.4.4) means that the spikes of $(i-2)$th and the ith neurons overlap in time. Next, we introduce the number

$$T^0_S = \sum_{i=1}^{N} \xi^0_{i,\mathrm{mod}(i-2,N)}, \tag{3.4.5}$$

which asymptotically coincides with the expected period of passage of wave excitation on the ring. Assume that

$$T^0_S < T_2, \tag{3.4.6}$$

i.e., the period of passing of the wave of excitation is less than the neuron's own activity. Next, suppose that

$$T_S - \xi^0_{i,\mathrm{mod}(i-2,N)} > T_R \quad (i = 1, \ldots, N), \tag{3.4.7}$$

i.e., during time required for the return of the wave to the neuron, the neuron comes out of the refractory state. The given conditions (3.4.4) through (3.4.7) are completely analogous to the requirements in Sect. (3.3.4) through (3.3.7) in Sect. 3.3. In addition to the preceding, we assume that for any $i = 1, \ldots, N$

$$\sum_{j=1}^{m} \xi^0_{\mathrm{mod}(i+2j,N),\mathrm{mod}(i+2j-2,N)} + T_1 < T_R, \tag{3.4.8}$$

where, remember, $N = 2m + 1$. The latter inequality means that during spike of the ith neuron, the $(i+1)$th neuron is in a refractory state.

Now as the synaptic coefficients $g_{i,\mathrm{mod}(i-2,N)}$ $(i = 1, \ldots, N)$ we select:

$$g_{i,\mathrm{mod}(1-2,N)} = \frac{T_2 - T^0_S}{\xi_{i,\mathrm{mod}(i-2,N)}} \tag{3.4.9}$$

With such a choice of specified weights, regardless of the values $g_{i,\mathrm{mod}(i-1,N)}$ $(i = 1, \ldots, N)$ the system of equations (3.4.1) through (3.4.3) has the attractor consisting of the waves of excitation, which, propagating along the ring structure, successively generate the spikes of the $1, 3, \ldots, N, 2, 4, \ldots, N$-1st neurons. The numbers $\xi_{i,\mathrm{mod}(i-2,N)}$ $(i = 1, \ldots, N)$ asymptotically coincide with the time intervals between the beginnings of pulses. All of this follows directly from the results of previous Sects. 3.2 and 3.3 of this chapter.

We now turn to the attractor of the first type, when the excitation wave propagates in natural ascending order of the numbers of neurons. Generally speaking, there can be several such the attractors depending on the values of the synaptic coefficients. We choose (if possible) the remaining disposal coefficients $g_{i,i-1}$ in the system of equations so that it has the attractor, which has certain properties. For the incoming solutions, the spikes of neighbouring (according to the total umber neurons) overlap in time, but the spikes of neurons passing through a number of neurons no longer possess this property.

To derive asymptotic formulas, we need the value of the latent period of the response of the neuron to the sequence of two spikes. By analogy with Sect. 3.1, the problem reduces to the analysis of equation:

$$\dot{u}_1 = \lambda \left[\begin{array}{c} -1 - f_{\mathrm{Na}}(u) + f_K(u(t-1)) + \\ + \alpha H(u)(g_1 V(t - t_* - \xi_1) + g_2 V(t - t_*)) \end{array} \right] u. \qquad (3.4.10)$$

Here the functional $H(u)$ is given by expression (3.1.5) and the function $V(t) = 1$ at $t \in [0, T_1]$ and $V(t) = 0$ for $t \notin [0, T_1]$. Furthermore, $t_* - \xi_1 (\xi_1 > 0)$ and t_* are the moments of the beginning of the first and second pulses. Finally, $u(s) = \varphi(s) \in S$ for $s \in [-1, 0]$, where S is the class of functions introduced in Sect. 1.6 of this chapter.

Latent period $Q(t_*, g_1, g_2, \varphi)$ is said to be the difference between the first positive moment of time when $u(t)$ crosses a single value with positive speed (i.e., the beginning of a new pulse of the neuron) and the moment t_* of the beginning of the second external pulse. The value $Q(t_*, g_1, g_2, \varphi)$ is obtained by integrating Eq. (3.4.10) by the method of steps. We write out the given value only for the particular case we are interested in.

Let

$$0 < \xi_1 < T_1, \quad t_* - \xi_1 > T_R, \qquad (3.4.11)$$

where T_R is asymptotic meaning of the refractory period. In addition, let

$$\tau < t_* < \tau + \xi_1(1 + g_2), \qquad (3.4.12)$$

where $\tau = T_2 - T_1(1 + g_1 + g_2)$. Then

$$Q(t_*, g_1, g_2, \varphi) = \frac{T_2 - t_* - g_1 T_1}{1 + g_2} + o(1). \qquad (3.4.13)$$

Note that from formulas (3.4.11) through (3.4.13) easily follows inequality:

$$\xi_1 + Q(t_*, g_1, g_2, \varphi) > T_1. \tag{3.4.14}$$

It is also useful to restate the previous statement in the following form. Let inequality (3.4.11) hold, the latent period satisfy inequality (3.4.14), and the condition $Q(t_*, g_1, g_2, \varphi) < T_1$ hold. Then its value is calculated by formula (3.4.13).

Using the value of the latent period, we can write the finite-dimensional part of mapping defining the attractor. Let the spike of the first neuron occur at zero time, and let $\xi_i (i = 2, \ldots, N)$ be time intervals between the starts of spikes of $(i - 1)$th and the ith neurons at the initial wave of excitation. Next, let ξ_1' be the time interval over which the spike of the Nth neuron induces the spike of the first neuron and $\xi_1'(i = 2, \ldots, N)$ be intervals between the starts of spikes of $(i - 1)$th and the ith neurons at the next cycle of the wave excitation.

Make a number of a priori assumptions. Let

$$0 < \xi_i < T_1 \quad (i = 2, \ldots, N),$$
$$0 < \xi_i' < T_1 \quad (i = 1, \ldots, N),$$
$$\xi_i + \xi_{i+1} > T_1 \quad (i = 2, \ldots, N - 1),$$
$$\xi_N + \xi_1' > T_1,$$
$$\xi_i' + \xi_{i+1}' > T_1 \quad (i = 1, \ldots, N - 1).$$

This means that the spikes of the ith and $(i+1)$th neurons overlap in time; however, we do not observe overlapping of the spikes of the 1-st and i + 1st neurons. Next, suppose that

$$\sum_{j=2}^{N} \xi_j < T_2, \quad \sum_{j=i}^{N} \xi_j + \sum_{j=1}^{i-2} \xi_j' < T_2, \quad (i = 3, \ldots N), \quad \sum_{j=1}^{N-1} \xi_j' < T_2.$$

These inequalities guarantee that the wave from each neuron returns to it for a time period that is shorter than the period of the neuron's own activity.

Finally, assume that

$$\sum_{J=2}^{N-1} \xi_J > T_R,$$

$$\sum_{j=3}^{N} \xi_j > T_R,$$

$$\sum_{J=i+1}^{N} \xi_J + \sum_{j=1}^{i-2} \xi_j' > T_R, \quad (i = 4, \ldots, N - 1),$$

$$\sum_{J=1}^{N-2} \xi_j' > T_R.$$

These inequalities guarantee that at the second cycle of the passage of the excitation wave, at the moment of start of spike of the ith neuron, $(i+2)$th neuron is already out of the refractory state.

In view of the a priori conditions formulated at the second cycle of the excitation wave, formula (3.4.13) for the latent period of the respond is valid in the problem of action on the neuron of the two previous neurons. As a result, we obtain the following relations:

$$\xi_1' = \left(\left(T_2 - g_{1,N-1}T_1 - \sum_{j=2}^{N} \xi_j \right) \middle/ (1+g_{1,N}) \right) + o(1). \tag{3.4.15}$$

$$\xi_2' = \left(\left(T_2 - g_{2,N}T_1 - \sum_{j=3}^{N} \xi_j - \xi_1' \right) \middle/ (1+g_{2,1}) \right) + o(1). \tag{3.4.16}$$

$$\xi_1' = \left(\left(T_2 - g_{i,i-2}T_1 - \sum_{j=i+1}^{N} \xi_j - \sum_{j=1}^{i-1} \xi_j' \right) \middle/ (1+g_{i,i-1}) \right) + o(1) \tag{3.4.17}$$

$$(i=3,\ldots,N-1), \xi_N' = \left(\left(T_2 - g_{N,N-2}T_1 - \sum_{j=1}^{N-1} \xi_j' \right) / (1+g_{N,N-1}) \right) + o(1), \tag{3.4.18}$$

which themselves represent the finite-dimensional part of mapping sought.

After discarding in (3.4.15) through (3.4.18) the term $o(1)$, we obtain a convergent iterative process. In fact, it is the method of the Seidel solution of a nonhomogeneous linear system with a symmetric positive definite matrix.

We try to dispose of the weight coefficients $g_{i,N}, g_{i,i-1}$ $(i=2,\ldots,N)$ so that in the neighbourhood of the fixed point, the a priori conditions, under which the formulas of mapping (3.4.15) through (3.4.18) were discharged, have been performed.

Let ξ_i^{00} $(i=1,\ldots,N)$ be the numbers satisfying conditions:

$$0 < \xi_i^{00} < T_1, \qquad \xi_i^{00} + \zeta_{\text{mod}(i-1,N)}^{00} > T_1 \quad (i=1,\ldots,N) \tag{3.4.19}$$

Write

$$T_S^{00} = \sum_{i=1}^{N} \zeta_i^{00} \tag{3.4.20}$$

and require that

$$T_S^{00} < T_2 - T_1 \max_{i=1,\ldots,N} g_{i,\text{mod}(1-2,N)}, \tag{3.4.21}$$

$$T_S^{00} - \zeta_i^{00} - \zeta_{\text{mod}(i-1,N)}^{00} > T_R \quad (i=1,\ldots,N) \tag{3.4.22}$$

If we consider that the numbers $g_{i,\mathrm{mod}(1-2,N)}$ $(i = 1, \ldots, N)$ are already chosen from the condition of the existence of the second attractor, which is determined by the values $\xi_i^0 (i = 1, .., N)$, then inequality (3.4.21) is a joint restriction on the choice of numbers ξ_i^0 and ξ_i^{00} $(i = 1, .., N)$. The same applies to inequalities (3.4.8) and (3.4.22).

Suppose we can choose numbers ξ_i^{00} $(i = 1, \ldots, N)$, satisfying the constraints (3.4.19) through (3.4.22). Put for $(i = 1, \ldots, N)$

$$g_{i,\mathrm{mod}(i-1,N)} = \left(T_2 - T_S^{00} - T_1 g_{i,\mathrm{mod}(1-2,N)}\right)/\xi_i^{00}. \qquad (3.4.23)$$

Mapping (3.4.15) through (3.4.18) with discarded term $o(1)$ has the vector $\xi_1^{00}, \ldots, \xi_N^{00}$ as a stable fixed point. If the initial values $\xi_i(i = 2, \ldots, N)$ are chosen close to ξ_i^{00} $(i = 2, .., N)$, then all the above a priori assumptions will be executed.

In this case, the system has the attractor consisting of consisting of excitation waves travelling in the natural order of ascending the numbers of neurons. At the same time, intervals between spikes of the $(i - 1)$th and the ith neurons are asymptotically close as $\lambda \to \infty$ to numbers ξ_i^{00}. This attractor coexists along with the one described previously.

3.5 Concluding Remarks

We can consider the problem of propagation of the waves of excitation along a homogeneous neural network, which has cellular structure. In such a network, each neuron (except for extreme) is connected with four neighbouring neurons. Homogeneity implies that the connections are the same and are bidirectional. The network architecture of nine neurons is shown in Fig. 3.3.

Figure 3.3 shows neurons in bold points, and links are the solid lines. The arrows show one of the possible paths of excitement. Let at zero moment of time be the

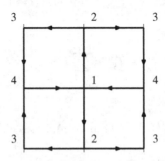

Fig. 3.3 Example of propagation of the wave excitation along a homogeneous network consisting of nine neurons

start the spike of the central neuron (no. 1). After some time, it induces ξ_1 simultaneous spikes of neurons marked with (no. 2), which in turn after some more time ξ_2 causes the spikes of angular neurons (no. 3). After time ξ_3, there spikes of the fourth neurons will start (no. 4). Finally, after time ξ_4 there will again begin spike of the central neuron (no. 1). In the stationary mode, $\xi_1 = \xi_2 = \xi, \xi_3 = \xi_4 = 0.5\xi$. The value ξ is determined by the synaptic weights. Because of the symmetry of links, the picture can be rotated by ninety degrees. There may be other, more complex, closed paths of conducting excitation.

References

Lorente de Nó, R. (1938). Analysis of the activity of the chains of internuncial neurons. *Journal of Neurophysiology, 1*, 207–244.

Eccles, J. (1966). *The physiology of synapses*. Moscow: Mir.

Rosenblatt, F. (1965). *Principles of neurodynamics*. Moscow: Mir.

Frolov, A. A., & Shulgina, G. I. (1983). Memory model based on the plasticity of inhibitory neurons. *Biophysics, 28*(3), 445–480.

Schade, J., & Ford, D. (1976). *Fundamentals of neurology*. Moscow: Mir.

Eckert, R., Randall, D., & Augustine, G. (1991). *Animal physiology* (vol. 1). Moscow: Mir.

Rabinovich, M. I., & Trubetskov, G. I. (1984). *Introduction to the theory of oscillations and waves*. Moscow: Nauka.

Anishchenko, V. S., Aranson, I. S., Postnov, D. E., & Rabinovich, M. I. (1986). Spatial synchronization and bifurcation of chaos in the chain of coupled oscillators. *Rep Acad Sci USSR, 28*(5), 1120.

Zaslavsky, G. M., & Sagdeev, R. Z. (1988). *Introduction to nonlinear physics*. Moscow: Nauka.

Dmitriev, A. S., & Kislov, V. Y. (1989). *Stochastic oscillations in radiophysics and electronics*. Moscow: Nauka.

Chua, L., & Yang, L. (1988a). Cellular neural networks: Theory. *IEEE Trans Circ Syst, 35*(10), 1257–1272.

Chua, L. O., & Yang, L. (1988b). Cellular neural networks: Applications. *IEEE Trans Circ Syst, 35*, 1273.

Osipov, G. V., Rabinovich M. I., & Shalfeev V. D. (1992). Dynamics of nonlinear synchronization networks (Vol. 2, p. 88). In *Proceedings of International Seminar "Nonlinear circuits and systems"*, June 16–18, Russia, Moscow.

Kashchenko, S. A. (1990a). Asymptotic analysis of the dynamics of the system of two coupled oscillators with delayed feedback. *Proceedings of Universities. A series of Radiophysics 33*(3), 308.

Kashchenko, S. A. (1990b). Spatially inhomogeneous structures in the simplest models with delay and diffusion. *Mathematical modelling, 2*(9), 49–69.

Zabrodin, J. M., & Lebedev, A. N. (1977). *Psychophysiology and psychophysics*. Leningrad: Nauka.

Lebedev, A. N. (1990). Human memory, its mechanisms and limits. *Research of memory* (pp. 104–118). Moscow: Nauka.

Lebedev, A. N., Mayorov, V. V., & Myshkin, I. Y. (1991). The wave model of memory. In A. V. Holden & V. I. Kryukov (Eds.), *Neurocomputers and attention, neurobiology, synchronisation and chaos* (Vol. 1, pp. 53–59). Manchester: Manchester University Press.

Kogan, A. B. (1967). *Problem of the movement of the nervous processes in the cortex of the brain*. Rostov-on-Don: RSU.

Kogan, A. B. (1979). *Functional organization of neural systems of the brain*. Leningrad: Nauka.

John, E. R. (1982). *Neurophysiological model of purposeful behaviour in neurophysiological mechanisms of behaviour* (pp. 103–128). Moscow: Nauka.

Kryukov, V. I. (1989). Attention and the principle of dominant by Ukhtomskii. In *International workshop "Neurocomputers and Attention"* (pp. 54–55), Moscow.

Kryukov, V. I. (1990). Do phase transitions take place in the brain or in a model? In *Neural networks: Theory and architecture* (pp. 95–116). UK: Manchester University Press.

Livanov, M. N. (1972). *Spatial organization of the processes of the brain*. Moscow: Nauka.

Livanov, M. N. (1989). *Spatiotemporal organization of potentials and systemic activity of the brain, in Selected works*. Moscow: Nauka.

Chapter 4
Model of Self-organization of Oscillations in the Ring System of Homogeneous Neural Models

According to the hypothesis of the wave encoding we are considering, data perceived by the brain are reflected and stored in the form of stable recurring combinations of spikes. One of the possibilities of packet formation of the desired structure is the adaptation of synapses causing interaction between neurons. Synapses themselves are quite plastic (Hebb 1949; Eccles 1966), and we can assume that there exists a mechanism of adjustment. In the theory of neural networks, this is a common assumption (Rosenblatt 1965; Dunin-Barkowski 1978); Frolov and Muravyov (1987, 1988). Modification of synapses takes time, and the wave packets are changeable formations. It seems that for adaptation of properties of the neural environment, there should be at least several cycles of the recurrence of the wave packet. Thus, the hypothesis of the wave coding, in our opinion, presupposes the existence of special memory for temporary packing of wave packets. We call it "short term," although the storage time can be prolonged. In our opinion, there are no synaptic changes in the neural environment when it is temporarily preserving wave packets. Memorizing is done by adjusting the phase oscillations, i.e., their self-organization. Note that in the classical models of neural networks (Rosenblatt 1965; Hopfield 1982), short-term and long-term memory do not differ. In contrast, biologists and psychologists conduct such classification (Pribram 1975).

Below is a model of a neural population that can sufficiently store different infinitely long wave packets. It represents a ring-oriented structure of neural modules containing excitatory and inhibitory elements. The division of neurons to excitatory and inhibitory states is due to the fact that each presynaptic neuron forms synapses of only one type (Eccles 1966, 1971): excitatory or inhibitory. It turns out (Mayorov 1994; Kashchenko and Mayorov 1995) that in the population within each module, part of the excitatory elements synchronizes the generation of spikes, and the remaining the pulses are repressed ineach remaining module. Temporary mismatches, uniquely determined by the number of inhibited excitatory neurons, occur among the groupings of spikes of neurons of neighboring modules. In modules, the composition and size of the groups of synchronously functioning excitatory neurons depends on the initial conditions. If we interpret the spatial distribution (i.e., patterns) of spikes in the module as an image, then the network can indefinitely store a given sequence of images. Both patterns themselves and their temporal distribution can be informative.

© Springer International Publishing Switzerland 2015
S. Kashchenko, *Models of Wave Memory*,
Lecture Notes in Morphogenesis, DOI 10.1007/978-3-319-19866-8_4

4.1   Model of Action of a Burst of Spikes on the Neuron

Here and throughout the chapter we talk about the interaction of neurons by way of chemical synapses. It is assumed that the dynamics of the membrane potential of an isolated neuron is described by Theorem 1.6.1. As done previously, we associate the beginning and end of a spike with moments of time when the membrane potential crosses the unit value with positive and negative speed, respectively. We believe that the effective time of action of a neurotransmitter coincides with the asymptotic duration of the spike of the presynaptic neuron. For the postsynaptic neuron, its spike and some time interval following it represent a zone of refractoriness or resistance to mediatory action, i.e., it is considered that part of the increased excitability is missing after a spike occurs. We assume that the duration of the refractory period represents two or three durations of spike.

We previously examined in sufficient detail the result of the action of the neurotransmitter in cases of one (Sects. 2.3 and 3.1) and two synapses (Sect. 3.4). To study the formations of neural modules, we must explore a more general situation.

Consider the neuron is under the synaptic action m of other neurons. In this case, the total system of Eqs. (3.1.1) for the neuronal population, where the synaptic interaction is modeled by (3.1.2) and (3.1.3), reduces to one equation:

$$\dot{u} = \lambda[-1 - f_{Na}(u) + f_K(u(t-1))] + \lambda\alpha H(u)\left[\sum_{k=1}^{m} g_k V(t - t_k)\right]. \qquad (4.1.1)$$

Here $f_{Na}(u)$ and $f_K(u)$ are positive and sufficiently smooth functions monotonically tending to zero as $u \to \infty$, the parameter $\lambda \gg 1$, and the value $\alpha = f_K(0) - f_{Na}(0) - 1 > 0$ given by formula (1.5.8). Thus, in the absence of an external action the conditions and conclusions of Theorem 1.6.1 are fulfilled. In (4.1.1), the function $V(t)$ is an indicator pointing to the presence of a neurotransmitter: $V(t) = 1$ for $t \in [0, T_1]$ and $V(t) = 0$ for $t \notin [0, T_1]$. Here T_1 is an asymptotic duration of spike. Furthermore, $t_k(k = 1, \ldots, m)$ is the start time of the release of the neurotransmitter on the kth synapse. In turn, g_k are the synaptic weights, which determine the contribution of the kth synapse in the dynamics of the membrane potential. Finally, in (4.1.1), $H(u)$ is a functional providing the presence of refractory period. As we have already noticed, it can be discharged in several ways. In this chapter, we will consider the following representation:

$$H(u) = \theta(1 - u(t)) \cdot \theta(1 - u(t - 1 - (1 - \varepsilon)T_1)) \\ \times \theta(1 - u(t - 1 - 2(1 - \varepsilon)T_1)), \qquad (4.1.2)$$

where $\theta(*)$ is function of Heaviside, and $0 < \varepsilon < 1$. Functional $H(u)$ vanishes during the time of the spike and during time $2(1 - \varepsilon)T_1$ after the spike. Thus, the duration of the refractory period, including the time of the spike, is $T_R = (3 - 2\varepsilon)T_1$. We assume that the refractory period is shorter than the period of the neuron's own activity, i.e., $T_R < T_2$ where T_2 by Theorem 1.6.1 asymptotically coincides with the period of its own activity. Chosen representation (4.1.2) for the functional $H(u)$ is fully consistent with the ideology described in Sect. 3.1, and it simultaneously simplifies the following calculations below.

According to the results of Sect. 2.3, if the synaptic weight is $g_k > 0$, then we consider the synapse as excitatory, and if $g_k < 0$ we consider the synapse to be inhibitory.

Introduce the notion of burst action, due to which this chapter owes its name. We enumerate synapses in the order of the beginning of the release of the neuro-transmitter on them, i.e., assume that $t_1 \leq t_2 \leq \ldots \leq t_m$. We call it "burst action" if the neurotransmitter appeared in the mth synapse earlier than ended its effective action released in the first synapse. This means that $t_m < t_1 + T_1$.

The choice of initial conditions is important for the description of the properties of the solutions of Eq. (4.1.1). Write $h = 2(1 - \varepsilon)T_1 + 1$, and introduce the set S_h, analogous to the set S. It consists of continuous functions $\varphi(s)$, on the interval $s \in [-h, 0]$ for which $\varphi(0) = 1$ and $0 < \varphi(s) \leq \exp(\lambda \alpha s / 2)$. Denote the solution of Eq. (4.1.1) with the initial conditions $u(s) = \varphi(s) \in S_h$ by $u(t, \varphi)$. Thus, the start of spike of neuron is initially confined to the zero moment of time.

Let the neuron carry on itself $m > 1$ of synapses of the excitatory type and all synaptic weights be the same: $g_k = g > 0 (k = 1, \ldots, m)$. Consider the situation when the burst of spikes acts on the neuron, which has come out of the refractory state. Then the following time relationships take place: $T_R < t_1 \leq t_2 \leq \ldots \leq t_m < T_2$ and $t_m < t_1 + T_1$. The presence of the excitatory neurotransmitter approximates the start t^{Sp} of the next spike of the neuron $t^{Sp} < T_2$.

We say that the neuron responds directly to bursts when its spike starts earlier than the disintegrated neurotransmitter generated by the first pulse, i.e., $t_m < t^{Sp} < t_1 + T_1$.

Introduce the values $\xi_k = t_k - t_{k-1} (k = 2, \ldots, m)$ of time mismatch of the beginning of the neurotransmitter's release in the kth and $(k-1)$th synapses. Denote by $Q = t^{Sp} - t_m > 0$ the delay of the start of spike of neuron with respect to the start of the last burst. By analogy with Sect. 3.1, call the value Q the "latent period." The condition under which the neuron responds directly to the burst action, in terms of the latent period, is as follows: $0 < Q < t_1 + T_1 - t_m$.

Lemma 4.1.1 *Let the neuron respond directly to the burst action. Then as $\lambda \to \infty$ holds asymptotic formula:*

$$Q = \left(T_2 - t_m - g \sum_{k=2}^{m} (k-1)\xi_k \right) / (1 + mg) + o(1), \qquad (4.1.3)$$

At the same time, $u(t^{Sp} + s, \varphi) \in S_h$ for $s \in [-h, 0]$.

The proof of Lemma (4.1.1) is carried out by the following scheme. The state of the neuron at moment of time t_1 is defined by formulas of Theorem 1.6.1, according to which $u(t_1, \varphi) = \exp[-\lambda\alpha(T_2 - t_1 + o(1))]$. For $t > t_m$, while $u(t, \varphi) \ll 1$ and $t < t_1 + T_1$, from Eq. (4.1.1) we immediately obtain:

$$u(t, \varphi) = \exp\left[-\lambda\alpha\left(T_2 - t_1 - g\sum_{k=1}^{m}(t - t_k) - (t - t_1) + o(1)\right)\right]. \qquad (4.1.4)$$

The moment t^{Sp} of the start of the spike (under condition $t_m < t^{\text{Sp}} < (t_1 + T_1)$) is defined from equation

$$T_2 - t_1 - g\sum_{k=1}^{m}(t - t_k) - (t - t_1) + o(1) = 0.$$

Introduced numbers $\xi_k (k = 2, \ldots, m)$ and Q allow transformation of the latter relationship to the form:

$$T_2 - t_m - g\left(mQ + \sum_{k=2}^{m}(k - 1)\xi_k\right) - Q + o(1) = 0,$$

from which follows equality (4.1.3). Inclusion of $u(t^{\text{Sp}} + s, \varphi) \in S_h$ follows from representation (4.1.4).

The conditions under which the neuron directly responds to external bursts can easily be written. From inequalities $t^{\text{Sp}} \equiv t_m + Q < t_1 + T_1$ and $t^{\text{Sp}} > t_m$, we obtain for the value of the latent period Q:

$$0 < Q < T_1 - \sum_{k=2}^{m}\xi_k. \qquad (4.1.5)$$

Using formula (4.1.3) for Q, the latter inequality and the relationship $t_1 > T_R$ (the burst acts on the neuron, which has come out of the refractory state), we obtain:

$$t_m > \max\left[\left(T_R + \sum_{k=2}^{m}\xi_k\right), T_2 - g\sum_{k=2}^{m}(k - 1)\xi_k - (1 + mg)\left(T_1 - \sum_{k=2}^{m}\xi_k\right)\right]. \qquad (4.1.6)$$

Thus, the burst must not start too early. Simultaneously, the burst cannot enter the synapses too late ($Q > 0$):

$$t_m < T_2 - g\sum_{k=1}^{m}(k - 1)\xi_k. \qquad (4.1.7)$$

We assume that the value of synaptic weight g is not too large and that the right-hand side of (4.1.7) is positive. Otherwise, the spike of the neuron occurs before the neurotransmitter begins to release on the mth synapse. This means that the bursts, the number of which is less than m., i.e., since the neuron passed into the refractory state during the last pulse, act on the neuron.

We can use the obtained formula (4.1.3) to predict the result of action of the next burst, i.e., for $t_k > t^{Sp}(k = 1,\ldots,m)$, replacing t_k with $t_k - t^{Sp}(k = 1,\ldots,m)$. In this regard, we consider the periodic burst action on the neuron. Initially let the burst, which has come on the neuron, have a start time t_m of the neurotranmitter's release in the mth synapse, thus satisfying inequalities (4.1.6) and (4.1.7). Later the same burst appears with period T. The spike of the neuron, as a response to the first burst, occurs at moment of time $t^{Sp} = t_m + Q$. With respect to this moment, the next release of the neurotransmitter in the mth synapse starts after time $T - Q$. We make an priori assumption that the neuron also responds directly to the second packet. Then the delay Q' of the new spike of the neuron with respect to the start of the release of neurotransmitter on the mth synapse is:

$$Q' = \left(T_2 - T + Q - g \sum_{k=2}^{m} (k-1)\xi_k \right) \Big/ (1 + mg) + o(1). \qquad (4.1.8)$$

the a priori assumption is satisfied if the value Q' satisfies inequality (4.1.5). From this follow the restrictions on the period T:

$$T_2 - mg \left(T_1 - \sum_{k=2}^{m} \xi_k \right) - g \sum_{k=2}^{m} (k-1)\xi_k < T < T_2 - g \sum_{k=2}^{m} (k-1)\xi_k. \qquad (4.1.9)$$

It is clear that in the future the mismatch of the start of the neurotransmitter's release on the mth synapse and on the spike of the neuron is also subject to (4.1.8). Discard in (4.1.8) the term $o(1)$. The obtained relationship can be viewed as the iterative process. It converges to the stationary point

$$Q^* = \left(T_2 - T - g \sum_{k=2}^{m} (k-1)\xi_k \right) \Big/ mg \qquad (4.1.10)$$

From the construction, it follows that the periodic burst action imposes its frequency on the neuron. Let the period of action T satisfies inequality (4.1.9). The initial spike of the neuron occurred at zero time $(u(s, \varphi) \in S_h)$. The minimum positive moment of time t_m of the start of the release of the neurotransmitter in the mth synapse is subordinated to inequalities (4.1.6) and (4.1.7). Then, as $t \to \infty$ and $\lambda \to \infty$, the spike of the neuron is delayed with respect to the next moment of the start of the release of the neurotransmitter in the mth synapse by value $Q^* + o(1)$. The form of the spike is given by the asymptotic equalities of Theorem 1.6.1 with the natural shift in time.

Bursts of spikes can quite differently influence the neuron if it carries on itself synapses of both the excitatory and inhibitory types. Let m_1 and m_2 $(m_1 + m_2 = m)$, respectively, be the number of excitatory and inhibitory synapses and all synaptic weights be the same in modulus: $|g_k| = g(k = 1, \ldots, m)$. The action is said to be inhibitory if $m_1 < m_2$. Introduce the set P_h^γ, consisting of continuous functions $\psi(s)$, on the interval $s \in [-h, 0]$ for which $0 < \psi(s) < \exp(-\lambda\gamma)$, where $\gamma > 0$. Fix moment of time $T_* > t_m + T_1 + h$.

Lemma 4.1.2 *Let $t_m + T_1 + h < t_* < T_*, \beta = gT_1(m_2 - m_1) - T_* > 0$ and for solution $u(t, \varphi)$ of Eq. (4.1.1) let the initial condition be $u(s, \varphi) \in P^\gamma$, where $\gamma < \alpha(T_* + gm_1T_1)$. Then as $\lambda \to \infty$ holds inequality $0 < u(t) < 1$ for $t \in [0, t_*]$, $u(t_* + s, \varphi) \in P_h^{\gamma_*} \subset P_h^\gamma$, where $\gamma_* = \gamma + \alpha\beta.$, is a valid inclusion.*

For the proof of Lemma 4.1.2, we use asymptotic equalities: $f_{\text{Na}}(u) = f_{\text{Na}}(0) + o(1), f_K(u) = f_K(0) + o(1)$ at $u \ll 1$. From Eq. (4.1.1) follows the representation $u(t, \varphi) = u(0, \varphi)\exp\lambda(\alpha + o(1))t$ at $t \in [0, t_1]$. Recall that the functions $V(t - t_k)(k = 1, \ldots, m)$ turn into one at $t \in [t_k, t_k + T_1]$, and otherwise are identically equal to zero. In view of this, from Eq. (4.1.1) immediately follows $0 < t < T_*$ the estimate for $0 < t < T_*$:

$$u(t, \varphi) \leq u(0, \varphi) \exp \lambda\alpha(t + gm_1T_1 + \delta) < u(0, \varphi) \exp \lambda\alpha(T_* + gm_1T_1),$$

where $\delta > 0$ is arbitrarily small. Since $u(0, \varphi) < \exp(-\lambda\alpha(T_* + gm_1T_1))$, then $u(t, \varphi) < 1$ as $\lambda \to \infty$. Furthermore, when $t_m + T_1 < t < T_*$, from Eq. (4.1.1) we obtain:

$$u(t, \varphi) = u(0, \varphi) \exp \lambda\alpha(t + gT_1(m_1 - m_2) + o(1)) \\ < u(0, \varphi) \exp \lambda\alpha(T_* + gT_1(m_1 - m_2)). \tag{4.1.11}$$

Hence follows the inclusion given in Lemma 4.1.2.

The meaning of Lemma 4.1.2 is quite simple. If initially the membrane is polarized so that the neuron, even under the action of excitatory synapses, can not generate spike until moment of time T_*, then the additional effect of the inhibitory synapses increases the degree of polarization of the membrane with respect to the initial one.

From Lemma 4.1.2 it follows that T-periodic inhibitory action, for which $T < gT_1(m_2 - m_1)$, can completely suppress the generation of pulses (i.e., inhibit the neuron). It is observed, for example, in the case when initially the membrane is quite strongly polarized: $u(s, \varphi) \in P_h^\gamma$, where $\gamma < \alpha(T + gm_1T_1)$.

4.2 Model of Action of the Burst of Spikes on the System of Two Neurons

Let on each of identical neuron act one and the same burst consisting of m-spikes incoming from the outside. In this case, neurons are interconnected by excitatory synapses. We consider all external synaptic weights to be the same as well as positive. By virtue of (3.1.1) through (3.1.3), the given neuronal formation is described by the system of equations:

$$\dot{u}_1 = \lambda \left[\begin{array}{l} -1 - f_{Na}(u_1) + f_K(u_1(t-1)) \\ + \alpha H(u_1) \left(g \sum_{k=1}^{m} V(t-t_k) + g_{1,2}\theta(u_2(t)-1) \right) \end{array} \right] u_1, \qquad (4.2.1)$$

$$\dot{u}_2 = \lambda \left[\begin{array}{l} -1 - f_{Na}(u_2) + f_K(u_2(t-1)) \\ + \alpha H(u_2) \left(g \sum_{k=1}^{m} V(t-t_k) + g_{2,1}\theta(u_1(t)-1) \right) \end{array} \right] u_2, \qquad (4.2.2)$$

Here u_1 and u_2 are the values of membrane potentials of the first and second neurons, respectively; $\theta(u_1(t)-1)$, $\theta(u_2(t)-1)$ are indicators of the neurotransmitter's presence released as the result of the spike of each neuron on the corresponding synapse of another neuron; $g_{1,2}$, $g_{2,1}$ are the weights of these synapses; and $g > 0$ are the synaptic weights of the external action. All other notations are described in the consideration of Eq. (4.1.1). In particular, the functional $H(u)$, guaranteeing the existence of refractory period, is given by formula (4.1.2).

We first consider the case when the second neuron does not influence the first one and the synaptic weight of action of the first neuron on the second neuron coincides with the weight of the external action: $g_{1,2} = 0$, $g_{2,1} = g$.

Let spike of the first neuron start at zero time and of the second at the time moment $\xi(0 < \xi < T_1)$, and $u_1(s) = \varphi_1(s) \in S_h$, $u_2(\xi+s) = \varphi_2(s) \in S_h$, i.e., initially the spikes are mismatched by value ξ. The first neuron is influenced by only the external burst. For the second neuron, the induced spike of the first neuron is added to the burst . We immediately note that by definition of burst action, $T_R + \xi < t_1 \le t_2 \le \ldots \le t_m < T_2$ and $t_m < t_1 + T_1$. Recall also that the neuron responds directly to the burst if its spike starts earlier than when the effective time T_1 of the neurotransmitter's action, which is released in response to the first burst, is over. Let t_1^{Sp} and t_2^{Sp} be the moments of the start of new spikes, respectively, of the first and second neurons and $\xi' = t_2^{Sp} - t_1^{Sp}$ be their new mismatch.

Lemma 4.2.1 *Let each of neurons responds directly to the burst action incoming to it. Then as $\lambda \to \infty$, holds representation:*

$$\xi' = \xi/(1 + g(m+1)) + o(1), \qquad (4.2.3)$$

thus, we have the inclusions $u_1(t_1^{Sp} + s) \in S_h$ and $u_2(t_2^{Sp} + s) \in S_h$.

The proof of Lemma 4.2.1 is as follows. According to Lemma 4.1.1, we determine the time moment of the start of spike of the first neuron, $t_1^{Sp} = t_m + Q_1$, where $Q_1 = Q$, and the value Q is calculated by formula (4.1.3). Because the first neuron responds directly to the burst action, $t_1 < t_1^{Sp} < t_1 + T_1$. Thus, the spike of the first neuron is related to the burst acting on the second neuron. By virtue of Lemma 4.1.1, we determine the moment of the start of the spike of the second neuron, $t_2^{Sp} = t_1^{Sp} + \xi'$, where $\xi' = Q$. The value Q is still calculated by formula (4.1.3), in which we should replace m to $m+1$, denote $t_{m+1} = t_1^{Sp} - \xi$, and write $\xi_{m+1} = Q_1$. As a result, we obtain:

$$\xi' = \left(T_2 - \left(t_1^{Sp} - \xi \right) - g \sum_{k=2}^{m} (k-1)\xi_k - mgQ_1 \right) / (1 + (m+1)g) + o(1).$$

The given expression reduces to (4.2.3) if we use formula (4.1.3) for the value Q_1. Inclusions $u_1(t_1^{Sp} + s) \in S_h$ and $u_2(t_2^{Sp} + s) \in S_h$ follow directly from Lemma 4.1.1.

The first neuron responds directly to the burst action if moment of time t_m of the beginning of the release of neurotransmitter in the mth synapse, which is adjacent to it, satisfies inequalities (4.1.6) and (4.1.7). The conditions under which the second neuron responds directly to the burst action are also written out on the basis of inequalities (4.1.6) and (4.1.7), which in this case take the form:

$$t_1^{SP} - \xi > T_2 - g \sum_{k=2}^{m} (k-1)\xi_k - gmQ_1 - (1 + (m+1)g)\left(T_1 - \sum_{k=2}^{m} \xi_k - Q_1 \right),$$

$$t_1^{SP} - \xi < T_2 - g \sum_{k=2}^{m} (k-1)\xi_k - gmQ_1.$$

Inequalities can be simplified by using formula (4.1.3) for the value Q_1. As a result, we obtain:

$$0 < \xi < (1 + (m+1)g)\left(T_1 - \sum_{k=2}^{m} \xi_k - Q_1 \right), \qquad (4.2.4)$$

where Q_1 is given by (4.1.3). Thus, the second neuron responds directly to the burst if the initial mismatch ξ of the start of the spikes of neurons is not too large. Note that the new mismatch ξ' of neuron spikes also satisfies the resulting inequality.

Let us now consider the problem of a T-periodic action of external burst on the system of two neurons, each of which influences the other with the weight of external influence, i.e., we assume that in Eqs. (4.2.1) and (4.2.2) that $g_{1,2} = g_{2,1} = g$. In the asymptotic integration of the system (4.2.1) and (4.2.2), we use the described previously (Sect. 3.2) the method of choice for determining the initial conditions for each of the equations in their own time. We give the moments of the spikes' starts and, as known from biological reasons, the prehistory of evolution of the membrane potentials on the time intervals preceding the spikes. On the basis of this information, the solution of the system is uniquely constructed.

Let $u_1(s) = \varphi(s) \in S_h$ and $u_2(\xi + s) = \varphi(s) \in S_h$, i.e., initially confine the spike of the first neuron to zero moment of time and spike of the second neuron to the time moment ξ. We consider that $0 < \xi < T_1$. In the time interval $t \in [0, \xi]$, the value $u_2(t) < 1$, and therefore $\theta(u_2(t) - 1)) = 0$ and Eq. (4.2.1), can be integrated. The asymptotic approximation is constructed with the help of Theorem 1.6.1. At $t > \xi$ for system (4.2.1) and (4.2.2), we obtain the usual Cauchy initial problem.

For solution of system (4.2.1) and (4.2.2), we construct an asymptotic approximation on any finite interval. According to biological meaning, each of the neurons acts on the other neuron only during the generation of its spike. Simultaneously the neuron itself during its spike and in time interval of h duration is refractory to any external action. The second neuron generates its first spike in the time interval $[\xi, \xi + T_1 + o(1)]$, to which the first neuron is refractory (in Eq. (4.2.1) the functional $H(u_1) = 0$). Thus, Eq. (4.2.1), in any case, is integrated independently of Eq. (4.2.2) before the beginning of the generation of a new spike by the second neuron. We find the moment t_1^{Sp} of the start of the spike by Lemma 4.1.1 as the solution of the problem of the external action of the burst on the first neuron. The spike of the first neuron is added to the burst, thus acting on the second neuron. We also find the moment t_2^{Sp} of the start of the spike of the second neuron by Lemma 4.1.1. As a result, we return to the initial situation if we count the time from the moment t_1^{Sp} of the start of the spike of the first neuron. It is easy to see that we actually solve the problem that we have already studied regarding the external burst action on the system of two neurons in the case when the second neuron does not act on the first one. If the neurons directly respond to burst action, the new mismatch ξ' of the start of the neurons' spikes can be determined by formula (4.2.2) of Lemma 4.2.1. It remains to write out the conditions under which the respond of the neurons is immediate.

Let at the first cycle of external burst action, at the moment time t_m of the beginning of the mth pulse, satisfy inequalities (4.1.6) and (4.1.7). Assume that for the period T of external action, the inequality is satisfied. Under these conditions, the first neuron will respond directly to T, i.e., a periodic burst action. Denote the initial delay of the start of the second neuron's spike with respect to the first one through $\xi(0)$, and through $\xi(k)(k = 1, 2, \ldots)$ are the subsequent values of this quantity depending on the number of cycles of external influence. Assume that for $\xi(0)$ inequality (4.2.4) is satisfied, and $0 < \xi(0) < T_1$. Then the second neuron will directly respond to periodically received bursts, which include the spike of the first neuron. Applying Lemma 4.2.1, we obtain:

$$\xi(k) = \xi(k-1)/(1 + g(m+1)) + o(1) \qquad (k = 1, 2, \ldots).$$

As a result, because of the $\lambda \to \infty$ and $k \to \infty$ mismatch of the start the neurons' $\xi(k) \to o(1)$, i.e., over time, the neurons will function almost synchronously. The delay of the start the neurons' spikes with respect to the beginning of the mth pulse of the external burst is $Q_* + o(1)$, where the value Q_* is given by expression (4.1.10).

We discussed the role of external burst action in the model of organization of oscillations on the example of two neurons. The construction is transferred to the case of the system of an arbitrary number of elements that are interconnected by an excitatory action and synapses that are equal in weight. If the initial mismatch of the start of the spikes is not large, then over time the pulses of neurons are almost synchronized. The external periodic action imposes its –period on the system.

Physiologists (Livanov 1972, 1989; Osipov et al. 1992; Singer et al. 1976) performed experiments (on both animals and humans) on the effects on the action on the brain of periodic light signals. According to experimental data, they observed a synchronization of the activity of the neuronal ensembles and the phenomenon of imposing frequency.

4.3 Architecture and Equations of Neural Network with Modular Organization

A neural network is a population of similar neurons that are functionally related to each other. It is extremely difficult to clarify the structure of relations in the nervous system of living organism because each neuron forms synapses on thousands of other neurons. According to Blum et al. (1988), one of the possible ways of organizing neural networks hierarchically is as follows. A neural population is divided into levels, between which there is definite dependence. Each level of neurons (or many levels) accepts signals from many (or all) of the neurons of the previous level. In its turn, this level of neurons acts on many (or all) neurons at the next level. Problems of the interaction between elements of one level with another are not discussed here. According to Blum et al. (1988), the hierarchical organization of structures is responsible only for the initial processing of information.

The viewpoint of Blum et al. (1988) is in harmony with the ideas of Edelman and Mountcastle (1981), Edelman (1987) about the principle of modular organization of the brain when neurons split into bands (mini-columns) that are dynamically linked to each other. To explain the mechanism of the formation of bands, Edelman and Mountcastle developed the selective theory. Groups of neurons are fixed as a result of selection. The point of view of the constancy of groups (as well as the principles of selection) was criticized by Crick (1990). According to a remark by Crick, it is conceivable that part of the neurons in groups is replaced by the other neurons.

In the model considered below, modules, although are fixed, play the role of reservoirs from which, as a result of an action, can release groups of synchronously functioning neurons. An important place is occupied by the inhibitory neurons, the spikes of which suppress pulses of the elements related to them.

The functional significance of inhibitory neurons has been proved by Eccles (1971) (currently there are difficulties in the identification of inhibitory and excitatory elements). According to Eccles (1971), it seems that inhibitory components are not uncommon. Apparently they must be part of any neural association. In Edelman and Mountcastle's theory, inhibitory neurons are also an integral part of neural associations.

To avoid misunderstanding, it should be noted that in the papers of listed well-known physiologists, the ring organization of neural modules is not considered. For us the composition of modules and the possible hierarchical principle of connection between them is important . At the same time, morphological ring neural formations, caused in the brain by its own structure, have long been identified (Lorente de Nó 1938). Therefore, their simulation is of interest.

Let us turn to the model of functioning of neurons organized in a ring-modular structure. Consider the network consisting of N neural modules (associations). Let each of modules contain n excitatory and q inhibitory neurons. Consider that within each module any excitatory neuron has a synaptic influence on any other neuron. In turn, each inhibitory neuron acts only on the excitatory elements but does not have access to the inhibitory ones.

Assume that the neural modules form a ring-oriented structure, in which communications between the modules are arranged as follows. Each excitatory element of the ith module acts on any excitatory neuron of the $(i+1)$th module, but it has no synaptic endings on the inhibitory neurons of the $(i+1)$th module. At the same time, associations with numbers $i + kN(k = 0, \pm 1, \ldots)$ are identified. The parameters of all neurons and the absolute values of synaptic weights are the same.

The structure of the excitatory and inhibitory connections in the population is consistent with biological concepts (Eccles 1966, 1971): The excitation process is more global than the process of inhibition.

Let us enumerate the excitatory and inhibitory elements separately by a pair of indices (i,j), where i is number of module, and j is number of the neuron. The synaptic weights, when reflecting the influence of excitatory elements, are positive and for inhibitory elements are negative. Denote by $u_{i,j}$ and $v_{i,m}$ the values of the membrane potentials of the (i,j)th and (i,m)th of the excitatory and inhibitory neurons, respectively. The system of Eqs. (3.1.1), where the synaptic interaction is given by (3.1.2) and (3.1.3), for the considered population takes the form:

$$
\dot{u}_{i,j} = \lambda \left[
\begin{array}{l}
-1 - f_{Na}\left(u_{i,j}\right) + f_K\left(u_{i,j}(t-1)\right) \\
+ \alpha g H\left(u_{i,j}\right)\left(\sum_{r=1}^{n}[\theta\left(u_{i-1,r}-1\right) + \theta\left(u_{i,r}-1\right)] - \sum_{r=1}^{q}\theta\left(u_{i,r}-1\right)\right)
\end{array}
\right] u_{i,j},
$$

$$(4.3.1)$$

$$
\dot{u}_{i,m} = \lambda\left[-1 - f_{Na}\left(v_{i,m}\right) + f_K\left(v_{i,m}(t-1)\right) + \alpha g H(v_{i,m})\sum_{r=1}^{n}\theta\left(v_{i,r}-1\right)\right] v_{i,m},
$$

$$(4.3.2)$$

where $i = 1,\ldots,N; \quad j = 1,\ldots,n; m = 1,\ldots,q$. The functional $H(*)$, providing the presence of the refractory period, is defined by formula (4.1.2), and g is the total in absolute magnitude value of the synaptic weights. Functions of the form $\Theta(u-1)$ are indicators of the presence of the neurotransmitter, which is released as a result of the spike of the corresponding neuron. The number $\alpha = f_K(0) - f_{Na}(0) - 1 > 0$, i.e., the speed parameter, is $\lambda \gg 1$.

Let us state the hypothesis about the possible, manner of organization of oscillations in the ring structure of neural modules. Let for some reason in one of the modules, with a slight mismatch in time, begin the spikes of excitatory elements. A created burst first stimulates the spikes of the inhibitory neurons, which suppress the pulses of the excitatory neurons of this module and are not included in the initial group. Second, the given burst, acting on excitatory neurons of the next module, causes spikes of some of them. These are the neurons, the membranes of which are relatively less polarized. Thus, the initial situation is repeated for the neuron module following the initial one. As a result, the wave of spikes cyclically spreads on the neural structure. In any module, each wave passage begins with the pulses of excitatory spikes and ends with the spikes of inhibitory neurons. The wave represents the unity of the two components of excitation and inhibition.

4.4   Algorithm of Asymptotic Integration of the System of Equations of a Neural Network with Ring-Modular Organization

We assume that the number of neurons of inhibitory type in modules satisfying the condition $\beta_0 = gT_1(q-2n) - T_2 > 0$, is sufficiently large. In the asymptotic integration of system (4.3.1) and (4.2.3) the above-described above methodological procedure is used to specify the initial conditions for the equations at different moments of time. We will carry out the countdown of modules with some arbitrarily fixed number k. We first define the initial conditions for the equations describing the excitatory neurons. Inside the ith module ($i = k, \ldots, k+N-1$), choose $n_i \leq n$ of the first excitatory neurons. Denote by $t_{i,j}$ the moments of the start of their spikes. Let $t_{i,j+1} \geq t_{i,j}$ ($i = k, \ldots k+N-1$; $j = 1, \ldots, n_i$), $t_{i+1,1} > t_{in_i}(i = k, \ldots, k+N-2)$, i.e., the spikes start in order of ascending numbers of neurons within a module, and the spike of the first excitatory neuron of the module follows the spike of the last of the chosen excitatory neurons in the previous module. We consider that inside modules, spikes form bursts: $t_{i,j} \leq t_{i,1} + T_1(j = 2, \ldots, n_i; i = k, \ldots k+N-1)$. As the initial functions for Eq. (4.3.1), we take $u_{i,j}(t_{i,j}+s) = \varphi_{i,j}(s) \in S_h$ at $i = k \ldots, k+N-1$; $j = 1, \ldots, n_i$. For the rest of the equations of excitatory neurons we $u_{i,j}(t_{i,1}+s) = \varphi_{i,j}(s) \in P_h^{\gamma_0}$ at $i = k, \ldots, K+N-1$; $j = n_i+1, \ldots, n$. Here the number $\gamma_0 = \alpha(T_2 + 2nT_1) > 0$. The initial functions for the latter equations are exponentially small as $\lambda \to \infty$ according to the definition of class $P_h^{\gamma_0}$ (recall that $\varphi(s) \in P_h^{\gamma_0}$ if $0 \leq \varphi(s) \leq \exp(-\lambda\gamma_0)$).

Let us turn to specifying the initial conditions for the equations of inhibitory neurons. Denote the moments of the start of their spikes through $\tau_{i,j} (i = k, \ldots, k+N-1; \quad j = 1, \ldots, q)$. Consider that $\tau_{i,j} > t_i n_i$, but $\tau_{i,j} < t_{i,1} + T_1$. This corresponds to the case when in the ith module' sinhibitory neurons directly respond to burst of the first n_i excitatory elements. Let the initial functions for Eq. (4.3.2) satisfy the conditions $v_{i,j}(\tau_{i,j} + s) = \varphi_{i,j}(s) \in S_h$ at $i = k, \ldots, k+N-1; \quad j = 1, \ldots, q$. We note an important fact: Spikes of inhibitory elements in the ith module have time to complete before the first n_i excitatory neurons come out of the refractory state. In fact, before anyone else, the state of refractoriness ends at the first neuron. The moments of the end of the spikes of inhibitory neurons are close to numbers $t_{i,j} + T_1$. It remains to note that the following inequality holds: $t_{i,j} + T_1 < t_{i,1} + 2T_1 < t_{i,1} + T_R$.

Remember that the structure of the interaction of neurons is that each of Eqs. (4.1.3) and (4.1.2) for $u_{i,j}$, and $v_{i,j}$ can be integrated independently of the others during generation of a spike by the corresponding neuron in the time interval of duration h after the moment the pulse terminates. We called given time interval the "period of refractoriness" T_R. Simultaneously, functions $u_{i,j}$ and $v_{i,j}$, describing (i,j)th excitatory and inhibitory neurons, respectively, are included in equations of other neurons only on the intervals of time where $u_{i,j} \geq 1$ and $v_{i,j} \geq 1$. These are intervals of the generation of pulses, respectively, by the (i,j)th excitatory and inhibitory neurons.

We extend the definition of continuous initial functions for the equations of excitatory neurons of the latter $(k+N-1)$th module on the intervals of time $t \in [t_{k,1}, t_{k+N-1,j}]$ $(j = 1, \ldots, n_{k+N-1})$, respecting the condition $u_{k+N-1}(t) < 1$. By the choice of moments of time $\tau_{k,j}$, the spikes of inhibitory elements of the kth module do not act on the first excitatory neuron, i.e., functions $v_{k,j}(t)$ $(j = 1, \ldots, n)$ are not included in equation for $u_{k,1}$. The equation for $u_{k,1}$ is integrated on the interval $t \in [t_{k,1}, t_{k+N-1,1}]$ independently of the others. For $t > t_{k+N-1,1}$, this represents the problem considered previously of external action on the neuron of burst generated in this case by the excitatory elements of the $(k+N-1)$th module. We make the a priori assumption that the first neuron of the kth module responds directly to the burst action. Then Lemma 4.1.1 allows us to find the moment of time $t'_{k,1}$ of the start of its new spike while $u_{k,1}\left(t'_{k,1} + s\right) \in S_h$.

Note further that the spikes of inhibitory neurons of the kth module do not act on its second excitatory element. The equation for the second excitatory element of the kth module is the problem of the external action of burst of excitatory neurons of the $(k+N-1)$th module, to which the spike of the first excitatory neuron additionally joins. This case is considered previously in Lemma 4.2.1. If we make the a priori assumption that the second neuron responds directly to the burst action, then moment of time $t'_{k,2}$ of the start of its new spike is defined by formula (4.2.3), and at the same time $u_{k,2}\left(t'_{k,2} + s\right) \in S_h$. We emphasize that the second neuron does not act on the first one. The situation for equations describing the excitatory neurons in the kth module with numbers $j = 3, \ldots, n_k$, is analogous to the case of the second neuron.

In turn, for all of parts of Eqs. (4.3.2) describing the inhibitory neurons of the kth module, there is a problem of burst action on the part of the first n_k excitatory element of this module. If we make the a priori assumption that all of these neurons respond directly to the burst action, then the moments of the start of spikes $\tau'_{k,j}(j = 1, \ldots, q)$ are defined by Lemma 4.1.1; at this, $v_{k,2}\left(\tau'_{k,j} + s\right) \in S_h$.

It should be noted that under the a priori assumptions, new spikes of the first n_k excitatory elements of the kth module generate a burst. The spikes of inhibitory neurons have time to complete before the given excitatory elements come out of the refractory state.

Let us turn to equations of the excitatory neurons of the kth module with numbers $j > n_k$. In the time interval $t \in [t_{k,1}, t'_{k,1}]$ excitatory neurons of the kth and the $(n + N - 1)$th modules and inhibitory neurons of the kth module act on the given neurons. The total number of acting excitatory elements is $n_k + n_{k+N-1} < 2n$. Recall that the number q of inhibitory neurons satisfies the conditions $\beta_0 = gT_1(q - 2n) - T_2 > 0$. As a result, the excitatory neurons of the kth module with numbers $j > n_k$ are summarily under the inhibitory influence, and $u_{k,j}\left(t'_{k,1} + s\right) \in P_h^{\gamma_0 + \alpha\beta_0}$. This can be proven by using Lemma 4.1.2 or can easily be deduced directly from Eq. (4.3.1).

By the described method in system (4.3.1) and (4.3.2), we sequentially consider the equations for neurons belonging to the modules with numbers $i = k+1, \ldots, k+N-1$. After the analysis of equations, $(k+N-1)$ we return to the initial situation of the $(k+N-1)$th module. First, the calculated functions do not satisfy system of Eqs. (4.3.1) and (4.3.2). Starting from the moment of the full complete traversal of the ring structure, when the $(k+N-1)$th module will be considered $(k+N-1)$, this method gives a classical solution of the system. Using the algorithm, we can construct the solution of system (4.3.1) and (4.3.2) on any finite time interval.

Write out formulas for the moments $t'_{k,i}(i = 1, \ldots, n_k)$ and $\tau'_{k,i}(i = 1, \ldots, q)$ of the start of the spikes of the excitatory and inhibitory neurons of the kth module. Using Lemmas 4.1.1 and 4.2.1, we obtain:

$$t'_{k,1} = t_{k-1,n_{k-1}} + \left[\begin{array}{c} T_2 - \left(t_{k-1,n_{k-1}} - t_{k,i}\right) \\ -g \sum_{r=2}^{n_{k-1}} (r-1)\left(t_{k-1,r} - t_{k-1,r-1}\right) \end{array} \right] \Big/ (1 + gn_{k-1}) + o(1),$$

$$(4.4.1)$$

$$t'_{k,j} = t'_{k,j-1} + [t_{k,j} - t_{k,j-1}]/(1 + g(n_{k-1} + j - 1)) + o(1), (j = 2, \ldots, n_k), \quad (4.4.2)$$

$$\tau'_{k,j} = t'_{k,n_k} + \left[\begin{array}{c} T_2 - \left(t'_{k,n_k} - \tau_{k,j}\right) - \\ -g \sum_{r=2}^{n_k} (r-1)\left(t'_{k,r} - t'_{k,r-1}\right) \end{array} \right] \Big/ (1 + g n_k) + o(1), (j = 1, \ldots, q).$$

$$(4.4.3)$$

These formulas allow us to calculate the moments of the start of the spikes of neurons occurring in succession. During the transition to the $(k+1)$th module in expressions (4.4.1) through (4.4.2), we should write $t_{k,j} = t'_{k,j}$ $(j = 1, \ldots, n_k)$, where $t'_{k,j}$ are previously found start moments of the spikes of neurons of the kth module.

4.5   Dynamics of the Neural Network with Ring-Modular Organization

From the algorithm, it follows that one of the manners of the organization of oscillations in the modular structure is the cyclic propagation of the wave of spikes. We choose the first module as the initial one. We call the process of the consecutive generation of spikes by the neurons of modules, from the first up to the Nth, the "cycle of the wave propagation." Let zero cycle of wave propagation, started by the spike of the first excitatory neuron of the first module, begin at moment of time $t_w = t_{1,1} = 0$. Consider the vectors:

$$\bar{t} = \left(t_{1,1}, \ldots, t_{1,n_1}, t_{2,1}, \ldots, t_{2n_2}, \ldots, t_{N,1}, \ldots t_{N,n_N}\right),$$
$$\bar{\tau} = \left(\tau_{1,1}, \ldots, \tau_{1,q} \ldots, \tau_{2,1}, \ldots, \tau_{N,1}, \ldots \tau_{N,q}\right),$$

consisting of the moments of the start of spikes of neurons at zero cycle. We consider that

$$t_{i,j} \le t_{i,j+1} \le t_{i,1} + T_1 (i = 1, \ldots, N; \quad j = 1, \ldots, n_i - 1), \qquad (4.5.1)$$

$$t_{i,n_i} \le t_{i+1,j} \le t_{i,1} + T_1 (i = 1, \ldots, N-1; \quad j = 1, \ldots, n_{i+1}), \qquad (4.5.2)$$

$$t_{i,n_1} + T_R < T_{N,1}; \quad t_{1,1} + T_2 > t_{N,n_N} \qquad (4.5.3)$$

$$t_{i,n_i} \le \tau_{i,j} \le t_{i,1} + T_1 \quad (i = 1, \ldots, N; \quad j = 1, \ldots, q), \qquad (4.5.4)$$

i.e., spikes of the neuron form described above burst at the initial cycle. Moreover, by virtue of (4.5.3), spikes of uninhibited neurons of the latter Nth module start after the uninhibited excitatory neurons of the first module come out of the refractory state but before the time the spikes can spontaneously generate.

From the initial functions the vectors, we form $\bar{\varphi} = \left(\varphi_{1,1}, \ldots \varphi_{1,n}, \ldots, \varphi_{N,1}, \ldots \varphi_{N,n}\right)$, $\bar{\psi} = \left(\psi_{1,1}, \ldots \psi_{1,q}, \ldots, \psi_{N,1}, \ldots \psi_{N,q}\right)$. Here $\varphi_{i,j} \in S_h$ $(i = 1, \ldots, N;$

$j = 1, \ldots, n_i)$, $\varphi_{i,j} \in P_h^{\gamma_0}$, where $\gamma_0 = \alpha(T_2 + 2nT_1) > 0$ $(i = 1, \ldots, N;$ $j = n_i + 1, \ldots, n)$. Finally, $\psi_{i,j} \in S_h$ $(i = 1, \ldots, N;$ $j = 1, \ldots, q)$. Denote by W the set consisting of sets $t, \bar{t}, \bar{\varphi}, \psi$, that satisfies the above- enumerated conditions. We are interested in the solutions of system (4.3.1) and (4.3.2) with the initial conditions of W, i.e., those solutions for which $u_{i,j}(t_{i,j} + s) = \varphi_{i,j}(s)$ $(i = 1, \ldots, N;$ $j = 1, \ldots, n)$, $v_{i,j}(t_{i,j} + s) = \psi_{i,j}(s)$ $(i = 1, \ldots, N;$ $j = 1, \ldots, q)$.

According to formula (4.4.1), we calculate the moment t'_W of the beginning of the next cycle of wave passage. This moment coincides with the start of the new spike of the first excitatory neuron of the first module. At the start of the new cycle, we will count the moments of the start of spikes of excitatory and inhibitory neurons of the modules from their start time t'_W and denote, respectively, through $t'_{i,j}(i = 1, \ldots, N;$ $j = 1, \ldots, n_i)$ and $\tau'_{i,j}(i = 1, \ldots, N;$ $j = 1, \ldots, q)$. At the initial time, the spikes of neurons start at the moments $t'_W + t'_{i,j}$ and $t'_W + \tau'_{i,j}$. Recurrently using Formulas (4.4.1) through (4.4.3), we obtain:

$$t'_W = t_{N,n_N} + \left[T_2 - t_{N,n_N} - g \sum_{r=2}^{n_N} (r-1)(t_{N,r} - t_{N,r-1}) \right] \bigg/ 1 + g n_N + o(1), \quad (4.5.5)$$

$$t'_{1,1} = 0, \qquad\qquad\qquad (4.5.6)$$

$$t'_{1,j} = t'_{1,j-1} + (t_{1,j} - t_{1,j-1})/(1 + g(n_N + j - 1)) + o(1)(j = 2, \ldots, n_1), \quad (4.5.7)$$

$$t'_{i,1} = t'_{i-1,n_{i-1}} + \left[\begin{array}{c} T_2 - \left(t'_W + t'_{i-1,n_{i-1}} - t_{i,1} \right) \\ -g \sum_{r=2}^{n_{i-1}} (r-1)\left(t'_{i-1,r} - t'_{i-1,r-1} \right) \end{array} \right] \bigg/ (1 + g n_{i-1}) + o(1) \quad (4.5.8)$$

$$(i = 2, \ldots, N),$$
$$t'_{i,j} = t'_{i,j-1} + (t_{i,j} - t_{i,j-1})/(1 + g(n_{i-1} + j - 1)) + o(1) \qquad (4.5.9)$$
$$(i = 2, \ldots, N; \quad j = 2, \ldots, n_i),$$

$$\tau'_{i,j} = t'_{i,n_i} + \left[\begin{array}{c} T_2 - \left(t'_W + t'_{i,n_i} - \tau_{i,j} \right) \\ -g \sum_{r=2}^{n_i} (r-1)\left(t'_{i,r} - t'_{i,r-1} \right) \end{array} \right] \bigg/ (1 + g(n_i) + o(1) \quad (4.5.10)$$

$$(i = 1, \ldots, N; \quad j = 1, \ldots, q).$$

We emphasize that in view of Lemmas 4.1.1 and 4.2.1, we have the following inclusions:

$$u_{i,j}\left(t'_{i,j}+s\right)=\varphi'_{i,j}(s)\in S_h \quad (i=1,\ldots,N; \quad j=1,\ldots,n_i), \tag{4.5.11}$$

$$v_{i,j}\left(\tau'_{i,j}+s\right)=\psi'_{i,j}(s)\in S_h \quad (i=1,\ldots,N; \quad j=1,\ldots,q), \tag{4.5.12}$$

The a priori conditions of the applicability of formulas (4.1.1) through (4.1.3) lie in the fact that neurons respond directly to burst action. At the same time, bursts occur after the neurons come out of the refractory state, but they do not have time to generate their own spikes spontaneously. As a result, in addition to (4.5.1) through (4.1.2) indirect restrictions occur on the components of the vectors $\bar{t}$ and $\bar{\tau}$. These are inequalities (4.5.1) through (4.1.2), in which $t_{i,j}$ and $\tau_{i,j}$ change to $t'_{i,j}$ and $\tau'_{i,j}$, respectively:

$$t'_{i,j}\leq t'_{i,j+1}\leq t'_{i,1}+T_1 \quad (i=1,\ldots,N; \quad j=1,\ldots,n_i-1), \tag{4.5.13}$$

$$t'_{i,n_i}\leq t'_{i+1,j}\leq t'_{i,1}+T_1 \quad (i=1,\ldots,N-1; \quad j=1,\ldots,n_{i+1}), \tag{4.5.14}$$

$$t'_{1,n_1}+T_R<t'_{N,1}; \quad t'_{1,1}+T_2>t'_{N,n_N}, \tag{4.5.15}$$

$$t'_{i,n_i}\leq \tau'_{i,j}\leq t'_{i,1}+T_1 \quad (i=1,\ldots,N; \quad j=1,\ldots,q), \tag{4.5.16}$$

Furthermore, it is necessary to require:

$$t_{N,n_N}\leq t'_W+t'_{1,j}<t_{N,1}+T_1 \quad (j=1,\ldots,n_1), \tag{4.5.17}$$

$$t_{i+1,n_{i+1}}+T_R<t'_W+t'_{i,1} \quad (i=1,\ldots,N-1), \tag{4.5.18}$$

$$t_{i+1,1}+T_2<t'_W+t'_{i,n_i} \quad (i=1,\ldots,N-1). \tag{4.5.19}$$

We restrict the initial class W for all of these conditions to be simultaneously performed. Along with the class W, we introduce its projection—set W_1—consisting of finite-dimensional vectors $\bar{t},\bar{\tau}$.

Formulas (4.5.5) through (4.5.12) define the operator $\Pi:W\to W$, which describes the consecutive clock cycles of the passage of the wave of spikes around the ring system of the neural modules. Expressions (4.5.5) through (4.5.10) define its finite-dimensional and (4.5.11) and (4.5.12) its infinite-dimensional components. The main part of the finite-dimensional element is obtained from (4.5.5) through (4.5.10) if we substitute expression (4.5.5) into (4.5.6) through (4.5.10) and discard the term o (1). This represents the mapping $\Pi_1:W_1\to W_1$.

Below we write out inequalities connecting numbers n_i,T_R,T_2 and q, under the performance of which the mapping Π_1 has a stable fixed point. Denote this by $(\bar{t}_*,\bar{\tau}_*)$. From the fact of the existence of a fixed point, it follows that in the class W

there are the initial conditions of the system (4.3.1) and (4.3.2) forming the attractor. For these solutions, at each cycle of the passage of the wave, moments of the start of spikes satisfy the relations: $t_{i,j} = t_{i,j}^* + o(1)$ $(i = 1, \ldots, N;$ $j = 1, \ldots, n_i)$, $\tau_{i,j} = \tau_{i,j}^* + o(1)$ $(i = 1, \ldots, N;$ $j = 1, \ldots, q)$, respectively, for the excitatory and inhibitory neurons. We cannot study the detailed structure of the attractor. However, we can assert that periodic solution belongs to it. The answer to the question of the stability of the periodic solution is unknown.

To prove the existence of fixed point mapping Π_1, we introduce new variables. For the initial cycle of each cycle of the wave passage denote by $\xi_{i,1} = t_{i,1} - t_{i-1,n_{i-1}} > 0$ $(i = 2, \ldots, N)$ the delay of the start of the spike of the first excitatory element of the ith module with respect to the start of spike of the latter n_{i-1}th uninhibitory excitatory neuron of the $(i-1)$th module. Let $\xi_{i,j} = t_{i,j} - t_{i,j-1} > 0$ $(i = 1, \ldots, N;$ $j = 2, \ldots, n_i)$ be the time interval between the start of spikes of the $(j-1)$th and the jth excitatory neurons of the ith module and $\eta_{i,j} = \tau_{i,j} - t_{i,n_i} > 0$ $(i = 1, \ldots, N;$ $j = 1, \ldots, q)$ be the mismatch of the start of the spikes of the n_ith excitatory and jth inhibitory neurons, respectively. Denote by $\xi_{1,1}' = t_W' - t_{Nn_N} > 0$ the time interval between the start of spike of the latter n_N uninhibited excitatory neuron of the latter Nth module and and the beginning of the next cycle of wave passage. Let $\xi_{i,j}', \eta_{i,j}'$ be the values of the described variables at the new cycle of the passage of the wave.

The introduced variables allow us to write the mapping Π_1 in the form:

$$\xi_{1,1}' = \left[T_2 - \sum_{v=2}^{n_1} \xi_{1,v} - \sum_{\mu=2}^{N} \sum_{v=1}^{n_\mu} \xi_{\mu,v} - g \sum_{v=2}^{n_N} (v-1)\xi_{N,v} \right] / (1 + g n_N), \quad (4.5.20)$$

$$\xi_{1,j}' = \xi_{1,j}/(1 + g(n_N + j - 1)) \quad (j = 2, \ldots, n_1), \quad (4.5.21)$$

$$\xi_{i,1}' = \left[T_2 - \sum_{v=2}^{n_i} \xi_{i,v} - \sum_{\mu=i+1}^{N} \sum_{v=1}^{n_\mu} \xi_{\mu,v} - \sum_{\mu=1}^{i-1} \sum_{v=1}^{n_\mu} \xi_{\mu,v}' - g \sum_{v=2}^{n_{i-1}} (v-1)\xi_{i-1,v}' \right] / (1 + g n_{i-1}),$$
$$(i = 2, \ldots, N)$$

$$(4.5.22)$$

$$\xi_{i,j}' = \xi_{i,j}/(1 + g(n_{i-1} + j - 1)) \quad (i = 2, \ldots, N; \quad j = 2, \ldots, n_i), \quad (4.5.23)$$

$$\eta_{i,j}' = \left[T_2 - \left(\sum_{\mu=i+1}^{N} \sum_{v=1}^{n_\mu} \xi_{\mu,v} - \eta_{i,j} + \sum_{\mu=1}^{i} \sum_{v=1}^{n_\mu} \xi_{\mu,v}' \right) - g \sum_{v=2}^{n_i} (v-1)\xi_{\mu,v}' \right] / (1 + g n_i),$$
$$(i = 1, \ldots N; \quad j = 1, \ldots, q).$$

$$(4.5.24)$$

It is easy to see that for the study of the convergence of the iterative process (4.5.20) through (4.5.24), it suffices to consider the iterative process for the variables $\xi_{i,1}(i = 1, \ldots, N)$, defined by formulas (4.5.20) and (4.5.22), in which we should write $\xi_{i,j} = \xi'_{i,j} = 0$ for $j = 2, \ldots, n_i$. The resulting iterative formulas can be rewritten in the following form:

$$(1 + gn_N)\xi'_{i,1} + \sum_{j=2}^{N} \xi_{j,1} = T_2,$$

$$\sum_{j=1}^{i-1} \xi'_{j,1} + (1 + gn_{i-1})\xi'_{i,1} + \sum_{j=i+1}^{N} \xi_{j,1} = T_2 (i = 2, \ldots, N-1),$$

$$\sum_{j=1}^{N-1} \xi'_{j,1} + (1 + gn_{N-1})\xi'_{N,1} = T_2.$$

These relationships represent the Seidel method of the solution of the linear system with a symmetric positive definite matrix. It is known to be convergent. Denote the limit point by $(\xi_1^0, \ldots, \xi_N^0)$. It is determined from the system:

$$(1 + gn_{n-1})\xi_i^0 + \sum_{j \neq 1} \xi_j^0 = T_2 (i = 1, \ldots, N),$$

which is easily solved:

$$\xi_i^0 = T_2 \bigg/ \left[n_{i-1}\left(g + \sum_{j=1}^{N} n_j^{-1}\right)\right] (i = 1, \ldots, N) \qquad (4.5.25)$$

Returning to the mapping Π_1, given in the form (4.5.20) through (4.5.24), we can assert that the iterative process, starting from some neighbourhood of the point $\xi_{i,1} = \xi_i^0$ $(i = 1, \ldots, N)$, $\xi_{i,j} = 0$ $(i = 1, \ldots, N; \ j = 2, \ldots, n_i)$, $\eta_{i,j} = \xi_{i+1}^0$ $(i = 1, \ldots, N; \ j = 1, \ldots, q)$, converges to it.

The obtained fixed point Π_1, in terms of the initial variables $\bar{t}, \bar{\tau}$, must belong to the domain of its definition W_1, i.e., it must hold inequalities (4.5.21) through (4.1.2) and (4.5.13) through (4.5.19). These conditions in terms of variables ξ_i^0 acquire a very simple form:

$$0 < \xi_i^0 < T_1,$$

$$T_R < \sum_{j=1}^{N} \xi_j^0 - \xi_i^0 < T_2 (i = 1, \ldots, N).$$

Using the explicit form of the variables, we obtain:

$$0 < T_2 / \left[n_{i-1} \left(g + \sum_{j=1}^{N} n_j^{-1} \right) \right] < T_1 (i = 1, \ldots, N), \qquad (4.5.26)$$

$$T_R < \left(T_2 \sum_{k \neq i} n_k^{-1} \right) / \left((g + \sum_{j=1}^{N} n_j^{-1} \right) < T_2 (i = 1, \ldots, N), \qquad (4.5.27)$$

The conducted construction guarantees the existence of an invariant domain at operator Π, which is introduced in formulas (4.5.5) and (4.5.12) presented previously. For study by us of the system of Eqs. (4.3.1) through (4.3.2), this proves the existence of the attractor [100, 63], which consists of solutions with a number of specific properties.

Statement 4.5.1 *Distinguish in each module to n_i of excitatory neurons so that inequalities (4.5.26) and (4.5.27) hold. We assume the number q of inhibitory neurons in each module is sufficiently large: $\beta_0 = gT_1(q - 2n) - T_2 > 0$. Then as $\lambda \to \infty$, the system (4.5.26) and (4.5.27) has the attractor, which consists of solutions with the following properties. The time intervals between the starts of spikes of the chosen excitatory neurons within each module differ by a value $o(1)$. The intervals between the starts of spikes of excitatory neurons of the $(i-1)$th and the ith modules are close to numbers ξ_i^0, which is given by formula (4.5.25). Within the ith module, spikes of inhibitory neurons are delayed with respect to the pulses of excitatory elements by a value close to ξ_{i+1}^0. The remaining excitatory elements do not generate spikes.*

The described attractors consist of solutions, which can be naturally called the "waves of excitement." These waves do not exhaust all the attractors of system (4.3.1) and (4.3.2). For example, there are possible situations when several waves simultaneously (one after another) propagate around the ring module system.

In the above-formulated Statement 4.5.1, we propose the solution to the problem of the synthesis of the neural system, in which [depending on the initial state (one can specify an external action)] the waves of a predetermined structure propagate. On one hand, when choosing the number of uninhibitory excitatory neurons in modules, we can adjust the time intervals between the beginnings of the bursts of neighbouring modules. On the other hand, the spatial structure of uninhibitory excitatory elements within the modules can be informative. For the theory of neural networks, these problems are important and are related to the issues of the simulation of short-term memory. According to the wave hypothesis [84] about the nature of memory, information is recorded and stored in the brain as stable, undamped waves of neural activity. The considered neural network is capable of storing the wave patterns by self-oscillation without the adaptation of synaptic weights.

4.6 Conclusion

On the model of the ring structure of neural modules, we demonstrate the fundamental possibility of storing information in a dynamic way. The spatial distribution of spikes (patterns) in the module can be interpreted as an image (i.e., visual). The ring system of neural modules can store infinitely long given sequence of images. It does not require the adaptation of synapses to store the new sequence that allows us to consider the construction as well as the phenomenon itself as a model of short-term memory.

Certainly, it is too simplistic to represent that information must be stored in the brain as in the photograph. If different dynamics of the system correspond to various classes of the initial conditions, then steady-state oscillatory modes (attractors) can serve as codes of perceived information associated with the initial conditions. It remains to note that the number of steady-state modes in a given network of neural modules is large.

If you choose a module and monitor its activity, a specific spatial pattern of spikes of neurons will periodically appear and disappear. We can determine the time when the pattern appears according to the increase of total neural activity. There arises the hypothesis that in the brain, by locally analyzing in some domain the total electrical activity, we can also observe the updating and camouflage of patterns. This hypothesis was tested experimentally by analyzing the so-called EPs generated by light stimuli. Indeed, we detected cyclically recurring zones of neural activity caused by differences of stimuli (colors) from each other. On the contrary, no differences were observed in the intervals between these zones. The corresponding results will be discussed in Chap. 9 to follow.

Note that in the model it is quite insignificant as to where the neural modules are actually located. They can intertwine with each other, and they can simultaneously be in completely different areas of the brain. In the latter case, it is easy to take into account the delay in the conduction of excitation. Corollaries of the model will not change.

References

Blum F., Leiserson A., & Hofstefer L. (1988). *Brain, mind and behaviour*. Moscow: Mir.

Crick, F. (1990). Neural Edelmanism. *TINS, 12*, 240–248.

Dunin-Barkowski, V. L. (1978). *Information processes in neural structures*. Moscow: Nauka.

Eccles J. 1966. *The physiology of synapses*. Moscow: Mir.

Eccles J. 1971. *Inhibitory pathways in the central nervous system*. Moscow: Mir.

Edelman G. M., & Mountcastle V. B. 1981. *The mindful brain*. Moscow: Mir.

Edelman G. M. 1987. Neural Darwinism in Basic Books.

Frolov, A. A., & Muravyov, I. P. (1987). *Neural models of associative memory*. Moscow: Nauka.

Frolov, A. A., & Muravyov, I. P. (1988). *Information characteristics of neural networks*. Moscow: Nauka.

Hebb D. O. 1949. *The organization of behaviour: neuropsychological theory.* N. Y.: J. Wiley and Sons: 335.

Hopfield, J. J. (1982). Neural networks and physical systems with emergent collective computational abilities. *Proceedings of the National Academy of Sciences, 79*(8), 2554–2558.

Kashchenko, S. A., & Mayorov, V. V. (1995). Wave structures in ring systems from homogeneous neural modules. *Reports of the Russian Academy of Sciences, 342*(3), 318–321.

Livanov, M. N. (1972). *Spatial organization of the processes of the brain.* Moscow: Nauka.

Livanov, M. N. (1989). *Spatiotemporal organization of potentials and systemic activity of the brain, in selected works.* Moscow: Nauka.

Lorente de Nó, R. (1938). Analysis of the activity of the chains of internuncial neurons. *Journal of Neurophysiology, 1,* 207–244.

Mayorov, V. V. (1994). Structure of the oscillations in the ring system of homogeneous neural modules. *Modelling and Analysis of Information Systems Yaroslavl, 2,* 50–77.

Osipov, G. V., Rabinovich M. I., & Shalfeev V. D. 1992. Dynamics of nonlinear synchronization networks. *Proceedings of International Seminar "Nonlinear circuits and systems,* June 16–18. Russia: Moscow, 2: 88.

Pribram K. (1975). *Languages of the Brain.* Moscow: Mir.

Rosenblatt F. (1965). *Principles of neurodynamics.* Moscow: Mir.

Singer, W., Tretter, F., & Cynader, M. (1976). The effect of reticular stimulation on spontaneous and evoked activity on the cat visual cortex. *Brain Research, 102*(1), 71–90.

Chapter 5
Model of Adaptation of Neural Ensembles

The problem of choice of synaptic weights in neural systems is called the "problem of learning." There exist two fundamentally different approaches to its solution. According to one of them, the weights are calculated outside the neural network and then imported to the synapses. The finding of the weights is usually reduced to the solution of some optimization problem. In the case of classical neural networks, we often minimize the total standard deviation of the desired and the actual state of the outputs of neurons.

In Sect. 3.3 for the ring neural structure, we obtain the values of synaptic weights, which guarantee the existence of the attractor of predetermined structure. Thus, the problem of external training has been already resolved.

Direction, associated with the import of weights, is widespread, but the modeling of biological systems does not seem natural. In accordance with the second approach to the training system, synaptic weights are adjusted during operation of the neural network. The direction goes back to the ideas of Hebb (1949), which was developed by Rosenblatt (1965). Some investigators (Dunin-Barkowski 1978; Kohonen 1980, 1984; Frolov and Muyarov 1987, 1988; Frolov 1993) considered the presence of the mechanism as of adaptation an inherent attribute of the neural system. We agree with this opinion.

The problem of adaptation is not easy with regard to dynamic environments. However, we are able to solve it for the ring neural structures. In the process of adaptation, the central place is given to the axoaxonic synapse. By means of this, the substitution of the signal incoming into the neuron from the previous ring of the neuron to the standard (training) takes place. Thus, the model reveals one of the possible functional appointments of axoaxonic contact.

5.1 Model of Adaptation of Individual Neurons

Consider an oriented (for simplicity) ring structure consisting of N identical excitatory neurons. In it, each preceding neuron acts on the subsequent neuron by chemical synapse. The first neuron follows the Nth neuron. Let the synaptic weights be chosen (Sect. 3.3) so that the start of spike of the ith neuron is delayed with respect

© Springer International Publishing Switzerland 2015
S. Kashchenko, *Models of Wave Memory*,
Lecture Notes in Morphogenesis, DOI 10.1007/978-3-319-19866-8_5

to the start of spike of the $(i-1)$th neuron by value $\xi_i^0 + o(1)$, where $0 < \xi_i^0 < T_1$, and T_1 is asymptotic duration of the spike (it coincides with the effective time of action of neurotransmitter). Call the given ring a "reference" or "standard" ring. We similarly the neurons belonging to this ring "reference neurons."

In addition to the neurons of the reference ring, we also consider N neurons, which are said to be adaptive. Each ith adaptive neuron is postsynaptic for the ith and $(i-1)$th reference neurons. The synaptic weight of action of the ith reference neuron on the ith adaptive neuron is equally fixed. In turn, the synaptic weight of action of the $(i-1)$th reference neuron on the ith adaptive neuron changes in the process of functioning of the network. The purpose of adaptation is to synchronize the functioning of the ith standard and the ith adaptive neurons.

We can give this problem a biological interpretation. The spike of the single reference neuron is a relatively weak signal, which can be masked in the neuronal noise. Capturing of the adaptive neuron by the reference neuron leads to increased signal. It is clear that several (many) adaptive neurons can be arranged to the ith reference neuron. This will further strengthen the signals. The wave structure manifests and stands out from the neural noise.

Denote the membrane potentials of the reference neurons by u_i and the adaptive neurons as w_i. The system of Eq. (3.1.1) for the neuronal population, where the synaptic interaction is given by (3.1.2) and (3.1.3), in this case takes the form:

$$\dot{u}_i = \lambda\left[-1 - f_{Na}(u_i) + f_K(u_i(t-1)) + \alpha g_{i,i-1} H(u_i)\theta(u_{i-1} - 1)\right]u_i, \qquad (5.1.1)$$

$$\dot{w}_i = \lambda\left[\begin{array}{c} -1 - f_{Na}(w_i) + f_K(w_i(t-1)) \\ + \alpha H(w_i)\left[G_{i,i-1}\theta(u_{i-1} - 1) + g\theta(u_i - 1)\right] \end{array}\right], \qquad (5.1.2)$$

where $i = 1, \ldots, N$, whereas number $i = 0$ is identified with $i = N$. Assume that the positive functions $f_{Na}(u)$ and $f_K(u)$ tend monotonically to zero as $u \to \infty$, the number $\alpha = f_K(0) - f_{Na}(0) - 1 > 0$, and the speed parameter is $\lambda \gg 1$. The functional $H(*)$, which provides the presence of refractory period, is defined by formula (4.1.2), i.e.,

$$H(u) = \theta(1 - u(t))\theta(1 - u(t - 1 - (1 - \varepsilon)T_1)) \\ \cdot \theta(1 - u(t - 1 - 2(1 - \varepsilon)T_1)), \qquad (5.1.3)$$

where $0 < \varepsilon < 1$. The functional $H(u)$ vanishes at the time of spike and during time $2(1 - \varepsilon)T_1$ after the spike. The duration of the refractory period, including the time of the spike, is $T_R = (3 - 2\varepsilon)T_1$. We assume that the refractory period is shorter than the period of the neuron's own activity, i.e., $T_R < T_2$, where T_2, according to Theorem 1.6.1, asymptotically coincides with the period of its own activity.

In (5.1.1) and (5.1.2), functions of the form $\theta(u - 1)$ are indicators of the presence of the neurotransmitter released from the spike of the corresponding neuron. Synaptic weights $g_{i,i-1}$ of action of the $(i-1)$th neuron on the ith neuron in reference ring are calculated by formulas (3.3.8):

$$g_{1,N} = \left(T_2 - \sum_{i=1}^{N}\xi_i^0\right)\bigg/\xi_1^0, \quad g_{i,i-1} = \left(T_2 - \sum_{i=1}^{N}\xi_i^0\right)\bigg/\xi_i^0, \quad (i=2,\ldots,N).$$

$$(5.1.4)$$

Here the numbers ξ_i^0 satisfy the conditions:

$$0 < \xi_i^0 < T_1, \quad T_2 - \sum_{i=1}^{N}\xi_i^0 > 0, \quad \sum_{i=1}^{N}\xi_i^0 - \xi_j^0 > T_R$$

for all $j = 1,\ldots,N$.

Furthermore, in (5.1.1) and (5.1.2), g is the general fixed value of the synaptic weight of action of the ith reference on the ith adaptive neuron. Finally, $G_{i,i-1}$ are synaptic weights of action of the $(i-1)$th reference neuron on the ith adaptive neuron. These weights will change in the process of the functioning of the neural network.

The choice of synaptic weights $g_{i,i-1}$ by formula (5.1.4) according to Theorem 3.3.1 ensures that in the standard ring is installed in the oscillatory mode, in which the start of the spike of the ith neuron is delayed with respect to the start of the spike of the $(i-1)$th neuron by the value $\varepsilon_0 + o(1)$. Forming bursts spikes of the $(i-1)$th and the ith standard neurons act on each ith adaptive neuron act. The period of action is:

$$T_\nu = \sum_{i=1}^{N}\xi_i^0 + o(1) < T_2.$$

The problem of this kind of action was studied in detail in Sect. 4.1. If the synaptic weight g is not small, and the weight $G_{i,i-1}$ is not too large, then (Lemma 4.1.1) the start of the spike of the ith adaptive neuron is delayed with respect to the start of the spike of the ith standard neuron. The response is immediate: The spike of the ith adaptive neuron occurs before the spike of the $(i-1)$th standard neuron is completed. If we increase the synaptic ratio $G_{i,i-1}$, then the value of the delay of the start of the spike of the ith adaptive neuron, with respect to the start of the spike of the ith standard neuron, decreases. The idea of fine tuning is based on this fact.

It is convenient to perform modification of the synaptic weight $G_{i,i-1}$ on the interval of time duration $T_{Ad} < T_R$ after the start of the spike of the ith adaptive neuron. The corresponding synapse is not used on this interval, because the neuron is refractory. Simultaneously, it is convenient to perform modification of the weight

$G_{i,i-1}$ only if the spike of the ith standard neuron is observed on this interval (there can be no external signals incoming to the ith adaptive neuron, and the question of fine tuning does not arise).

The functional

$$H_v(w_i, u_i) = \theta(w_i(t) - 1)\theta(w_i(t - T_{Ad}) - 1)\theta(u_i(t) - 1) \qquad (5.1.5)$$

at the moment of time t becomes one only if there the spikes are simultaneously observed of the ith standard and ith adaptive neurons and at the same time the spike of the ith adaptive neuron started not later than moment of time $t - T_{Ad}$, where $T_{Ad} < T_1$. Under other conditions, the functional H_v takes the value zero (naturally, here w_i, u_i are functions describing the dynamics of membrane potentials). We assume that the modification of synaptic weight $G_{i,i-1}$ occurs on time intervals where the functional H_v takes a single value.

Next, we introduce the functional

$$H_{Sp}(u) = \int_{t-T_1}^{t} \theta(u(\tau) - 1)d\tau\theta(u(t) - 1). \qquad (5.1.6)$$

During the spike of the neuron, it specifies the time that has elapsed since the start of the spike; otherwise, it takes the value zero. In the time interval when the of the ith reference and the ith adaptive neurons are simultaneously observed, the value of the difference $H_{Sp}(w_i) - H_{Sp}(u_i)$ coincides with the mismatch of the start of spikes of the given neurons, i.e., with phase difference of oscillations. We assume that the speed of change of synaptic weight $G_{i,i-1}$ where the change is possible, is in proportion to this difference. As a result, we obtain an equation that describes the process of modification of synaptic weight:

$$\dot{G}_{i,i-1} = q^* \left(H_{Sp}(w_i) - H_{Sp}(u_i)\right)H_v(w_i, u_i), \qquad (5.1.7)$$

where $q^* > 0$, and the functional H_v and H_{Sp} are given by formulas (5.1.5) and (5.1.6). Thus, the system of Eqs. (5.1.1), (5.1.2), and (5.1.7) is a model of adaptation.

Choose and fix the ith adaptive neuron arbitrarily. Let at the zero moment time and at moment of time ξ_i^0 be the start the spikes, respectively, of the $(i-1)$th and the ith standard neurons. Assume that the start of the spike of the ith adaptive neuron is delayed with respect to the start of the spike of the ith standard neuron by the value $\eta_i > 0$. In the time interval $t \in \left[\xi_i^0 + \eta_i, \xi_i^0 + \eta_i + T_{Ad}\right]$, there is a change of the value of synaptic weight $G_{i,i-1}$. For the simplicity of calculations, we assume that $\eta_i + T_{Ad} < T_1$, i.e., the process of modification of the synapse is terminated before the end of the spike of the ith standard neuron. Suppose that prior to the adaptation $t < \xi_i^0 + \eta_i$ the synaptic rate had the value $G_{i,i-1}^0$. It is easy to see that from Eq. (5.1.7) for $t > \xi_i^0 + \eta_i + T_{Ad}$ until the start of new spike of the adaptive neuron, the synaptic weight takes the value

$$G_{i,i-1} = G_{i,i-1}^0 + q\eta_i, \tag{5.1.8}$$

where $q = q^* T_{Ad}$.

The wave of spikes propagates around the ring structure independently of the adaptive neurons. Let at the initial cycle of the passage of the excitation wave the number $\eta_i > 0$ be the mismatch of the start of the spikes of the ith adaptive and the ith standard neurons and $G_{i,i-1}$ be the value of the synaptic weight after the initial fine tuning, which flows during the spike of the ith adaptive neuron. From Lemma 4.1.1 we define η_i' as the new delay of the start of the spike of the adaptive neuron at the next cycle of wave passage. Let $G_{i,i-1}'$ be the value of the synaptic weight after modification that occurred during the spike of the adaptive neuron at a cycle of passage of the wave. This can be found on the basis of formula (5.1.8). As a result, for the values η_i and $G_{i,i-1}'$ we obtain the following relationships:

$$\eta_i' = \left(T_2 - \sum_{j=1}^{N} \xi_j^0 + \eta_i - G_{i,i-1}\xi_i^0 \right) \bigg/ \left(1 + G_{i,i-1} + g \right), \tag{5.1.9}$$

$$G_{i,i-1}' = G_{i,i-1} + q + \eta_i'. \tag{5.1.10}$$

Here we do not take into account the terms $o(1)$. We regard formulas (5.1.9) and (5.1.10) as the iterative process reflecting consecutive clock cycles of the adaptation process. In this capacity, formulas are applicable if they fulfill the conditions under which they are written out: $\xi_1^0 + \eta_i' < T_1$ (the condition of the immediate response) and $\xi_1^0 + \eta_i' < T_{Ad}$. Restrictions are satisfied if the synaptic weight is

$$g > 1 + \left(T_2 - \sum_{j=1}^{N} \xi_j^0 \right) \bigg/ \left(T_1 - \xi_i^0 \right) \tag{5.1.11}$$

This estimate follows from (5.1.9) (it is too high).

It is obvious that mapping (5.1.9) and (5.1.10) has the fixed point $\eta_i = 0$, $G_{i,i-1} = g_{i,i-1}$, where the number $g_{i,i-1}$ is the corresponding synaptic weight in the standard ring calculated by formula (5.1.4). We investigate the problem of convergence of the iterative process, which we consider a priori in the area:

$$0 \le G_{i,i-1} \le g_{i,i-1}, 0 \le \eta_i < T_1 - \xi_i^0.$$

Introduce new variables $z = \eta_i/\xi_i^0$ and $y = g_{i,i-1} - G_{i,i-1}$. We rewrite the mapping (5.1.9) and (5.1.10) in the form:

$$z' = (z + y)/(1 + g_{i,i-1} + g - y),   \tag{5.1.12}$$

$$y' = y - q\xi_i^0 z'.   \tag{5.1.13}$$

It should be considered subject to the constraints: $0 \leq z < (T_1 - \xi_i^0)/\xi_i^0 \equiv z_0$, $0 \leq y \leq g_{i,i-1}$. We single out on the plane (y, z) the triangle $M : 0 \leq y \leq g_{i,i-1}$, $0 \leq z \leq (z_0/g_{i,i-1})y$. Let the synaptic weight g satisfy the condition (5.1.11) and the coefficient q satisfy inequality

$$0 < q < \left[\xi_i^0\left(z_0/g_{i,i-1} + 1\right)\right]^{-1} \equiv \left[(T_1 - \xi_i^0)/g_{i,i-1} + \xi_i^0\right]^{-1}.   \tag{5.1.14}$$

Then the given triangle is invariant for mapping (5.1.12) and (5.1.13). We explain the reasoning.

If the point $(y, z) \in M$, then from (5.1.12) it follows that

$$0 \leq z' \leq \frac{z_0/g_{i,i-1} + 1}{1 + g} y.   \tag{5.1.15}$$

By virtue of given inequality from (5.1.13), we obtain:

$$y \geq y'\left[1 - q\xi_i^0 \frac{z_0/g_{i,i-1} + 1}{1 + g}\right]y.   \tag{5.1.16}$$

The required restrictions immediately follow from this and from condition (5.1.14) $0 \leq y' \leq g_{i,i-1}$.

Furthermore, from (5.1.14) and (5.1.16), we deduce the inequality

$$y \leq y'(1 + g)/g,$$

according to which, from (5.1.15), it follows that

$$0 \leq z' \leq y'\left(z_0/g_{i,i-1} + 1\right)/g.$$

If the synaptic weight g satisfies inequality

$$g > 1 + g_{i,i-1}/z_0,   \tag{5.1.17}$$

then the required restriction $0 \leq z' \leq (z_0/g_{i,i-1})y'$ is performed. Inequality (5.1.17) is easily reduced to form (5.1.11).

The problem of convergence of the iterative process (5.1.12) and (5.1.13) in the triangle M is simple. Let $y(k), z(k) (k = 0, 1, \ldots)$ be a sequence of iterations. Then, from formula (5.1.13), it follows that $0 \leq y(k) \leq \mu y(k - 1)$, where

$$0 < \mu = 1 - \frac{q \xi_i^0}{1 + g_{i,i-1} + g} < 1.$$

In turn, from inequality (5.1.5) follows the estimate $0 \leq z(k) \leq Cy(k - 1)$ (C is constant). It is obvious that $y(k) \to 0, z(k) \to 0$ at $k \to \infty$.

Return to the original mapping (5.1.9) and (5.1.10). If the initial point of the iterative process given by mapping satisfies inequalities

$$0 \leq G_{i,i-1} < g_{i,i-1}, \quad 0 < \eta_i < (T_1 - \xi_i^0)(g_{i,i-1} - G_{i,i-1})/g_{i,i-1}, \tag{5.1.18}$$

then the sequence of iterations $G_{i,i-1}(k) \to g_{i,i-1}, \eta_i(k) \to 0$ at $k \to \infty$.

This is the scheme of adaptation and its almost complete justification, for the completion of which it is necessary to make an important and rather complex note. Mapping (5.1.9) and (5.1.10) was obtained at an priori condition $\eta_i' > 0$, and we did not take into account terms of order $o(1)$. Because in the process of adaptation $\eta_i(k) \to 0$ at $k \to \infty$, then the a priori condition can be impaired. It is conceivable that at $\eta_i < 0$, the process of adaptation can diverge $\eta_i(k) \nrightarrow 0, G_{i,i-1}(k) \nrightarrow g_{i,i-1}$ at $k \to \infty$. Let us discuss this problem.

The value η_i' under condition $\eta_i' < 0$ is written on the basis of Lemma 4.1.1. The corresponding expression can be combined with formula (5.1.9) and has the following form:

$$\eta_i' = \left(T_2 - \sum_{j=1}^{N} \xi_j^0 + \eta_i - G_{i,i-1}\xi_i^0 \right) \bigg/ \left(\begin{array}{c} 1 + G_{i,i-1} \\ + g\theta \left(T_2 - \sum_{j=1}^{N} \xi_j^0 + \eta_i - G_{i,i-1}\xi_i^0 \right) \end{array} \right),$$

$$\tag{5.1.19}$$

where $\theta(*)$ is the function of Heaviside, and the terms of order $o(1)$ are omitted.

There arises the question of convergence of the iterative process defined by formulas (5.1.19) and (5.1.10) in the case when the value η_i is close to zero, and $G_{i,i-1}$ is little different from $g_{i,i-1}$ [the number of $g_{i,i-1}$ is given by (5.1.4)]. As mentioned previously, we write in (5.1.19) and (5.1.10) that $\eta_i = z\xi_i^0$ and $G_{i,i-1} = g_{i,i-1} - y$. Then make the substitution $z \equiv z, y = x - z$. By M_1 denote the image of the triangle M on the plane (x, z). Below we are particularly interested in the part M_1' of the triangle M_1, which is designated by the inequalities:

$$0 \leq x \leq g_{i,i-1}, \quad 0 \leq z \leq x(z_0/g_{i,i-1})/(1 + z_0/g_{i,i-1}).$$

Mapping (5.1.19) and (5.1.10) for the variables (x, z) takes the form:

$$x' = \left[1 + \frac{1 - q\xi_i^0}{1 + g_{i,i-1} + z - x + g\theta(x)}\right]x - z, \qquad (5.1.20)$$

$$z' = \frac{x}{1 + g_{i,i-1} + z - x + g\theta(x)}. \qquad (5.1.21)$$

It is necessary to investigate the behavior of the iterative process defined by mapping (5.1.20) and (5.1.21), the initial point of which belongs to a small neighborhood of zero. We denote the sequence of iterations through the $x(k), z(k)$. Let $x(0) > 0$. If the point $(x(0), z(0)) \in M_1$, we immediately note that the sequence of iterations tends to zero. Let $(x(0), z(0)) \notin M_1$. Then two variants are possible. According to the first variant, the sequence has the property $x(k) > 0 (k = 1, 2, \ldots)$. As we study the small neighborhood of zero, then we can linearize mapping (5.1.20) and (5.1.21) if $x > 0$:

$$x' = \left[1 + \frac{1 - q\xi_i^0}{1 + g_{i,i-1} + g}\right]x - z, \qquad (5.1.22)$$

$$z' = \frac{x}{1 + g_{i,i-1} + g}. \qquad (5.1.23)$$

Eigenvalues of this mapping (at small and positive values of q) are positive and less than one. The sequence of iterations $(x(k), z(k))$ of mapping (5.1.20) and (5.1.21) tends to zero.

According to the second variant, for some $k = k_1$ the sequence $(x(k), z(k))$ leaves the half-plane $x > 0$. From the form of mapping (5.1.20) and (5.1.21), it follows that $x(k_1) < 0$, $z(k_1) > 0$ and $x(k_1 + 1) < 0$, $z(k_1 + 1) < 0$. Again there are two variants. According to the first variant, the sequence of iterations for $k > k_1 + 1$ remains in the half-plane $x < 0$. We can again linearize mapping (5.1.20) and (5.1.21), but we do so under the condition $x < 0$. We obtain mapping (5.1.22) and (5.1.23), in which the parameter $g = 0$. Eigenvalues of the matrix are positive and less than one. The sequence of iterations $(x(k), z(k))$ of mapping (5.1.20) and (5.1.21) tends to zero.

In the second case, for $k = k_2$ the sequence $(x(k), z(k))$ leaves the half-plane $x < 0$. Then from the form of mapping (5.1.20) and (5.1.21), it follows that $x(k_2) > 0, z(k_2) < 0$. At the same time, for a finite number k_2 of iterations, members of the sequence cannot get away far from zero. From (5.1.20) and (5.1.21), it follows that $x(k_2 + 1) > 0$, $z(k_2 + 1) > 0$, and thus we have the estimate:

$$z(k_2 + 1) - \frac{z_0/g_{i,i-1}}{1 + z_0/g_{i,i-1}} x(k_2 + 1)$$

$$< \frac{xk_2}{1 + g_{i,i-1} + z(k_2) - x(k_2) + g} \left[1 - \frac{z_0/g_{i,i-1}}{1 + z_0/g_{i,i-1}} g \right] < 0. \tag{5.1.24}$$

In the derivation of the estimate, we use $z(k_2) < 0$ and inequality (5.1.17) for the weight g. From the inequality follows the inclusion $(x(k_2 + 1), z(k_2 + 1)) \in M_1' \subset M_1$. Triangle M_1 is invariant for mapping (5.1.20) and (5.1.21), and the iterative process within it tends to zero.

Thus, we have analyzed all of the cases and have shown that the sequence $(x(k), z(k)) \to 0$ at $k \to \infty$. As a result, we have established the convergence of iterations of mapping (5.1.19) and (5.1.10). For the model of neural populations, described by system (5.1.1), (5.1.2), and (5.1.7), this means that the adaptation process converges.

The above-mentioned adaptation began when the oscillations were established in the standard ring of neurons, (the components $u_i(t)$ of the solution belong to the attractor described in Theorem 3.3.1). The construction is also valid in the more general case.

Let $\xi_i(0) > 0 (i = 2, \ldots, N)$ be the time intervals between the starts of spikes of the $(i - 1)$th and the ith standard neurons in the initial cycle of the wave's passage around the ring structure. We denote by $\eta_i(0) > 0$ the delay of the start of spike of the ith adaptive neuron with respect to the start of spike of the ith standard neuron. Let $G_{i,i-1}(0)$ be the initial values of the synaptic weights of modifiable synapses. The values $\xi_i(k)$, $\eta_i(k)$ and $G_{i,i-1}(k)$ $(k = 1, 2, \ldots)$ have the same meaning, but they are related to subsequent cycles of wave propagation.

Lemma 5.1.1 *Let the numbers $\eta_i(0)$ and $G_{i,i-1}(0)$ satisfy inequalitities (5.1.8). There exists such $C_0 > 0$ that at $|\xi_i(0) - \xi_i^0| < C_0$ consequences $\xi_i(k) \to \xi_i^0 + o(1)$, $\eta_i(k) \to o(1)$, $G_{i,i-1}(k) \to g_{i,i-1} + o(1)$ at $k \to \infty$.*

Thus, we complete the study of the adaptation process of individual neurons. We note that we cannot answer the question of possible oscillations of the values $\xi_i(k)$, $\eta_i(k)$, $G_{i,i-1}(k)$, respectively, in the neighborhood of $\xi_i^0, 0, g_{i,i-1}$. From a practical point of view, this is not essential because the amplitude of such oscillations would be would be negligible for the parameter value $\lambda \to \infty$.

5.2   Model of Adaptation of the Ring Neural Structure

Assume that we generate a sequence of spikes as a result of some neural processes. Let ξ_i^0 be time intervals between the beginning of the $(j - 1)$th and the jth pulses. The duration of each spike is close to T_1. They overlap in time: $\xi_j^0 < T_1$. Assume that the sequence is periodically repeated: $\xi^0_{j+N} \equiv \xi^0_j$. We can consider that the period is $N > 4$. Call the given sequence of spikes "reference" or "standard."

Consider the oriented ring of N neurons, in which the ith element acts the pulse of the $(i-1)$th neuron and the spike with the number $j = i + kN$ $(k = 0, 1, 2, \ldots)$ from the reference sequence. In the process of functioning of the ring structure. the synaptic weights of action of the $(i-1)$th neuron on the ith are modified. The purpose of adaptation is this: After disconnecting external signals, the ring structure must generate a sequence of spikes identical to the reference sequence.

As mentioned previously, the neurons belonging to the ring are considered adaptive. The reference can be the sequence of spikes generated by sensory structures of the brain, or, for example, some ring formation. In the latter case, we talk about the reproduction of identical rings. We will return to this question later in the text.

Earlier in Sect. 2.3, a generally accepted classification of chemical synapses, depending on their location, was given. Among the types we mentioned were the axosomatic and axoaxonic synapses. The Axoaxonic synapse plays a central role in the process of adaptation. In it, the end of the axon of the presynaptic neuron forms a contact on the axon of the postsynaptic neuron. The given postsynaptic axon is able to carry the excitation to the axosomatic synapse, which is located directly on the body (soma) of the third neuron. As a result, the latter is influenced by the first two neurons. For them, part of the axon after axoaxonic contact is a common terminal. We assume that it carries out excitation regardless of which neuron spike went through the site of axoaxonic contact. If the spike has passed along the common terminal, then the neurotransmitter is released in its terminal, i.e., the axosomatic synapse.

We will speak about the exciting type of neurotransmitter and, hence, of excitatory synapse. Let $u(t)$ and $w(t)$ bee current values of the membrane potentials of neurons, the common terminal of which terminates with a given axosomatic synapse. As the indicator of the effective time of action of neurotransmitter, we can take the function $\theta(u(t) + w(t) - 1)$.

Note that the axon has the property of refractoriness. For some time after passage of the nerve pulse, passage of a new spike is impossible. The phenomenon is easily taken into account in the model. As a result, the formula for the indicator of the effective period of the neurotransmitter's action becomes more complex. For example, for the indicator we can take the function $\theta(u(t)H(w(t - T_1)) + w(t)H(u(t - T_1)) - 1$, where the functional $H(*)$, providing refractoriness, is given by (4.1.2). However, in this case all of the following construction and conclusions are retained.

We describe the architecture of the neural system carrying out the process of adaptation. Let the sequence of spikes with intervals ξ^0_i between them generate some reference-oriented neural ring. In the adaptive ring, the preceding $(i-1)$th neuron acts on the subsequent ith neuron through the axosomatic synapse. The $(i-1)$th reference neuron forms a contact at the appropriate axon. Through the axosomatic synapse, it also controls the ith adaptive neuron. For simplicity, we assume that all of the weights of the axosomatic action of reference neurons on the adaptive neurons are the same.

We denote by u_i and w_i $(i = 1, \ldots, N)$ the membrane potentials, respectively, of standard and adaptive neurons. The system of Eqs. (3.1.1) for the considered neural population (synaptic interaction is given by formulas (3.1.2) (3.1.3)) takes the form:

$$\dot{u}_i = \lambda\left[-1 - f_{Na}(u_i) + f_K(u_i(t-1)) + \alpha g_{i,i-1} H(u_i)\theta(u_{i-1} - 1)\right]u_i, \qquad (5.2.1)$$

$$\dot{w}_i = \lambda\left[\begin{array}{c} -1 - f_{Na}(w_i) + f_K(w_i(t-1)) \\ + \alpha H(w_i)\left[G_{i,i-1}\theta(u_{i-1} + w_{i-1} - 1) + g\theta(u_i - 1)\right]\end{array}\right]w_i, \qquad (5.2.2)$$

where $i = 1, \ldots, N$, whereas the number $i = 0$ is identified with $i = N$. In (5.2.2), the function $\theta = u_{i-1} + w_{i-1} - 1$ is the indicator of the effective interval of action of the neurotransmitter in the synapse, by which the common terminal of the $(i - 1)$th standard and adaptive neurons terminates on the body of the ith adaptive neuron. The other entire notation is described in detail in the previous section for system (5.1.1) and (5.1.2). We emphasize that only the synaptic weights $g_{i,i-1}$ are chosen in accordance with formulas (5.1.4). As a result, the reference neural ring is capable of generating a cyclic sequence of spikes with specified mismatches $\zeta_i^0 + o(1)$ regarding the time of their start.

The idea of adaptation is based on the following simple considerations. Imagine that the spikes of each adaptive neuron are delayed with respect to the spikes of the corresponding reference neuron. Then only the $(i - 1)$th and ith reference neurons act on the ith adaptive neuron, which follows from the properties of common terminal. In the previous section, the mechanism of the modification of synaptic weights $G_{i,i-1}$ was discussed. According to Lemma 5.1.1, over time the values $G_{i,i-1}$ tend to numbers $g_{i,i-1}$. In this case, the ith adaptive and the ith standard neurons will function almost synchronously. After modification, we can disable the reference ring, for example, to inhibit the appendages of the neurons going to the adaptive ring. However, the adaptive ring continues in an oscillatory mode, in which the spike of the ith neuron is delayed with respect to the spike of the $(i - 1)$th neuron by the value $\zeta_i^0 + o(1)$. This follows directly from Theorem 3.3.1. Note that now that the $(i - 1)$th adaptive neuron, instead of the standard neurons, acts equivalently on the ith adaptive neuron. In connection with this we call this adaptation scheme a "scheme with substitution."

To implement the plan, we assume that the synaptic weights $G_{i,i-1}$ satisfy differential Eq. (5.1.7). As a result, they change only on time intervals where the spikes of the ith adaptive and the ith reference neurons overlap. At each of these periods, the weight $G_{i,i-1}$ as a result changes by value proportional to the difference of the moments of the start of the spikes of the adaptive and reference neurons. Let us discuss the problem in more detail.

Let the oscillations in the reference ring be established. Around it propagates the wave of excitation, in which the spike of the ith neuron is delayed with respect to the spike of the $(i - 1)$th neuron by the value $\zeta_i^0 + o(1)$. We agree that every cycle of the passage of the wave opens with the spike of the first neuron and ends with the pulse of the Nth neuron. Consider some cycle of the passage of the wave. Denote by $\eta_i (i = 1, \ldots, N)$ the difference between the moments of the start of spikes of the ith adaptive and the ith reference neurons in this cycle. Let $G_{i,i-1}$ be the values of the synaptic weights after modification. Denote by η_i' and $G_{i,i-1}'$ the analogous values for the next cycle of the wave passage.

We calculate η_i' with the help of Lemma 4.1.1. Thus, it is necessary to consider a number of cases depending on the signs η_i' and η_{i-1}'. In turn, the modified synaptic weights $G_{i,i-1}'$ are defined by formulas (5.1.10). As stated previously, we believe that the time T_{Ad} of adaptation at each cycle is not very long: $\eta_i + T_{Ad} < T_1$. As a result, we obtain the following relationships for the values η_i' and $G_{i,i-1}'$:

$$
\eta_i' = \left(T_2 - \sum_{j=1}^{N} \xi_j^0 + \eta_i - G_{i,i-1}\left(\xi_i^0 + |\eta_{i-1}'|\theta(-\eta_{i-1}')\right) \right) \Big/ \Big(1 + G_{i,i-1}
$$

$$
+ g\theta\left(T_2 - \sum_{j=1}^{N} \xi_j^0 + \eta_i - G_{i,i-1}\xi_i^0 \right) \Big),
$$
(5.2.3)

$$
G_{i,i-1}' = G_{i,i-1} + q\eta_i'.
$$
(5.2.4)

Here $i = 1, \ldots, N$, and at $i = 1$ we should assume $\eta_0' = \eta_N$. Parameter $q = q^* T_{Ad}$ where $g^* > 0$ is the coefficient from Eqs. (5.1.7). In formulas (5.2.3) and (5.2.4) are omitted the terms of order $o(1)$.

Mapping, as given by formulas (5.2.3) and (5.2.4), obviously has the fixed point $\eta_i = 0$, $G_{i,i-1} = g_{i,i-1}$. As previously mentioned, the question arises about the convergence of the iterative process, which is given by the mapping.

We assume that the synaptic weight $g > 0$ of the axosomatic action of the ith reference on the ith adaptive neuron satisfies condition (5.1.11) (is relatively not small) and that the parameter $q > 0$ satisfies inequality (5.1.14) (i.e., is not large). Let $\eta_i \geq 0$, $G_{i,i-1} \geq 0$ and inequalities (5.1.18) hold. In this case, the relationships (5.2.3) and (5.2.4) pass to formulas (5.1.9) and (5.1.10) (i.e., the mapping is split into the components with numbers i). In the previous section, it was shown that the domain released by inequalities (5.1.18) is invariant for the mapping and that iterations converge to the desired fixed point. Thus, the process of adaptation converges.

It should be noted that certain difficulties arise. Formulas (5.2.3) and (5.2.4) are written out with accuracy to the terms of order $o(1)$. Is it conceivable that at small η_i and $g_{i,i-1} - G_{i,i-1}$, the iteration process leaves the invariant domain and begins to diverge. Consequently, for strictness it is necessary to study the mapping as a whole at least for small values η_i and $g_{i,i-1} - G_{i,i-1}$. Unfortunately, a number of difficulties arise that are typical for no smooth mappings. We conducted a computer analysis of mapping. It turns out that it converges in a wide range of parameters and initial conditions. Moreover, we have not found the examples of divergence.

Until now, we have assumed that the adaptation process began when the oscillations of the reference ring had already been established. The construction is also valid in the more general case.

Let $\xi_i(k)$ be the delay of the spike of the ith neuron with respect to the spike of the $(i-1)$th neuron in the reference ring at the kth cycle of wave propagation. By $\eta_i(k)$ denote the mismatch of the moments of the start of the spikes of the ith adaptive and the ith standard neurons in the kth cycle. Finally, let $G_{i,i-1}(k-1)$ be the values of the synaptic weights modified on the previous tact.

Theorem 5.2.1 *Let iterative process (5.2.3) and (5.2.4) converge in some fixed neighbourhood of the point $\eta_i = 0$, $G_{i,i-1} = g_{i,i-1}$ and the numbers $\eta_i(0)$ and $G_{i,i-1}(0)$ satisfy inequalities (5.2.4). There exists such $C_0 > 0$ that at $\left|\xi_i(0) - \xi_i^0\right| < C_0$ sequences $\xi_i(k) \to \xi_i^o + o(1)$, $\eta_i(k) \to o(1)$, $G_{i,i-1}(k) \to g_{i,i-1} + o(1)$ at $k \to \infty$.*

The question about the details of the behavior of the quantities $\xi_i(k)$, $\eta_i(k)$, $G_{i,i-1}(k)$ in the neighborhood, respectively, of the values $\xi_i^o, 0, g_{i,i-1}$ remains open. From an application point of view, it is not of particular interest because of the small potential of possible oscillations.

The described process of adaptation and Theorem 5.2.1 essentially complete the representations about the properties of ring neural systems. It turns out they can not only store the wave packets but can adjust to remember them. In this case, the nature of the reference sequence of spikes is absolutely not important. It can be generated by the sensory structures of the brain. Thus, it is permissible to consider the scheme of adaptation as the model of storing information provided in the wave form.

In the case when adaptation occurs under the influence of the reference ring, a copy of the ring is created. As noted, the influence of the reference ring after adjustment can be blocked. The axoaxonic synapses can be located on the axon appendages running to the adaptive ring. The latter, perhaps, are the endings of the axons of the inhibitory neurons generating spikes with sufficiently high frequency. In his work, Eccles (1971) always paid great attention to inhibitory pathways. These inhibitory pathways are also part of the model (Frolov and Shulgina 1983).

It is important that the neural rings are able to self-reproduce, which is a common property of all living things. Von Neumann was the first to raise the question about the problem with self-replicating systems. He posed the problem of synthesis of the machine capable to build its exact copy from a standard set of parts. For the point of classical neural networks, the problem is discussed, for example, in the work of Vvedensky and Ershov (1992). The authors of this work introduce into the model the mechanism rhythmically and synchronously changing the thresholds of neurons. It is possible to make copies in the process of exchanging information between the neural systems.

The adaptation scheme with replacement is not unique. We can propose a simpler neural system for adaptation without the involvement of axoaxonic synapses. In it, the ith reference neuron acts only on the ith adaptive neuron by means of an axosomatic synapse. The weights of the given action are the same for all pairs of neurons. The neural population is described by system (5.2.1), in which we should replace in the equation for the ith adaptive neuron $\theta(u_{i-1} + w_{i-1})$ to $\theta(w_{i-1} - 1)$. We construct the analogue of mapping (5.2.3) and (5.2.4) for mismatches of η_i spikes of the adaptive and reference neurons and the modified synaptic weights $G_{i,i-1}$. Computational experiments have shown that iterations of mapping converge. However, satisfactory analytical researches could not be held. It appears that the scheme has a lower reliability.

References

Dunin-Barkowski, V. L. (1978). *Information processes in neural structures*. Moscow: Nauka.

Eccles, J. (1971). *Inhibitory pathways in the central nervous system*. Moscow: Mir.

Frolov, A. A. (1993). Structure and functions of learning neural networks. *Neurocomputer as the basis of thinking computers* (pp. 92–100). Moscow: Nauka.

Frolov, A. A., & Shulgina, G. I. (1983). Memory model based on the plasticity of inhibitory neurons. *Biophysics, 28*(3), 445–480.

Frolov, A. A., & Muravyov, I. P. (1987). *Neural models of associative memory*. Moscow: Nauka.

Frolov, A. A., & Muravyov, I. P. (1988). *Information characteristics of neural networks*. Moscow: Nauka.

Hebb, D. O. (1949). *The organization of behaviour: Neuropsychological theory* (p. 335). New York: Wiley.

Kohonen, T. (1980). *Associative memory*. Moscow: Mir.

Kohonen, T. (1984). *Self-organization and associative memory*. Berlin: Springer.

Rosenblatt, F. (1965). *Principles of neurodynamics*. Moscow: Mir.

Vvedensky, V. L., & Ershov, A. A. (1992). Connection of electrical phenomena in the brain with information processes in model neural networks in *Results of science and technology*. In *Physical and mathematical models of neurl networks. Model networks and living organisms* (Vol. 4, pp. 3–37). Moscow: Inst. VINITI.

Chapter 6
Model of the Neural System, Synchronizing the Wave Packets

Synchronization (and desynchronization) in neural structures is an essential condition of the normal functioning of the brain. It has been considered, for example, that distinct rhythms of slow brain activity occur as a result of synchronization Elul 1972; Guselnikov and Suping 1968). Violation of these rhythms can indicate pathology.

The synchronization of oscillations plays an important role in processing information. Livanov (1972, 1989) discovered that excitation from one domain of the brain to the other is conducted better if the oscillations in the domains are synchronous and there is a certain phase shift. Investigations have been continued by Lebedev (1990); see also Zabrodin and Lebedev (1977), who, on a qualitative level, proposed the theory of wave representation of information in the brain. This theory is based on a number of assumptions. The main ones are as follows:

- All images in the brain are coded by waves of undamped periodic neural activity (wave patterns).
- If the phase difference between two identical waves is less than some critical value, then the waves must be synchronized.
- The identity of the same image is recognized at the time when the crests of the waves coincide.

The second assumption is very important. If we know the period of the wave (Lebedev considers that it coincides with the period of α-rhythm) and the critical difference of phases, then it is possible to estimate the degree of diversity of the waves (the amplitudes are considered to be the same). Furthermore, according to Lebedev, the material carrier of each wave is some neural ensemble (not fixed). We estimate the number of neurons in the ensemble, and thus the order of the number of neurons in the brain is determined. We can find a number of different types of oscillations and, consequently, the information characteristics of the brain. We introduce the notion of the alphabet and deduce a formula for the amount of memory.

Accepted by Lebedev, the second assumption is based on experimental data, but it is still a hypothesis. However, the formula for the amount of memory can be tested experimentally. It turns out that the theoretical predictions agree well with experimental results (Lebedev 1990).

© Springer International Publishing Switzerland 2015
S. Kashchenko, *Models of Wave Memory*,
Lecture Notes in Morphogenesis, DOI 10.1007/978-3-319-19866-8_6

The third assumption seems natural: At the moment of coincidence of crests of patterns, we observe a larger value of the total quantitative characteristics of the system. The event can be discovered, for example, with the help of some threshold mechanism. It turns out that we have also found indirect evidence (Lebedev 1990) for the third assumption.

It is obvious that the considered assumptions suggest the existence of some mechanisms of synchronization. Their simulation is of great interest. Below we propose a system of two coupled neural rings that synchronizes wave packets. Relations between the rings are homogeneous and unaddressed. Each neuron of one ring potentially equally influences all neurons of another ring. This is dictated by the fact that the information about the numbering of neurons in the rings is hidden to the outside observer (the numbering depends on the ring's arbitrariness). Despite the unaddressed relations, the synchronization of oscillations develops over time for identical rings.

Note that in the works of Lebedev, we can talk about waves of so-called "slow activity" (these are easy to register). However, a correlation between slow activity and pulse potentials has been shown in a number of studies (Elul and Adey 1966; Adey 1967).

6.1   Architecture and the Neural Network Model

Consider two identically oriented ring structures, each of which contains N simulation neurons. Let the conditions of Theorem 3.3.1 be fulfilled and the waves of spikes propagate around the rings. In general, the waves differ by phases. There is the task of synthesizing such a system of interaction between neurons of different rings so that over time the waves become synchronized.

It would seem that the easiest variant of this task is to connect the neurons corresponding to each other (the ith neuron of the first ring with the ith neuron of the second ring) by the excitatory synapses. In this case, we can construct mapping describing the passage of consecutive wave cycles. As already stated, synchronization occurs over time. The difficulty lies in the fact that it is unknown which neurons in the rings correspond to each other because their numbering depends on our arbitrariness: We operate by trial and error. Consistently choose the number $k = 1, \ldots, N - 1$ and declare that the ith neuron of one ring and the $(i + k)$th neuron of the other ring correspond to each other. There exists such value k that there the synchronization of oscillations will be observed. The scheme assumes the existence of some complex external control device that seems quite unnatural.

We can avoid assuming the existence of an external control if we consider that the principle of the unaddressed action is implemented during the process of synchronization. This means that each neuron of one ring can influence any neuron of another ring. However, the potential of possible action is realized only when there is a coincidence in time of certain events. Therefore, the model is based on the synapse, which we call "correlative." The idea of such synapse is found in the literature [22] on neural modeling and does not contradict neurophysiological data.

The correlative synapse is the communication channel, and it has a minimum of two inputs. For simplicity, let us say that there are indeed two inputs. The first input is activating and leads the synapse to a state of readiness. If in the state of activation the spike comes to the second input, then the correlative synapse works. This influences the evolution of the membrane potential of the neuron on which it is located. Assumptions about the character of influence are the same as for the chemical synapse. We consider that the time of activation of the correlative synapse coincides with T_1 (asymptotic duration of spike) and denote the effective time of its action by h_s. In the model, we assume that $h_s < T_1$. Introduce time delay h_0 of the activating input. Let it be larger than the effective time of action of the synapse: $h_s < h_0$. As the correlative synapse, complexes of synapses can act on the neuron body. Simultaneously, its role can be played by the external neuron as detector.

Denote by u and v the membrane potentials of the neurons acting, respectively, on the activating and starting inputs of the correlative synapse. The functional

$$H_{Co}(u, v) = \theta(u(t - h_0) - 1)\theta(v(t) - 1)\theta(1 - v(t - h_s)) \qquad (6.1.1)$$

is the indicator of the period of the effective time of action of the correlative synapse. It takes a single value on those time intervals of duration of h_s, where there are two events. The spike of the neuron connected to activating input, started earlier, at least for the time h_0 (delay of activation), but since its start has passed no more than $T_1 + h_0$. Simultaneously, no time longer than h_s. has passed since the moment of the start of spike of the neuron connected to the start input. Otherwise, the functional takes zero value. If some neurons have equal access to the starting input of the correlative synapse, then in the formula for the indicator we substitute the sum of the corresponding membrane potentials.

Excitatory correlative synapses are used in the construction of the system of interaction between the ring structures, thus ensuring the synchronization of oscillations. We choose for certainty the first neural ring. The architecture of connections for the second ring is symmetrical. Around the ring propagates the wave of spikes caused by the $(i - 1)$th neuron acting on the ith neuron. It is reasonable to carry out an adjustment of phase of the ith neuron on the interval of action. We assume that the correlative synapse located on the ith neuron is activated by the $(i - 1)$th neuron. All of the neurons of the second ring (the unweighted sum of the corresponding membrane potentials at the input is taken) have equal access to the starting input of the given synapse. Thus the principle of unaddressed action is implemented.

Denote by u_i and v_i $(i = 1, \ldots, N)$ the membrane potentials of neurons, respectively, of the first and second rings. The system of equation (3.1.1) for the neuronal population is being upgraded. In formulas (3.1.2) and (3.1.3) the effective time of action for correlative synapses is given by the functional (6.1.1). We arrive at the system of equations:

$$\dot{u}_i = \lambda \left[\begin{array}{l} -1 - f_{\mathrm{Na}}(u_i) + f_{\mathrm{K}}(u_i(t-1)) \\ + \alpha H(u_i) \left[g_{i,i-1}\theta(u_{i-1}(t)-1) + gH_{\mathrm{Co}}\left(u_{i-1}, \sum_{j=1}^N v_j\right) \right] \end{array} \right] u_i, \qquad (6.1.2)$$

$$\dot{v}_i = \lambda \left[\begin{array}{l} -1 - f_{\mathrm{Na}}(v_i) + f_{\mathrm{K}}(v_i(t-1)) \\ + \alpha H(v_i) \left[g_{i,i-1}\theta(v_{i-1}(t)-1) + gH_{\mathrm{Co}}\left(v_{i-1}, \sum_{j=1}^N u_j\right) \right] \end{array} \right] v_i, \qquad (6.1.3)$$

where $i = 1, \ldots, N$, thus the number $i = 0$ is identified with $i = N$. We consider that the positive functions $f_{\mathrm{Na}}(u)$ and $f_{\mathrm{K}}(u)$ monotonically tend to zero as $u \to \infty$, the number $\alpha = f_{\mathrm{K}}(0) - f_{\mathrm{Na}}(0) - 1 > 0$, and the speed parameter is $\lambda \gg 1$. Therefore, the conclusion of Theorem 1.6.1 is valid for the equations for each of the neurons. The functional $H(*)$, defined by formula (4.1.2), vanishes during the spike and within time $h = 2(1 - \varepsilon)T_1 (0 < \varepsilon < 1)$ after the spike. Synaptic weights $g_{i,i-1}$ of action of the $(i - 1)$th neuron on the ith neuron in the rings are given by formula (3.3.8):

$$g_{i,i-1} = \left(T_2 - \sum_{i=1}^N \xi_i^0 \right) \bigg/ \xi_i^0 > 0. \qquad (6.1.4)$$

As a result, the system of equations of each of the rings has an attractor consisting of solutions where the spike of the ith neuron is delayed with respect to the spike of the $(i - 1)$th neuron by a value close to ξ_i^0.

In the system of equations (6.1.2) and (6.1.3), g is common for all correlative synapses' weight of action. We assume that the value h_0 of the activation delay and the effective time of action h_s of the correlative synapses are subject to inequalities $h_0 + h_s < \min \xi_i^0$. In this case, as already noted, $h_s < h_0$.

From a purely mathematical point of view, there is the problem of investigation of the dynamics of two coupled oscillators in the neighborhood of the family of attractors, the trajectories of which have a rather complex structure but are asymptotically close to cycles. Below we show that in the considered situation there exists an attractor, in which the oscillations of generators (rings) occur almost simultaneously. The given attractor of the system of equations (6.1.2) and (6.1.3) is called the "attractor of synchronous oscillations."

6.2 The Existence of the Attractor of Synchronous Oscillations

We have already described in detail the method of asymptotic study of the system of equations (6.1.2) and (6.1.3) in Sects. 3.2, 3.3, and 4.4. At initial cycle of the passage of the excitation wave, the moments of the start of the spikes of neurons are given a priori. Let these be the numbers $t_{i,j}(i = 1, \ldots, N; j = 1, 2)$, respectively, for the ith neuron of the jth ring. We consider that $0 \le t_{1,j} < t_{2,j} < \cdots < t_{N,j} < T_2$ $(j = 1, 2)$. The initial conditions for the system of equations (6.1.2) and (6.1.3) are given separately: $u_i(t_{i,1} + s) = \varphi_{i,1}(s) \in S_h$, $v_i(t_{i,2} + s) = \varphi_{i,2}(s) \in S_h$, $(i = 1, \ldots, N)$. Recall that S_h is the set of continuous functions φ_s on the interval $s \in [-h, 0]$ $(h = 2(1 - \varepsilon)T_1)$, for which $\varphi(0) = 1$ and $0 < \varphi(s) \le \exp(\lambda \alpha s/2)$ at $s \in [-h, 0]$. The system of equations (6.1.2) and (6.1.3) is recurrently integrated. Formulas are asymptotically simplified if we assume that the parameter $\lambda \to \infty$. There are the moments $t'_{i,j}(i = 1, \ldots, N; \quad j = 1, 2)$ of the start neurons' spikes at the new cycle of wave propagation. At the same time $u_i\left(t'_{i,1} + s\right) = \varphi_{i,1}(s) \in S_h$, $v_i\left(t'_{i,2} + s\right) = \varphi_{i,2}(s) \in S_h$ allows the algorithm to continue.

If we count the start of each cycle at each cycle of the wave's passage, for example, from the start of spike of the first neuron of the first ring, then arises the mapping $t_{i,j} \to t'_{i,j}$, $\varphi_{i,j} \to \varphi t'_{i,j}$. It is approximately constructed. If the main (independent of λ) part of the finite-dimensional constituent of mapping has a stable fixed point $t^0_{i,j}$, then we can make a conclusion about the existence of the attractor of the system of equations (6.1.2) and (6.1.3). In the attractor, at each cycle of the waves' passage, the moments of the start of spikes are close to numbers $t^0_{i,j}$.

We should note that at different combinations of moments in time $t_{i,j}(i = 1, \ldots, N; j = 1, 2)$ of the start of the neurons' spikes at the initial cycle, we obtain different formulas for the time of the start of spikes at the next cycle. The final mapping $t_{i,j} \to t'_{i,j}$ is continuous, but it turns out to be stitched from parts. It is difficult to consider all the variants of the behavior of iterations of mapping analytically. We restrict ourselves to the case when the starts of spikes $t_{i,1}$ and $t_{i,2}$ $(i = 1, \ldots, N)$ of a pair of ith neurons at some point are not too different. In other words, we consider the completion of the synchronization process and show the existence of the attractor of synchronous oscillations.

Let each cycle of the passage of the wave be open with the spike of the first neuron of the first ring. Denote by $\xi_i(i = 2, \ldots, N)$ the time between the start of spikes of the $(i - 1)$th and ith neurons of the first ring at the initial cycle of wave propagation. Let $\eta_i = t_{i,2} - t_{i,1}$ $(i = 1, \ldots, N)$ be the mismatch of the start of spikes of the ith neurons of the first and the second rings in the initial cycle. Denote by ξ'_1 the time interval between the start of the spikes of the Nth and the first neurons of the first ring (it is calculated). The values $\xi'_i(i = 2, \ldots, N)$, $\eta'_i(i = 1, \ldots, N)$ have the same value, but they are related to the next cycle of wave passage.

Lemma 6.2.1 *There exists such independent of λ value $C_0 > 0$ that at $\left|\xi_i - \xi^0_i\right| < C_0(i = 2, \ldots, N)$, $|\eta_i| < C_0(i = 1, \ldots, N)$ and $\lambda \to \infty$ are valid asymptotic formulas:*

$$\eta_1' = \left(g_{1,N}\eta_N + \eta_1\right)/\left(1 + g_{1,N} + g\right) + o(1) \tag{6.2.1}$$

$$\xi_1' = \left(T_2 - \sum_{m=2}^{N} \xi_m\right)\bigg/\left(1 + g_{1,N}\right) + \theta\left(-\eta_1'\right)g\eta_1'/\left(1 + g_{1,N}\right) + o(1), \tag{6.2.2}$$

$$\eta_i' = \left(g_{i,i-1}\eta_{i-1}' + \eta_i\right)/\left(1 + g_{i,i-1} + g\right) + o(1), \quad (i = 2, \ldots, N), \tag{6.2.3}$$

$$\xi_1' = \left(T_2 - \sum_{m=i+1}^{N} \xi_m + \sum_{m=1}^{i-1} \xi_m'\right)\bigg/\left(1 + g_{i,i-1}\right) \tag{6.2.4}$$

$$+ \theta\left(-\eta_i'\right)g\eta'/\left(1 + g_{i,i-1}\right) + o(1), \quad (i = 2, \ldots, N-1) \tag{6.2.5}$$

$$\xi_N' = \left(T_2 - \sum_{m=1}^{N-1} \xi_m\right)\bigg/\left(1 + g_{N,N-1}\right) + \theta\left(-\eta_N'\right)g\eta_N'/\left(1 + g_{N,N_1} + o(1)\right). \tag{6.2.6}$$

In the proof of Lemma 6.2.1, we should consider two cases for each pair of the ith neurons of the first and second rings. The spike of neuron of the first ring at a new cycle of wave passage either leads the spike of the neuron of the second ring or is delayed with respect to such. For the rest of the cycle, the construction is carried out similarly to the proof of Lemma 4.1.1. This construction is simple but rather cumbersome.

Using Lemma 6.2.1, we can study consecutive cycles of the passage of the excitation wave. Let $\xi_i(k)$ and $\eta_i(k)$ $(i = 1, \ldots, N)$ be the values of the described variables at the kth cycle. From formulas (6.2.1) and (6.2.3), it follows that $\eta_i(k) \to o(1)$ at $k \to \infty$. If in formulas (6.2.2), (6.2.4), and (6.2.5) we discard the terms containing η_i' and the terms $o(1)$, then we obtain the iterative process, the convergence of which is justified in Sect. 3.3, Chap. 3. This solution of linear systems with symmetric positive definite matrix is the method of Seidel. The limit point of the given process is $\xi_1^0, \ldots, \xi_n^0$, which can be checked by direct substitution attracting formula (6.1.4) for the weights $g_{i,i-1}$.

From Lemma 6.2.1 immediately follows

Lemma 6.2.2 As $\lambda \to \infty$ there exists such independent of λ value $C_0 > 0$ that at $\left|\xi_i(0) - \xi_i^0\right| < C_0 (i = 2, \ldots, N)$, $\left|\eta_i(0)\right| < C_0 (i = 1, \ldots, N)$ with an increase of the number k of cycle of the passage of the wave time intervals $\xi_i(k)(i = 1,\ldots,N)$ between the starts of spikes of the $(i-1)$th and ith neurons of the first ring tend to $\xi_i^0 + o(1)$, and the mismatches $\eta_i(k)(i = 1,\ldots,N)$ of the start of spikes of the ith neurons of the first and second rings tend to $o(1)$.

From Lemma 6.2.2 follows the statement about the existence of an attractor of synchronous oscillations for the system of equations (6.1.2) and (6.1.3).

Statement 6.2.1 *As $\lambda \to \infty$, the system of equations (6.1.2) and (6.1.3) has the attractor consisting of solutions for which the spikes of the ith neurons $(i = 1,\ldots,N)$ of the first and the second rings start in an asymptotically small time period. The spikes of the ith neurons in each of the rings are delayed with respect to the spikes of the $(i-1)$th neurons by a value asymptotically close to ξ_i^0. All of the solutions included in the attractor have a common asymptotic behavior on any finite time interval. The attractor has a periodic solution.*

Thus, the neural structure of the connected rings of the system, described by Eqs. (6.1.2) and (6.1.3), actually performs synchronization of the wave packets, which are identical but ith respect to phases. The problem of synchronization is important in many problems. As already noted, it is of particular importance in modeling neural networks. The synchronization of wave packets can be interpreted (Lebedev 1990) as the identification of images, for which they are codes. Obviously, as a result of synchronization, the peak values of the total signal increase. By this event, we can identify the coincidence of wave packets. Unfortunately, it is not analytically possible to determine the size of the basin of attraction of the attractor of synchronous oscillations. Therefore, it is safe to say only that there is partial solution of the problem of comparison of the waves generated by ring neural formations.

References

Adey, V. R. (1967). Organization of brain structures in terms of communication and storage of information. *Modern problems of electrophysiology of the central nervous system* (pp. 324–340). Moscow: Nauka.

Elul, R. (1972). The genesis of the EEG. *International Review of Neurobiology, 15*(2), 227–242.

Elul, R., & Adey, W. R. (1966). Nonlinear relationship of spike and waves in cortical neurons. *The Physiologist, 8*, 98–104.

Guselnikov, V. I., & Suping, A. J. (1968). *Rhythmic activity of the brain.* Moscow: Moscow State University Press.

Lebedev, A. N. (1990). Human memory, its mechanisms and limits. *Research of memory* (pp. 104–118). Moscow: Nauka.

Livanov, M. N. (1972). *Spatial organization of the processes of the brain.* Moscow: Nauka.

Livanov, M. N. (1989). *Spatiotemporal organization of potentials and systemic activity of the brain: Selected works.* Moscow: Nauka.

Zabrodin, J. M., & Lebedev, A. N. (1977). *Psychophysiology and psychophysics.* Leningrad: Nauka.

Chapter 7
Model of the Neural System with Diffusive Interaction of Elements

This chapter discusses the problems of dynamics of a population of neurons with an electric interaction between them. We encounter electrical synapses in the nervous system relatively less than we do chemical synapses. The electrical transmission of signals between excitable cells was demonstrated for the first time by Furshpan and Potter in 1959 in experiments on the giant axons of crayfish. Soon electrical excitation transfer was found in the central nervous system, in heart muscle, and in receptor cells (Eckert el al. 1991). Because during electrical transmission the current goes directly from cell to cell, electrical synapses have a high degree of reliability. Neurophysiologists consider that electrical synapses play an obvious role in the synchronization of neuronal activity (Pappas and Waxman 1973; Eckert et al. 1991) when several nerve cells are affected by the excitation. As noted by Pappas and Waxman (1973), electrical contacts are often found in neuronal nuclei pacemakers of some fish species, and they apparently serve the same purpose.

According to the concepts of Sect. 2.2, neurons interacting with each other by way of electrical synapses are described by the system of equations with difference diffusion. Below we consider populations consisting of identical neurons with homogeneous couplings (diffusion coefficients are the same for all neurons). It turns out that the latent period for diffusively coupled neurons is very stable. Recall that the latent period is the period of time between the beginning of an action and the start of a neuron's spike. Thus, the simulation results of electrically connected neurons are consistent with generally accepted notions of biological safety.

In the first section of this chapter, we consider the features of oscillations in local networks where each neuron is connected only with their nearest adjacent neurons. Using asymptotic method, we prove that oscillatory modes exist in which spikes of neurons pass through the same time period (in the asymptotic sense).

Note the following: The discovery of the semiconducting properties of electrical contacts (Pappas and Waxman 1973) and other biological data make the problem of the dynamics of the network with inhomogeneous diffusion meaningful. We can transfer almost all of the above-mentioned construction for networks with chemical synapses to the case of neurons with electrical interaction. The ideology is the same; changes occur only in the formulas. Difficulties arise only for ring populations of neural modules because electric interaction is excitatory in nature.

© Springer International Publishing Switzerland 2015
S. Kashchenko, *Models of Wave Memory*,
Lecture Notes in Morphogenesis, DOI 10.1007/978-3-319-19866-8_7

In the second section of this chapter, a computer analysis of the homogeneous fully connected neural network with electrical interaction is presented. Initially, full connectivity may seem surprising. However, computer reconstruction of nerve tissue and fiber shows that both the interlacing and contacts of cells are rather complex in nature (Schade and Ford 1976). The existence of "noise," i.e., complex (nonsynchronous) modes, for a population of ten neurons is numerically demonstrated. Note that they are present along with the homogeneous attractor.

The fact of the existence of nonsynchronous oscillations is very important. If we act on the neural system, which is in "noise mode" created by the sequence of external pulses, then a wave packet arises, which carries on itself the features of the external action. It exists for some time and then washes out. The phenomenon can be interpreted as the short-term memorization and subsequent forgetting of sensory information. It has been observed that the "more complex" the mode noise, the more adequately the wave packet reflects the external action.

The given phenomenological consequence of the model was transferred as a hypothesis onto the human brain. The experimental results are discussed in detail in the next chapter. We statistically establish as valid the idea that the "more complex" an individual's electroencephalogram signal, the greater the amount of the individual's short-term memory.

Furthermore, the form of the wave patterns in the model again leads us to the conclusion that there are time zones where the structure of the oscillations in a neural ensemble is specific (i.e., they have characteristic features). At the same time, there are nonspecific zones with weakly expressed differences. The property of human perception of stimuli has also been studied experimentally and is discussed in Chap. 9.

7.1   Oscillations in Diffusion-Related Systems of Equations Simulating Local Neural Networks

A local network is a population of similar neurons, each of which is connected only with nearby adjacent neurons. The term "local" indicates the communication architecture, which is one of the attributes characterizing the network. The next feature is the type of elements. Different types of neural elements are considered in mathematical models and in physical implementations of local structures, e.g., McCulloch-Pitts neurons, operational amplifiers, and Wiener neurons. Kashchenko and Mayorov (1995b) considered the network in which the formal neurons are described by Hutchinson equations.

Usually the operational amplifiers (Chua and Yang 1988a, b) with S-shaped features are preferred. Network data can easily be implemented in hardware. Mathematically they are described by systems of ordinary differential equations. In the phase space, states of equilibrium, cycles, and tori can exist. Chaotic behavior of the trajectory is also possible. However, in the case of symmetric relations, any trajectory of the system tends to some state of equilibrium (Hopfield 1982, 1984; Chua and Yang 1988a, b). At a sufficiently high slope, the output currents of the amplifiers are close to saturation.

The behavior of symmetric local networks from the diffusionally connected neurons under consideration is much more complex (Kashchenko et al. 1995c). Note that "diffusionally connected elements" is another characteristic that points to the mechanism of interaction.

7.1.1   Oscillations in the System of Two Neurons

Consider two identical neurons that are diffusionally connected to each other. As stated previously, the dynamics of their membrane potentials u_1 and u_2 is described by the system of equations:

$$\dot{u}_1 = \lambda[-1 - f_{Na}(u_1) + f_K(u_1(t-1))]u_1 + d(u_2 - u_1), \qquad (7.1.1)$$

$$\dot{u}_2 = \lambda[-1 - f_{Na}(u_2) + f_K(u_2(t-1))]u_2 + d(u_1 - u_2), \qquad (7.1.2)$$

where d is the coefficient of difference diffusion. We consider that it is compatible with the value $\lambda(\lambda \gg 1)$ and is small: $d = \exp(-\lambda\sigma)$ where $\sigma > 0$. Assume that the positive functions $f_{Na}(u)$, $f_K(u)$ monotonically tend to zero as $u \to \infty$ and the number $\alpha = f_K(0) - f_{Na}(0) - 1 > 0$. As a result, the conclusions of Theorem 1.6.1 are valid for each neuron's equation. According to this theorem, an important role in the asymptotic formula for solving the isolated neuron's equation is played by the numbers:

$$\alpha_1 = f_K(0) - 1 > 0, \quad \alpha_2 = f_{Na}(0) + 1 > 0.$$

Recall that the ascending and descending parts of the spike are approximated by exponentials with multipliers in exponents $\lambda\alpha_1$ and $(-\lambda)$. Immediately after the spike, within a certain unit of time, the process develops exponentially with a multiplier in the exponent $(-\lambda\alpha_2)$. This is followed by slow evolution of the membrane potential with the multiplier in the exponent $\lambda\alpha$.

In the study of system (7.1.1) and (7.1.2), we must consider a number of cases and subcases. To reduce their number, we draw some biological considerations. Analysis of experimental data (see Ochs 1969) shows that the descending part of the spike is longer than the ascending part. The longest part is from the minimum point until the start of the next spike. Using this and formulas for α, α_1, α_2, it is easy to obtain that $0 < \alpha < 1$ and $1 < \alpha_1 < \alpha_2 < \alpha_1 + 1$.

A methodical procedure, which is described in Sect. 3.2, is used for the asymptotic integration of the system (7.1.1) and (7.1.2). The initial conditions for the equations are set individually for each neuron in its initial moment of time.

As mentioned previously, S is the class of continuous functions $\varphi(s)$ on the interval $s \in [-1,0]$, for which $\varphi(0) = 1$ and $0 < \varphi(s) \le \exp(\lambda \alpha s/2)$.

Let $u_1(s) = \varphi_1(s) \in S$ and $u_2(\xi + s) = \in S$ for $s \in [-1,0]$ where $0 < \xi < T_2$ and T_2 is the asymptotic value of the period of spike generation by the isolated neuron given by formula (1.6.4). Initially the spikes of neurons are mismatched by the value ξ. Ignoring the term $\exp(-\lambda \sigma) u_2(t)$, we can integrate Eq. (7.1.1) on the interval $t \in [0, \xi]$ and find the asymptotics $u_1(\xi)$ for $\lambda \to \infty$. At $t \ge \xi$ for system (7.1.1) and (7.1.2), we have the ordinary Cauchy problem, which is solved by the method of steps. At each step for the principal terms of the asymptotics as $\lambda \to \infty$, we obtain the triangular system of equations, which is integrated. Asymptotics can be constructed on any finite time interval. Denote by t_2^1 and t_2^2 the starts of the new spikes of the first and second neurons, respectively. The principal terms of asymptotics t_2^1 and t_2^2 as $\lambda \to \infty$ are uniquely determined by ξ and are independent of $\varphi_1(s)$ and $\varphi_2(s)$. Assume that $u_1(t_2^1 + s) = \varphi_1'(s) \in S$ and $u_2(t_2^2 + s) = \varphi_2'(s) \in S$. As a result, we return to the initial situation where the mismatch of the start of spikes of neurons is $\xi' = t_2^2 - t_2^1$.

Thus arises the continuous mapping $\prod : (\xi, \varphi_1(s), \varphi_2(s)) \to (\xi', \varphi_1'(s), \varphi_2'(s))$. In it, a one-dimensional constituent is represented in the form $\xi' = f(\xi) + o(1)$. If the main part of this mapping, in the asymptotic sense, $\xi' = f(s)$, has a stable fixed point ξ^*, then for system (7.1.1) and (7.1.2) we can draw a conclusion about the existence of the attractor. In it, the spike starts of the first and the second neurons are mismatched by value $\xi^* + o(1)$. The attractor contains a periodic solution.

Write $\sigma_1 = \alpha_1 + (\alpha_2 + \alpha_1 \alpha)/2$ and let $\sigma \in (0, \sigma_1)$. The function $f(\xi)$ is a linear spline. Note that the interval $(0, \sigma_1)$ falls into parts where the number of spline links differs and depends on σ_1. Nevertheless, as $\lambda \to \infty$ and for all $\sigma \in (0, \sigma_1)$ then the a priori assumption about the inclusion $\varphi_1'(s), \varphi_2'(s) \in S$ holds, and the mapping $\xi' = f(\xi)$ has, depending on the value σ, either one or two fixed points ξ^*. In all cases, the mapping in some neighborhood of the fixed point is constant, which guarantees its stability. In other respects, the form of the mapping is not of particular interest because the cycles are not detected.

For solutions that form the attractor, the asymptotic representations of spikes (i.e., the duration of the ascending part and descending one) are the same as in Theorem 1.6.1 for the case of the isolated neuron. Other formulas are cumbersome. We restrict ourselves to the most important results of the mismatch of spikes in time and around the minima $u_1(t), u_2(t)$. The latter is of interest because the interaction effect increases the minimum values. As a result, neurons generate spikes more commonly than they do in the free state. Recall that according to Theorem 1.6.1, the minimum value of the membrane potential of isolated neuron is $u^{\min} = \exp(-\lambda(\alpha_2 + o(1)))$. Speaking of moments of time, where there are minimum values we will count them from the start of the spike of the first neuron.

Write $\sigma_0 = \alpha_1 + \alpha_1 \alpha$. The fixed point ξ^* (or one of the points if there are two of them) of the mapping $\xi' = f(\xi)$ is given by formula:

$$\xi^* = \begin{cases} \sigma/\alpha_1 & \text{at } 0 < \sigma < \alpha_1 \\ 1 + (\sigma - \alpha_1)/\alpha & \text{at } \alpha_1 \le \sigma < \sigma_1 \end{cases}$$

Introduce one more value, which is dependent on σ:

$$\eta^* = \begin{cases} T - (\alpha_2/\alpha_1 - 1)\sigma/\alpha - \sigma/\alpha_1 & \text{at } 0 < \sigma \le \alpha_1 \\ \alpha_1 + 1 + (\alpha_1 - (\sigma - \alpha_1)/\alpha)/\alpha & \text{at } \alpha_1 \le \sigma < \sigma_0 . \\ \xi^* & \text{at } \sigma_0 \le \sigma < \sigma_1 \end{cases}$$

Depending on the value σ, we distinguish a number of cases: (1) $0 < \sigma < \alpha_1$, (2) $\alpha_1 < \sigma < \alpha_1 + \alpha(\alpha_1 - 1)$, (3) $\alpha_1 + \alpha(\alpha_1 - 1) < \sigma < \alpha_1(1 + \alpha) \equiv \sigma_0$, (4) $\sigma_0 \equiv \alpha_1(1 + \alpha) < \sigma < \alpha_1 + (\alpha_2 + \alpha_1\alpha)/2 \equiv \sigma_1$.

In the first three cases, the mapping $\xi' = f(\xi)$ has two stable fixed points; consequently, there exist two attractors at the system (7.1.1) and (7.1.2) as $\lambda \to \infty$. It is easy to understand that one attractor goes into the other if you change the numbering of the neurons. For solutions that form it, the start of the spike of the second neuron is delayed with respect to the start of the spike of the first neuron by the time $\xi^* + o(1)$. In turn, the time interval between the start of the spike of the second and the first neurons is $\eta^* + o(1)$ and $\eta^* > \xi^*$, i.e., the given interval is longer than the previous one. Thus, the spikes, which are adjacent in time, of each of the neurons are mismatched by the value $\xi^* + \eta^* + o(1)$.

Note that in the first case, the delay of the spike of the second neuron with respect to the first one is $\xi^* < 1$ and in the second and third cases is $1 < \xi^* < \alpha_1 + 2$. At the same time, in the first two cases, the maximum of the pulse of the second neuron is achieved on the descending part of the spike of the first neuron. In the third case, this maximum falls on a unit time interval, after the completion of which follows the spike of the first neuron.

In the first case, the minimum values of the membrane potentials of neurons are:

$$u_1^{min} = u_1(\alpha_1 + 2 + o(1)) \approx u^{min} \exp(\lambda\sigma(\alpha_2/\alpha_1 - 1)) > u^{min},$$
$$u_2^{min} = u_2(\alpha_2 + 2 + \xi^* + o(1)) \approx u^{min},$$

where $u^{min} = \exp(-\lambda(\alpha_2 + o(1))$ (the minimum value of the isolated neuron's membrane potential). Accordingly, in the second case these relations hold:

$$u_1^{min} = u_1(\alpha_1 + 2 + o(1)) \approx u^{min} \exp[\lambda[(\sigma - \alpha_1)/\alpha - \sigma + \alpha_2]] > u^{min},$$
$$u_2^{min} = u_2(\alpha_2 + 2 + \xi^* + o(1)) \approx u^{min}.$$

In the third case, the problem of minima is more complex (i.e., the case becomes separate subcases). For the membrane potential of the first neuron, the minimal value is achieved on the unit interval of time after the end of the spike and, necessarily, $u_1^{\min} > u^{\min}$. In one of the possible situations the minimum point is $t = \alpha_1 + 2 + o(1)$.

The increase of the minimum value of the first neuron's membrane potential is consistent in the biological sense: Action, incoming through the electrical contact from the side of the second neuron, does not allow the membrane's strong polarization after the completion of spike. This fact of the depolarizing action of synapses is found in the biological literature (Eckert et al. 1991; Tasaki 1971).

In the fourth case, $\sigma_0 \equiv \alpha_1(1 + \alpha) < \sigma < \alpha_1 + (\alpha_2 + \alpha_1\alpha)/2 \equiv \sigma_1$, the mapping $\xi' = f(\xi)$ has a unique stable fixed point (i.e., it is locally constant in its neighborhood). As a result, the above-described attractors of the system of equations. (7.1.1) and (7.1.2) are combined into one attractor. For its constituent solutions, the mismatch of the start of the spikes of the first and the second neurons is $\xi^* + o(1)$. The delay of the spike of the first neuron with respect to the spike of the second neuron has the same value: $\xi^* + o(1)$. Each of the neurons generates spikes through time $2\xi^* + o(1)$. The minimum values of the membrane potentials are $u_1^{\min} = u_1(\alpha_1 + 2 + o(1)) \approx u^{\min}$ and $u_2^{\min} = u_2(\alpha_1 + 2 + \xi^* + o(1)) \approx u^{\min}$. This attractor is naturally called "symmetric."

The conducted investigations are consistent with the biological data that the electrical interaction (diffusion links) helps synchronize the oscillations of two nerve cells. Local constancy of mapping $\xi' + f(\xi)$ suggests that synchronization (phase shift) occurs approximately every one cycle of oscillations. These results are obtained by the asymptotic method and are valid as $\lambda \to \infty$. For values of the parameter λ that are not to large, as considered in Sect. 2.2 by numerical example, the structure of the oscillations can be much more complex. The same effect is also observed for a fully connected network of more than two neurons, which will be shown in the next paragraph of this chapter.

Consider additionally the case when $\sigma_1 \equiv \alpha_1 + (\alpha_2 + \alpha_1\alpha)/2 < \sigma < \alpha_1 + \alpha_2 - \alpha$. In this situation, the main part $\xi' = f(\xi)$ of the mapping, in the asymptotic sense, $\prod$ has the following properties. Let $\xi_0 = 1 + (\sigma - \alpha_1)/\alpha$, $\xi_{00} = T_2 - \xi_0$. There exist such ξ_1 and ξ_2 ($\xi_1 < \xi_0 < \xi_{00} < \xi_2$) that $f(\xi) = \xi_0$ at $\xi \in [\xi_1, \xi_0]$, $f(\xi) = \xi$ at $\xi \in [\xi_0, \xi_{00}]$ and $f(\xi) = \xi_{00}$ at $\xi \in [\xi_{00}, \xi_2]$. Initially it seems that we can make a conclusion about the existence of two attractors. However, this is not so. The finite dimensional part of the mapping can be calculated more precisely:

$$\xi' \approx \xi - [\ln[\exp(-\lambda\alpha_2) + \exp(\lambda(\alpha\xi - \sigma - 1))]]/\lambda\alpha$$
$$+ [\ln[\exp(-\lambda\alpha_2) + \exp(\lambda(-\alpha\xi - \sigma + \alpha\alpha_1 + \alpha_1 + \alpha))]]/\lambda\alpha.$$

The right side of the given mapping, as $\lambda \to \infty$, converges pointwise to $f(\xi)$. There is one fixed point, $\xi^* = T_2/2$. It is stable, but the iterations converge slowly. Apparently, in this case there also exists a symmetric attractor. In any case, we can guarantee the presence at system (7.1.1) and (7.1.2) of the set of solutions for which, at asymptotically large intervals as $\lambda \to \infty$ time, the spikes of neurons alternate through time close to $T_2/2$.

   It should be noted that in all of the considered cases, at small positive values ξ the function $f(\xi) \equiv \xi$. To investigate the behavior of iterations of the mapping Π on these intervals, we should consider independent of λ any corrections to the function $f(\xi)$. Unfortunately, the corresponding formulas are very complex. Therefore, the domain of stability of the homogeneous attractor $(u_1(t) \approx u_2(t))$ is unknown.

7.1.2 Oscillations in the System of Three Neurons

Let us consider the behavior of a system of three identical neurons and give comments for the general case. Let the second neuron be connected with the first and third neurons in a diffusive manner. We say that the neurons are arranged linearly if the first and third neurons do not directly interact with each other. Let us say that the neurons lie on the circumference if the first and third neurons are connected with each other. In both cases, the membrane potentials $u_1(t)$ are described by the system of equations:

$$\dot{u}_i = \lambda[-1 - f_{Na}(u_i) + f_K(u_i(t-1))]u_i + d(u_{i+1} - 2u_i + u_{i-1}), \qquad (7.1.3)$$

where $i = 1, 2, 3$. For the system of N neurons, $i = 1, \ldots, N$. Consider that $\lambda \gg 1$ and $d = \exp(-\lambda\sigma)$, where $\sigma > 0$. In (7.1.3) for linearly arranged neurons $u_0 \equiv u_4 \equiv 0$ and for neurons lying on the circumference $u_0 \equiv u_3, u_4 \equiv u_1$. As stated previously, it turns out that as $\lambda \to \infty$, the asymptotic spike parameters are the same as for the free neuron.

   In the case of linearly arranged neurons at $0 < \sigma < \alpha_1 + (\alpha_2 + \alpha_1\alpha)/2 \equiv \sigma_1$, there exists the attractor for which the spike of the second neuron through time $\xi^* + o(1)$, after its start, generates spikes of the second and third neurons. The next spike of the second neuron starts at moment of time $\xi^* + \eta^* + o(1)$.

   Consider in detail the case $0 < \sigma < \alpha_1(1+\alpha) \equiv \sigma_0$. For given values σ, the attractor owes its existence to the fact that the minimum values of the membrane potentials of neurons satisfy the relationship: $u_2^{min} > u_1^{min} \approx u_3^{min}$. As a result, the spike of the second neuron starts earlier than it does in the absence of interaction. This influences the evolution of the membrane potentials of the first and third neurons, thus causing their induced (imposed) spikes. The attractor is the simplest example of the guiding center, in which the second neuron generates the wave. We call it the "principal."

   At values $\alpha_1(\alpha_2 - 1)/(\alpha_1 + \alpha_2 - 2) < \sigma < \alpha_1(1+\alpha) \equiv \sigma_0$, neither the first nor the third neuron act as principal. We explain the reason thusly. Assume that the first neuron is the principal. Let its spike start at time zero and the spikes of the second and third neurons start at close moments of time, respectively, to ξ^* and $2\xi^*$. Calculations show that the minimum values of membrane potentials satisfy the approximate relationships: $u_1^{min} > u_2^{min} \approx u_3^{min}$. As a result, during the next spike of the first neuron ($u_1(t) > 1$), the value $u_1(t)$ in the asymptotic sense is not part of the equation for the membrane potential $u_2(t)$ of the second neuron. The spike of the second neuron is not induced. A similar pattern is observed for the second and third neurons.

The relationship of the moments of the start of spikes is unstable. For example, let the initial time interval between the start of the spikes of the second and third neurons be larger than that between the start of the spikes of the first and second neurons. Then, over time, the start of the spike of the first neuron will be approximate to the start of spike of the second neuron. Then the spike of the first neuron will occur later than the spike of the second neuron. Steady oscillations of the system belong to the above-described attractor.

The situation is quite different if we assume that the value σ satisfies the inequality $0 < \sigma < \alpha_1(\alpha_2 - 1)/(\alpha_1 + \alpha_2 - 2)$, i.e., the interaction is strong enough. Then, along with the previous description, there exist two more attractors. In one of them, the first neuron is the principal. Its spike through time $\zeta^* + o(1)$ leads to the spike of the second neuron. Furthermore, the spike of the second neuron once more through time $\zeta^* + o(1)$ induces the spike of the third neuron. The process is repeated cyclically through time $\zeta^* + \eta^{**} + o(1)$ where it is essential that the value $\eta^{**} < \eta^*$. The given attractor owes its existence to the inequality $u_1^{\min} > u_2^{\min} > u_3^{\min}$, thus binding the minimum values of the membrane potentials. For the second attractor, the process of spike generation starts with the third neuron.

The construction is transferred to the case of arbitrary number N of linearly arranged neurons. Choose and fix a neuron with number k_0 and write $j_0 = \max(k_0, N - k_0)$. There exists such value $\sigma(j_0) > 0$ that at $0 < \sigma < \sigma(j_0)$ the system of equations (7.1.3) has an attractor, which is a guiding center with the k_0th principal neuron. The intervals between the starts of the spikes of the k_0th and ith neurons are close to the numbers $|i - k_0|\zeta^*$. Neighboring by virtue of time, the spikes of each of the neurons are separated by intervals $\zeta^* + \eta^{**}(j_0) + o(1)$, where $\eta^{**}(j_0) < \eta^*$. As stated previously, for the case of three neurons, the attractor exists due to the fact that minimum values $u_i^{\min}$ of membrane potentials of the ith neurons decrease with increasing the number $|i - k_0|$, i.e., with increasing distance from the principal neuron. Note that the numbers $\sigma(j)$ satisfy the inequality $\sigma(j) > \sigma(j+1)$.

At $\sigma_0 \equiv \alpha_1(1 + \alpha) < \sigma < \alpha_1 + (\alpha_2 + \alpha_1\alpha)/2 \equiv \sigma_1$ for the system of N, linearly arranged neurons, at time interval as asymptotically high as $\lambda \to \infty$, there exists an oscillatory mode of the following structure. The neurons are divided into two groups. The first group includes neurons with even numbers, and the second group includes neurons with odd numbers. Spikes of neurons of different groups begin almost simultaneously. The starts of the spikes of the neurons in different groups are mismatched by a value close to ζ^*. The question of the existence of the corresponding attractor remains unanswered. Difficulties of this analysis are caused by the fact that the external neurons are in an unequal position with respect to the internal neurons.

Let us return to the system of three neurons for the case in which they are located on the circumference. At $0 < \sigma < \sigma_1$, any neuron can be the principal neuron in the guiding center. The spike of the randomly selected principal neuron through time $\xi^* + o(1)$ causes spikes of the two neighboring neurons. The process of the generation of spikes is repeated cyclically after time interval $\xi^* + \eta^* + o(1)$. Along with this, at $\alpha_1 + \alpha(\alpha_1 - 1)/2 < \sigma < \alpha_1 + \alpha(\alpha_1 + \alpha\alpha_1 - \alpha)/(1 + 2\alpha)$, travelling waves sometimes occur. Depending on the direction of propagation, the spike of the jth neuron occurs earlier or is delayed with respect to the spike of the $(j + 1)$th neuron by time $\xi^* + o(1)$. The wave makes a full turn during the time $3\xi^* + o(1)$. At this interval, each neuron generates one spike. The main proof of the existence of these waves is that $\xi^* < \alpha_1 + 1 \equiv T_1$, where T_1 is the asymptotic duration of the spike. This relationship of times provides the direction of the excitation's propagation: The spike of the $(j + 1)$th neuron is initiated by the spike of the jth neuron if we talk about the wave moving in the direction of increasing numbers.

Consider the network of N neurons arranged in a ring. On the ring we will identify the neurons with numbers i and $i \pm N$. Let $0 < \sigma < \sigma(E(N/2))$, where $\sigma(*)$ is the value in the condition of the existence of the guiding center for the system N of linearly arranged neurons ($E(*)$ is the integer part of the number). Then each neuron can act as the principal. From this, the wave of spikes propagates in two ways. Assume that the principal is the k_0th neuron. Then the spikes of the $(k_0 - j)$th and $(k_0 + j)$th neurons $j \leq (E(N/2))$ occur almost simultaneously and are delayed with respect to the spike of the principal neuron by time close to $j\xi^* + o(1)$. The specifics of the propagation of counter-propagating waves around the ring is that they annihilate, i.e., destroy, each other. The annihilation point is at the opposite end of the diameter from the principal neuron (antipode). The determinant for the existence of the guiding center is that the minimum values of the membrane potentials of neurons decrease with increasing distance from the principal neuron.

Let $0 < \sigma < \alpha_1(1 + \alpha)$, and assume that there exist subintervals of change σ, on which the value ξ^* satisfies the condition:

$$(\alpha_1 + 1)/(N - 1) < \xi^* < (\alpha_1 + \alpha_2 + (\alpha_1 + 1)\alpha)/(\alpha_2 + \alpha(N - 1)),$$

if $0 < \sigma < \alpha_1$ and

$$(\alpha_1 + 1)/(N - 1) < \xi^* < (\alpha_1 + 1)(1 + \alpha)/(1 + \alpha(N - 1)),$$

if $\alpha_1 < \sigma < \alpha_1(1 + \alpha)$. Then, for σ values of these subintervals, there exist attractors of travelling waves. As for the case of three neurons, depending on the direction of propagation, the spike of the jth neuron occurs earlier than, or is delayed with respect to, the spike of the $(j + 1)$th neuron by a time asymptotically close to ξ^*. Full rotation occurs during $N\xi^* + o(1)$. On the given interval, each neuron generates a single spike. Such waves can be called "single-peaked" waves.

If the number of neurons in the ring is large enough $(N \geq 6)$, then one can observe two-peaked waves. We will discuss the wave propagating in the direction of increasing of numbers of neurons. For this, the spikes of the jth and $(j + m)$th $(3 \leq m \leq N - 2)$ neurons occur almost simultaneously. The start of the spike of the $(j + 1)$th neuron is delayed with respect to the start of the spike of the jth neuron by the value $\xi^* + o(1)$. The domains of the parameters, where there exist two-peaked waves, are highlighted in the following way. Let $N_1 = \min(m - 1, N - m)$ and $N_2 = \max(m - 1, N - m)$ and $0 < \sigma < \alpha_1(1 + \alpha)$. The value ξ^* must satisfy the condition:

$$(\alpha_1 + 1)/N_1 < \xi^* < (\alpha_1 + \alpha_2 + (\alpha_1 + 1)\alpha)/(\alpha_2 + \alpha N_2)$$

if $0 < \sigma < \alpha_1$ and

$$(\alpha_1 + 1)/N_1 < \xi^* < (\alpha_1 + 1)(1 + \alpha)/(1 + \alpha N_2)$$

if $\alpha_1 < \sigma < \alpha_1(1 + \alpha)$.

With large number of neurons, multi-peaked waves can occur.

In the case when $\alpha_1(1 + \alpha) < \sigma < \alpha_1 + (\alpha_2 + \alpha_1\alpha)/2$ and the number of neurons is even, the system of equations (7.1.1) has an attractor consisting of solutions of the following structure. Neurons are divided into two groups. The first group includes neurons with even numbers, and the second group includes neurons with odd numbers. Spikes of neurons within groups occur almost simultaneously. The spikes of neurons of different groups are mismatched by a value close to $\xi^* + o(1)$.

The described attractors hardly exhaust all the set of attractors in the ring system of $N > 3$ neurons. For example, several guiding centers, as well as their competition, can be observed. It is unclear how the competition is completed. The guiding center can be moved.

7.1.3   Some Structures of Oscillations in the Neural Network on a Flat Surface

Consider the lattice, i.e., a finite set of points Ω_N, on a plane in the domain Ω. Consider that its elementary cells are equilateral triangles. Any internal node is adjacent to six nodes. We introduce two-index numbering of nodes and through Z_{ij} denote the set of nodes adjacent to the (i,j)th. We will place neurons in the nodes. We assume that the neurons are connected diffusionally. Let each neuron interact only with adjacent neurons. For membrane potentials u_{ij}, we obtain the system of equations:

$$\dot{u}_{ij} = \lambda\left[-1 - f_{\mathrm{Na}}(u_{ij}) + f_{\mathrm{K}}(u_{ij}(t - 1))\right]u_{ij} + d \sum_{(m,k)\in Z_{ij}} (u_{mk} - u_{ij}), \qquad (7.1.4)$$

where $\lambda \gg 1$, $d = \exp(-\lambda\sigma)$. Let $\alpha_1 + (\alpha_1 - 1)\alpha/2 < \sigma < \alpha_1 + \alpha(\alpha_1 + \alpha\alpha_1 - \alpha)/(1 + 2\alpha)$. Waves of spikes can be propagated, i.e., can conduct excitation, through the elementary cells. The interaction of waves can lead to a regular structure of the oscillations.

Highlight the (i,j)th neuron. Assume that its spike started at time zero and that the spike of the neurons adjacent to it (i.e., with indices $(k,m) \in Z_{ij}$) started at moment of time $t_{k,m} > 0$. Let the set Z_{ij} be split into two parts, Z'_{ij} and Z''_{ij}, on the basis of: $t_{km} < \alpha_1 + 1$ for $(k,m) \in Z'_{ij}$ and $t_{km} > \alpha_1 + 1$ for $(k,m) \in Z''_{ij}$.

Write

$$\tau_1 = \max_{(k,m) \in Z'_{ij}} t_{km}, \quad \tau_2 = \min_{(k,m) \in Z''_{ij}} t_{km}$$

and consider that

$$\tau_2 < (\alpha_1 + 1 - \tau_1 + \alpha\alpha_1)/\alpha.$$

Then the next spike of the (i,j)th neuron starts at moment of time

$$t'_{ij} = \min_{(k,m) \in Z''_{ij}} t_{km} + \xi^* + o(1). \tag{7.1.5}$$

Consider one of the simplest variants of organization of oscillations on the plane related to the conduction of excitation through the closed circuit. Highlight three neurons that are adjacent to each other and that are internal in the lattice. Assign numbers to the neurons, and include each neuron in its own set G_1, G_2, G_3. These sets are shown in Fig. 7.1. They do not have common nodes and represent the intersection of spirals with multiple nodes Ω_N (i.e., the part of turns is broken). Introduce a new numbering of nodes (i,j) where the first index $i = 1,2,3$ indicates the neurons belonging to set G_i and the second index to the total number of neurons in the set. We will count the number of neurons in each of the sets G_i from the center point of the corresponding spiral, i.e., the originally selected neurons come first. Identify the sets G_i, G_{i+3}, G_{i-3}. For arbitrary node $i.j \in G_i$, we introduce the set $Z_{i,j}^{i-1} = G_{i-1} \cap Z_{i,j}$, where $i = 1,2,3$, i.e. is the totality of adjacent nodes belonging to the previous $i - 1$ set.

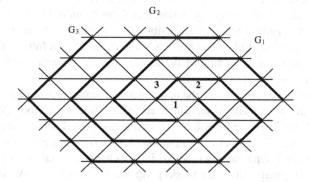

Fig. 7.1 Propagation of spikes along the elementary cells of the network consisting of nine neurons. Spikes of neurons marked with points follow through time close to ξ^*. The process is cyclically repeated

Choose and fix a sufficiently small, independent number $\varepsilon > 0$ of the parameter λ. Let the spike of the first neuron from the set G_1 start at moment of time $t_{1,1}$ and the spikes of the rest neurons start at moments of time $t_{1,j}$, where $|t_{1,j} - t_{1,1}| < \varepsilon$. Assume that the spikes of neurons from set $G_i (i = 2, 3)$ started at moments of time $t_{i,j}$, where $|t_{i,j} - i\xi^*| < \varepsilon$. Denote by $t'_{i,j}$ the moments of start of the following (nearest in time) spikes of neurons. From formula (7.1.5), it follows that

$$t'_{1,1} = t_{3,1} + \xi^* + o(1), \tag{7.1.6}$$

$$t'_{1,j} = \min_{(k,m) \in Z^3_{1,j}} t_{k,m} + \xi^* + o(1), \tag{7.1.7}$$

$$t'_{i,1} = t'_{i-1,1} + \xi^* + o(1), \tag{7.1.8}$$

$$t'_{i,j} = \min_{(k,m) \in Z^{i-1}_{i,j}} t'_{k,m} + \xi^* + o(1) \quad (i = 1, 2). \tag{7.1.9}$$

Count the time from the moment $t'_{1,1}$ of the start of the neuron's new spike with coordinates (1,1). Discard the terms $o(1)$. An iterative process that converges will occur. While checking the convergence, we employ the fact that for $j = 1, \ldots, 4$ is valid $Z^{i-1}_{i,j} = (i - 1, 1)$. As a result, neurons $(i, 1)(i = 1, 2, 3)$ impose their oscillations on the whole network.

The convergence of the iterative process ensures the existence as $\lambda \to \infty$ for the system of equations (7.1.4) of the attractor, which consists of solutions of the following structure. The spikes of neurons belonging to each of sets G_i start almost simultaneously. The start of the spikes of neurons from set G_i is delayed with respect to the start of the spikes of neurons from the set G_{i-1} by the value $\xi^* + o(1)$. Thus, in approximately equal time intervals, the spikes of neurons from sets $G_1, G_2, G_3, G_1, \ldots$ start. As a geometric object, each of the sets G_i is a spiral. The spatial-temporal structure of oscillations we can naturally call "reverberator."

In studying the structure of attractors for the case of the plane, it is important to plan the paths of the excitation (i.e., the waves of spikes), which is not easy to do. The problem can be facilitated by noting that from formula (7.1.5), we have the following. If the neuron receives an action (i.e., a spike) over time $\alpha_1 + 1$ after the start of its previous spike, then the new spike of the neuron follows through the fixed time ξ^* after the beginning of the action. The earlier action is less efficient (latent period is $\eta^* > \xi^*$).

In this regard, we can compare the neural network under consideration as well as the network of cellular automata, which are called Wiener neurons (Wiener and Rosenbluth [23]). These neurons function in discrete time. Each of them can be in one of three states: excitation (activity), refractoriness (resistance to action), or sleep (quiescent). If the Wiener neuron is active, it then changes its state to refractory, and after a cycle it changes its state to sleep. In the latter state, the Weiner neuron "pauses" for as long as there will be an excited neuron among its neighboring neurons. Note that Wiener developed the formal neuron based on experimental data of the functioning of heart tissue (i.e., cells that are diffusionally related).

There are two types of modes of functioning of the network of Wiener. In the first mode (trivial) over time no single excited neuron remains among all of the neurons. In the second mode, the set of Ω_N neurons is split into three disjoint nonempty subsets: G_1, G_2, G_3. Each neuron from the set G_i $(i = 1, 2, 3)$ has among its neighbours the neurons from the sets G_i, G_{i+1}. Among neurons at least one triple of neurons, which are adjacent to one another and belong to different subsets G_i, is necessarily present. For modes of the latter type, in consecutive discrete moments of time neurons belonging to the sets $G_1, G_2, G_3, G_1, \ldots$ become active. Note that these triads of neurons form closed paths of the conduction of excitation. Modes related to the conduction of excitation along the closed circuits are very diverse. For example, we can observe the system of reverberators, leading centers (without a principal neuron), different "textures", etc.

Using the network of Wiener, assume there is a split of the set Ω_N into subsets G_1, G_2, G_3. Let us return to our network. Let the spikes of neurons from the sets G_1, G_2, G_3 start, respectively, at moments of time that are close to zero, i.e., ξ^* and $2\xi^*$. For the moments of the start of the neuron spikes that will occur next, we can construct analogues of formulas (7.1.5). Let the iterative process, which describes the main mismatch of the start moments of the spikes, converge and the limit point, which is the spikes of neurons from G_i, be delayed with respect to the spikes from G_{i-1} by the value ξ^*. Then we can conclude that there exists the attractor, for which the spikes of neurons in each of the subsets G_i occur almost simultaneously and the spikes of neurons from the sets G_i and G_{i+1} mismatch by a value close to ξ^*.

Checking for convergence of the iterative process is specific in each case and is defined by mutual arrangement of the sets G_1, G_2, G_3. Complete analytical research is hardly possible. Without checking for convergence, we can assert that there exist the solutions of system (7.1.4) that behave as described on finite time intervals.

Pattern of oscillations related to the conduction of excitation along the closed circuits can be quite interesting. Consider an arbitrarily fixed curve γ on the lattice. Each point of this curve can be included into the elementary path of conduction of excitation.

Let the paths of the neighboring points of the curves have common nodes. Then (in any case, on a finite time interval) there exists an oscillatory mode, in which the spikes of neurons, located on γ, start practically synchronously. The spikes of neurons on γ that neighbor in time follow through the interval $3\xi^* + o(1)$. This is illustrated in Fig. 7.2. In this case the neurons located on γ are marked with points. Their synchronous spikes through time $\xi^* + o(1)$ lead to spikes of neurons depicted by circumferences. The latter, after another time $\xi^* + o(1)$, induce the spikes of neurons, which are marked with asterisks. Following this event, after time $\xi^* + o(1)$,the start spikes of neurons lying on γ start again.

Almost every curve can be repeated before the start of spikes. One example of the network size of 50 × 50 of neurons is shown in Fig. 7.3. Here k is discrete time with step ξ^* Black color is used to depict neurons whose starts of spikes are close to

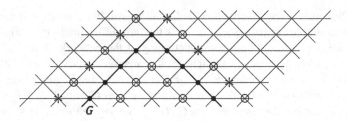

Fig. 7.2 Propagation of spikes along the elementary cells of the network. Spikes of neurons, marked with *points* (the curve G), circumferences, and *asterisks*, follow through time close to ξ^*. The process is cyclically repeated

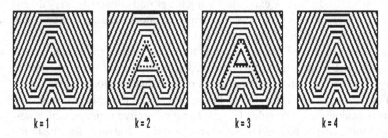

Fig. 7.3 Letter "A" on a network of 50×50 neurons is cyclically repeated in front of the wave of the spikes' start

$k\xi^*$. Cyclically through time $3\xi^* + o(1)$, the spikes of neurons located on a curve start, which is taken as the letter "A". They generate closed waves of the start of spikes.

Biological interpretation of the phenomena is connected with the hypothesis of wave coding of information in the brain (Lebedev 1990). The local network is able to store complex wave packets before their further processing. In this case, we do not require time modification of synaptic weights. Corresponding oscillatory modes organize themselves according to the initial conditions, which can be set along with the external action. Thus, the local network is a model of short-term memory based on the principles of the autowave. In the case when the described oscillatory modes exist only on the finite time interval, their destruction can be interpreted as the process of forgetting images.

7.2 Structure of the Oscillations in the Fully Connected Network of Diffusionally Interacting Neurons

Consider fully a connected network of N neurons where each element is diffusionally related to any other neurons. Communications are considered homogeneous. The neuronal population is described by the system of equations (7.1.4), in which the set of neurons adjacent to the (i,j)th neuron elements coincides with the whole set of elements, i.e., $Z_{i,j} \equiv \Omega_N$. As stated previously, let the parameter $\lambda \gg 1$, and the coefficient of difference diffusion be $d = \exp(-\lambda\sigma)$ where $\sigma > 0$.

We represent the set Ω_N of neurons as the union of two arbitrary disjoint nonempty subsets Ω_N^1 and Ω_N^2. Assume that $0 < \sigma < \alpha_1 + (\alpha_2 + \alpha_1\alpha)/2 \equiv \sigma_1$. Then system (7.1.4) has an attractor of the following structure. The spikes of neurons that belonging to each neuron of the set Ω_N^1 and Ω_N^2 start almost synchronously. The starts of the spikes of neurons from the set Ω_N^2 are delayed with respect to the starts of the spikes of neurons from the set Ω_N^1 by the value $\xi^* + o(1)$. The time interval between the spikes of neurons from the sets Ω_N^1 and Ω_N^2 is $\eta^* + o(1)$. The minimum values of the membrane potentials of neurons from the sets Ω_N^1 and Ω_N^2 are asymptotically the same as for the case of two neurons.

The existence of the attractor is can be proved simply enough by using the results presented in the first paragraph of this chapter. We add the following remarks. Let the spike of the neuron (from the set Ω_N^1) started at time zero. Assume that the group of spikes (generated by neurons from Ω_N^2) act on the neuron acts starting at or after—but not too long after—moment of time $\alpha_1 + 1$. Then the new spike of the selected neuron occurs over time $\xi^* + o(1)$ after the start of the first spike by the time pulse action of the group of neurons. Thus, due to the influence of many neurons from the set Ω_N^2, asymptotic behavior (in the main) for membrane potentials of neurons from Ω_N^1 is common. Completely analogously, neurons from Ω_N^1 establish uniform asymptotics for membrane potentials of the neurons from the set Ω_N^2 (while checking, we separately consider, as selected in the previous paragraph, the change intervals of the value σ).

We give one of the possible interpretations of the fact of the existence of oscillations of the described structure. Assume that the neurons are located on the plane. Then we can consider the set $\Omega_N^1 \subset \Omega_N$ as the "picture," and the set $\Omega_N^2 \subset \Omega_N$ as its negative image. A fully connected network at the cycle consistently presents the observer the image of a "picture" and its negative. Naturally, this example is greatly simplified because visual images are unlikely to be stored in the brain in the form of photographs.

Consider the case $\alpha_1 + (\alpha_1 - 1)\alpha/2 < \sigma < \alpha_1 + \alpha(\alpha_1 + \alpha\alpha_1 - \alpha)/(1 + 2\alpha)$. We represent the set Ω_N as the union of some three disjoint nonempty subsets $\Omega_N^1 \subset \Omega_N^2$, and The system of equations (7.1.4) has the attractor consisting of solutions for which spikes of neurons belonging to each of the sets $\Omega_N^1 \subset \Omega_N^2$, and Ω_N^3 occur almost synchronously. Spikes of neurons from the sets $\Omega_N^1, \Omega_N^2, \Omega_N^3, \Omega_N^1, \ldots$ start consistently through time $\xi^* + o(1)$.

Thus, the fully connected network actually organizes itself (without the adaptation of synaptic weights) into a ring structure of neural modules. The number of elements in the modules depends on the initial conditions (they can be set by external action). Visual interpretation of the phenomenon is analogous to that described previously.

Until now it was thought that the parameter $\lambda > 0$ is arbitrarily large. At values of λ that are not too large, there arises a number of effects that cannot be detected with the help of asymptotic methods. Let us turn to the numerical study of the fully connected network of N neurons. We assume that the conditions of Theorem 1.6.1 are fulfilled for the equation of each neuron. Specify the form of the nonlinear functions appearing in the equations assuming $f_{Na}(u) = R_1 \exp(-u^2)$ and $f_K(u) = R_2 \exp(-u^2)$. Numbering neurons and denoting their membrane potentials by u_i, we obtain the system of equations:

$$\dot{u}_i = \lambda\left[-1 - R_1 \exp\left(-u_i^2\right) + R_2 \exp\left(-u_i^2(t-1)\right)\right]u_i + d\sum_{j=1}^{N}\left(u_j - u_i\right), \quad (7.2.1)$$

where $i = 1,\dots N$. It is considered that $\alpha = R_2 - R_1 - 1 > 0$.

A system (7.2.1) for the network of ten neurons ($N = 10$) was studied numerically [95, 96] at the values of the parameters $\lambda = 3, R_1 = 1, R_2 = 2.2$. The corresponding value of the parameter $\alpha = 0.2$. Magnitude d was varied in a wide range of values. It was found that for large values d, the homogeneous periodic mode ($u_1 = u_2 = \cdots u_{10}$) of system (7.2.1) is stable. Its graph is shown above in Fig. 1.2 (it is also known as the graph of the solution of equation of the isolated neuron). The solutions are characterized by the presence of spikes. Note that the numerical calculation yields the value of the period of approximately 12, and from the Theorem 1.6.1 this value is approximately 13.2. Thus, even with this relatively small value of the parameter λ, the asymptotic formula for the period is in good agreement with the results of the experiment.

With a decrease of the value of the coefficient of difference diffusion d, the homogeneous solution either becomes unstable or the domain of its attraction strongly decreases. In this case, oscillatory modes of a new quality appear, the occurrence of which occurs nonlocally at relatively large values of d. For such solutions, the following points are true: (1) a much smaller biological period is typical, i.e., the middle interval between spikes of each of the neurons starting closely in time; and (2) the amplitude of the peaks is much larger. These are modes of self-organization.

Examples of the self-organization of modes are shown in Figs. 7.4 and 7.5. The modes respond to the value $d = 0.01$ and different initial conditions. Graphs of the initial conditions of the neurons in close proximity to each other are given in Fig. 7.4. In turn, Fig. 7.5 corresponds to the case when the neurons were divided into two groups of five neurons each. Within the groups the initial conditions differed little, and between groups they differed strongly. The above-listed modes of self-organization differ substantially. Graphs of neuronal activity, shown in Fig. 7.4, are located more or less uniformly; however, the oscillations of two neurons have almost merged, but another pair differs little. Quite a different situation is illustrated in Fig. 7.5, i.e., the graphs are not uniformly distributed. This is the consequence of the union of neurons into synchronously functioning groups. Such are the six groups shown: They are stable and are represented by one, two, or three neurons.

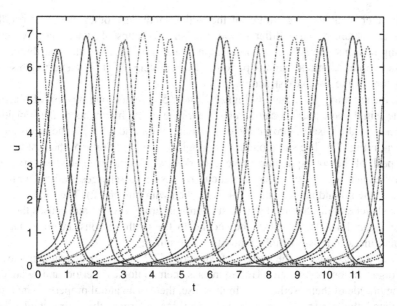

Fig. 7.4 Mode of self-organization as $\lambda = 3$, $R_1 = 1$, $R_2 = 2.2$, $d = 0.01$ and approximately the same initial conditions for $u_i(t), t \in [-1, 0] (i = 1, \ldots, 10)$

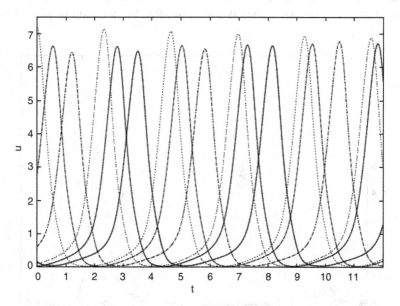

Fig. 7.5 Mode of self-organization as $\lambda = 3$, $R_1 = 1, R_2 = 2.2, d = 0.01$ and the initial conditions $u_1(t) \approx u_2(t) \approx \cdots \approx u_5(t) \neq u_6(t) \approx u_7(t) \approx \cdots \approx u_{10}(t), t \in [-1, 0]$

With other methods of grouping of the initial conditions, other modes of oscillations are obtained. They differ in the number of groups of synchronously functioning neurons, the size and composition of these groups, and the distribution of peaks in time. All modes are not periodic. Thus, system (7.2.1) has a very rich set of attractors of the complex structure.

In the situations illustrated in Figs. 7.4 and 7.5, a rather long process of blurring of the initial wave packets can be observed. One of the fragments of the transition process, in the case of proximity of the initial conditions of all the neurons, is shown in Fig. 7.6. The situation has a biological meaning. Spike of the first neuron (the peak of the corresponding curve is marked with numeral 1) acts as a stimulus for other neurons. There is an avalanche process of discharges starting with the spike of the second neuron (this peak is marked with numeral 2). The total effect of the avalanche of discharges on the first neuron is strong enough and causes the next spike (peak 1.1). At this time, the second to the tenth neurons are in a state of refractoriness (relative or full immunity to stimulus).

With the decrease in the system of equations (7.2.1), the coefficient d modes of self-organization degenerate: This increases their biological period and decreases the amplitude of their oscillations. In this case, there is a natural property: The more populated the group of synchronously working neurons, the more stable their biological period and lager their peak amplitudes.

Modes of self-organization of the type shown in Fig. 7.4 can be treated as neural noise. If we influence the neural network in noise mode by a sequence of pulses from the outside, then there occurs synchronization of the groups of neurons. Those

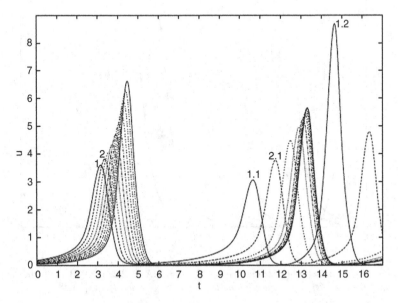

Fig. 7.6 Fragment of the transitional process to the mode of self-organization shown in Fig. 7.4 ($\lambda = 3$, $R_1 = 1$, $R_2 = 2.2$, $d = 0.01$)

neurons, for which at the moment of incoming outside pulses such an influence is sufficient to generate another pulse, more or less respond simultaneously. The package of the groups of spikes exist in the system for some time, called "evolutionarizing," and then "blurs." It seems to us that such a process can be treated as short-term storage and the subsequent forgetting of wave codes. In this case, the principle of non-addressed storage is realized because the structure of the wave packet, i.e., the spikes of which form it, rather than numbers of neurons, is most important.

The process of short-term memory was modeled by the system of equations (7.2.1) by adding all of the equations of the same term $\varphi(t)$, where $\varphi(t)$ is a sequence of rectangular pulses, to the right-hand sides. One of the results of modeling is shown in Fig. 7.7. The initial conditions for system (7.2.1) are chosen so that without external action the system is in the noise mode as shown in Fig. 7.4. The action in this case is the sequence of three pulses with s duration of 0.125 units and amplitude of 10 units each. Moments of time of the action in Fig. 7.7 are marked with the symbols $v1, v2, v3$. Under the action of the stimulus, three patterns of activity emerge; these are marked in the figure by numerals 1, 2 and 3. One neuron (marked C) "eluded" all three external pulses. Later the groupings are saved for some time, and the neuron labeled C induces the spike later (labeled $C.1$).

A long time has passed since the spike for the second neuron occurred. It is most "ready" for generation of the next pulse, which is marked on the curve as peak 2.1. The spike of the second neuron is not sufficient for stimulation of the first,

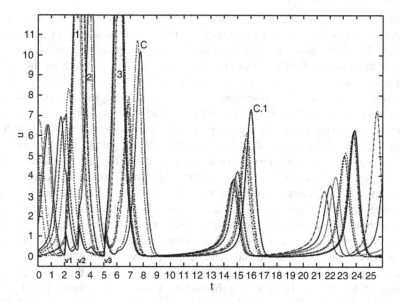

Fig. 7.7 Result of action on the noise mode of system (7.2.1) as $\lambda = 3, R_1 = 1, R_2 = 2.2, d = 0.01$ by a sequence of three pulses. Times of arrival of the pulses are marked with symbols $v1, v2, v3$

still-inhibited neuron, but it causes the avalanche process of the other neurons' pulses. This avalanche of spikes is a further reason for the generation of pulses by the first and then the second neurons.

Let us consider in more detail the results of the modeling of the action. Each external pulse has a depolarizing action on all other neurons. After the end of the action, the neurons are divided into two classes. Some graphs rise (are generated spikes), whereas others tend to a state of hyperpolarization. In the former case, enough time has passed since the last spike occurred, and the neurons are disinhibited. In the second case, the neurons have recently generated spikes and are in a state of refractoriness (resistance), so at the time of exposure they cannot start a new spike. Thus, in the life of the neural population, there are special moments when its elements "express their attitude" toward external influence: "yes" = generate spike, and "no" = the membrane of the neuron tends to a state of hyperpolarization. A third variant is also possible: External action has almost no effect on the neurons that generated spikes at the time of their arrival. The result of "voting" by each neuron is determined by the prehistory of its potential on the membrane.

Speaking about "special moments," it is appropriate to assume that the system has its own logic and is collective in nature. According to J. Hopfield, collectivism is the main feature of the neural structure.

In Fig. 7.7, the turning points of the graphs correspond to moments of time $v1, v2, v3$. At the point $v1$ formation of the first pattern of activity occurred, and at the points $v2$ and $v3$ formation of the second and third patterns occurred. The "free" neuron (with index C) really is not truly independent. Its spike is induced by the second pattern of activity.

We write these explanations, commenting on Fig. 7.7, but they do not differ from the arguments of the biologist examining the process of the arising of patterns of activity. This fact shows an obvious advantage over the proposed model.

Note that as a result of a sufficiently strong external action, i.e., after the signal with a long duration of effect on the amplitude, spikes are not observed in the system for a long time. This phenomenon is known in biological experiments as a "state of refractoriness" of the given system as a whole. The corresponding oscillatory mode becomes close to a uniform mode, the biological period of which is larger than the noise mode of self-organization. In the future, there occurs the above-described transitional process of destruction of synchronous oscillations. The system goes into noise mode and is again able to respond to outside action.

In numerical experiments, the richer the visual noise mode, the brighter and more adequate the response of the neural system. If we act on the population by some external action, which is located at the noise mode shown in Fig. 7.5 (i.e., the spikes of some of neurons are "stuck"), then there arise less obvious groupings of spikes. Given this, the phenomenological consequence suggested the idea of the experiments described in Chap. 8. The complexity of the rhythms of electrical activity of the human brain was estimated by number of neurons, i.e., by the correlation dimension. Then, using psychological tests, the number of numerals that an individual can remember nonrecurrently (i.e., the volume of short-term memory available for remembering a sequence of numbers) was determined. It was statistically significantly found that the higher the correlation dimension, the greater the amount of short term memory.

Let us discuss another phenomenological consequence. The groupings of spikes in Fig. 7.7 differ. However, in between bursts, there are no visual differences, i.e., the oscillations alternate specific (carrying attributes) and nonspecific zones. This gives rise to the hypothesis that this feature of the reaction of the neural system to an action can also be observed in the human brain. In Chap. 9, we describes the results of the analysis of evoked potentials of the brain in response to visual stimuli. It is shown that actually there are present time zones where evoked potentials for any two different stimuli statistically differ. These zones are separated by time intervals in which the differences are not detected.

We emphasize that these consequences are phenomenological in character. The model only helps us hypothesize about possible events in the organization and role of electric activity in the brain. The process of checking them is implemented experimentally.

References

Chua, L., & Yang, L. (1988a). Cellular neural networks: Theory. *IEEE Transactions on Circuits and Systems, 35*(10), 1257–1272.

Chua, L. O., & Yang, L. (1988b). Cellular neural networks: Applications. *IEEE Transactions on Circuits and Systems, 35*, 1273.

Eckert, R., Randall D., & Augustine, G. (1991). *Animal physiology* (Vol. 1). Moscow: Mir.

Hopfield, J. J. (1982). Neural networks and physical systems with emergent collective computational abilities. *Proceedings of the National Academy of Sciences, 79*(8), 2554–2558.

Hopfield, J. J. (1984). Neurons with gradual response have collective computational properties like those of two-state Neurons. *Proceedings of the National Academy of Sciences, 81*, 3088–3092.

Kashchenko, S. A., & Mayorov, V. V. (1995). Wave structures in cellular network from formal neurons of Hutchinson. *Radio Engineering and Electronics, 40*(6), 925–936.

Kashchenko, S. A., Mayorov, V. V., & Myachin, M. L. (1995). Oscillations in systems of equations with delay and differential diffusion, modelling local neural networks. *Reports of the Russian Academy of Sciences, 344*(3), 137–140.

Lebedev, A. N. (1990). Human memory, its mechanisms and limits. *Research of memory* (pp. 104–118). Moscow: Nauka.

Ochs, S. (1969). *Fundamentals of neurology.* Moscow: Mir.

Pappas, G., & Waxman, S. (1973). Ultrastructure of synapses. *Physiology and pharmacology of synaptic transmission* (pp. 7–30). Leningrad: Nauka.

Schade, J., & Ford, D. (1976). *Fundamentals of neurology.* Moscow: Mir.

Tasaki, I. (1971). *Nervous excitement.* Moscow: Mir.

Chapter 8
Pseudo-correlative Dimension of the Electroencephalogram and Its Relation to the Amount of Short-term Memory of the Human

Total electrical activity in the brain (large sections of it) is recorded as an electroencephalogram (EEG) based on activity recorded at the skin surface (i.e., scalp) of the head. The dynamics of membrane potentials of individual neurons and their associations is studied by the technique of microelectrode sensing. Corresponding experiments are either carried out on animals or accompany cranio-cerebral operations in the clinic. Temporal parameters of EEG oscillations and impulses of neural pools are completely different. Normally, in the EEG-dominant at-rest electrical activity for most people, there is the α-rhythm with an oscillation period of approximately 0.1 s. Recall that the length of the spikes of neurons is of the order of milliseconds. Therefore, the EEG reflects the waves of slow activity, and records that are registered with the microelectrode leads are, as a rule, pulse potentials.

There arises a question of the relationship of fast and slow potentials. Points of view on this problem at representatives of different schools are distinctly different and range from complete denial of relationship to its absolutization. At the time, Adrian (1934) correlated the EEG tracings with the impulses of neurons and explained total slow oscillations as the synchronization of impulse discharges of individual cells (see also Adrian and Moruzzi 1939). He put forward the hypothesis that the EEG is the result of simple summation of neural impulses. Recent research has found the connection of slow (spindly) cortical activity with pulse potentials (Arduini and Whitlock 1953; Brookchart and Zanchetti 1956; Golovchinsky 1965). The observations have been made in experiments on anesthetized animals. However, analogous experiments have shown that the flow of impulses with the deepening of anesthesia decreases sharply (Georgiev 1966; Atsev et al. 1968), and thus only the rhythms of slow electrical activity of the brain remain. Thus, the validity of the hypothesis of a simple summation is questioned. Moreover, Frost and Kellaway (1965) failed to find a clear relationship between slow and impulse potentials in their research. It is possible that the authors of the latter work just did not want to see this relationship because their results allowed the appropriate interpretation.

Observations of the agents of nerve cells and the animals identified a number of features of electrical activity of neurons. First, the levels of membrane potentials oscillate periodically (the frequency is understood in the biological sense). In this

© Springer International Publishing Switzerland 2015
S. Kashchenko, *Models of Wave Memory*,
Lecture Notes in Morphogenesis, DOI 10.1007/978-3-319-19866-8_8

case, there is always a slow component. Second, the cells generate impulses more likely close to the maximum of the wave of depolarization (see Elul and Adey 1966; Adey 1967; Pribram 1975). We should note that the given probability seems to be highly dependent on the experimental conditions. Third, we revealed an important fact. The flow of impulses in the cortex causes oscillations of surface macro potentials (Zhadin 1984; Gavrilov 1987). One could go on listing the arguments in favor of the existence of a relationship between the impulse and slow macro potentials (Livanov 1972; Shulgina 1978). At the same time, we can make counterarguments (see, for instance, Bernstein 1990, pp. 62–63). The waves of EEG really cannot be explained directly by the summation of pulsed discharges. There is too much of a difference between the duration of a single pulse and that of the wave. It should be recognized that there is currently no complete model explaining the relationship of slow and pulse brain potentials. The most plausible is represented by the following mechanism. Excitation, and especially inhibition, having come through synapses on neurons (Andersen and Eccles 1962), gives rise to local postsynaptic potentials. Which spread electrotonically (called "diffluence") on the surface of the soma, and branching of the axon and the dendritic tree. "Diffluence" goes along with attenuation, but the result is a relatively stable integral postsynaptic potential. The dendritic tree plays a special role where the postsynaptic potentials can exist during the time of the order of hundreds of milliseconds. Thus, there works some mechanism of averaging, the details of which are unclear. In this mechanism, for example, the conductivity of brain tissue, i.e., ordinary electric interaction, may play a role. The results of this chapter were obtained together with Myshkin and are described in the papers (Mayorov and Myshkin 1993a, b).

8.1   Pseudo-correlative Dimension and Its Calculation

The dynamics of electrical activity recorded from the surface of the scalp reflects the level of the functional state of the brain. This fact is not objectionable even though there is no (as already mentioned) common interpretation of the elec-troencephalogram (EEG) itself. The complexity and diversity of the EEG pattern and its frequency (in a certain sense) have led to the development of mathematical methods, statistical in nature, of analyzing brain biopotentials. There are several dozens of methods of mathematical processing of the EEG, but, as noted by Trush and Kornievsky (1978), there is no uniform classification and benchmarking. The choice of the method is determined, in our opinion, by two main factors:

- The investigators' views on the mechanisms of the origin of brain activity and the reasons for its rhythm.
- The specific goals and objectives facing the investigators.

In clinical practice, there are widespread spectral methods of analysis allowing the selection of typical frequency ranges, the identification of indexes, and the calculation of the energy of spectral components. These methods contribute much to our understanding of the processes of bioelectrical activity. However, their use has developed a certain relation to the EEG, which came to be regarded as "random process with a noise component." Recall that one of the problems of spectral methods is the analysis of such fine relations as the coherence of the periodic components of bioelectric activity of different brain areas. By virtue of this noise, this process is usually regarded as an undesirable artifact that masks the signal under investigation.

At the same time, it is clear that the stochastic nature of the EEG lies in the complex structure of interaction between different parts of the nervous system. There is a contradiction between the goals and methods of research: The method interprets too-complex organization of oscillations as noise due to the interaction, but determining this relationship is the purpose of the study. Functioning as a single entity, the brain shows its unique features, in particular, the phenomenon of information storage and extraction. At the same time, periodic electrical activity, according to the ideas of the phase-frequency method of encoding information in the brain, plays an important role (see Lebedev 1990). One hypothesis is that the system with certain irregular oscillatory modes has the ability to memorize and store information (see Sect. 7.2). Its ability will be determined by the range of allowable restructuring activity: The system functioning in a stable periodic mode as well as complete disorganization of periodic activity leads to a dramatic reduction of its information capacity. For proper functioning, the system must have variability of the level of activity in both the direction of synchronization and that of desynchronization, which we regard as a state of dynamic chaos, and connect with a variety of periodic modes.

A similar point of view was expressed by Adey (1974). Considering the brain as a noisy computer, the investigator believes that it should hold a sufficient level of independent activity of neural elements, thus providing a certain number of degrees of freedom for the system (Adey 1974; Soong and Stuart 1989). Failure to maintain the required level of independence, which occurs, for example, with epileptiform activity in a comatose state, constitutes a violation of regulatory mechanisms (see Jacome and Morilla-Pastor 1990). From this perspective, the presence of dynamical chaos is a necessary condition for the normal functioning of the brain.

The idea of using EEG to assess the measure of chaos and information capabilities of the brain has been made by several investigators (Elul 1967; Adey 1974), but adequate implementation was not proposed. Later appeared papers (Babloyantz et al. 1985; Babloyantz and Destexhe 1986; Başar et al. 1988, 1989; Başar 1990; Xu and Xu 1988; Soong and Stuart 1989; Belsky et al. 1993) reporting on EEG being investigated by methods of nonlinear dynamics. Let us briefly describe the method of research (Mayorov and Myshkin 1993a, b).

In the experiments, for example, in the process of work with the monopolarly allocated EEG, we deal with a one-dimensional signal $\varphi(t)$, which hopefully reflects the real physical process described by the flow in some phase space. If the flow were known, it would be natural to consider the attractor generating the observed signal $\varphi(t)$. In turn, as a characteristics of the attractor, we can take the value of the correlation dimension, which gives an idea of the complexity of the geometry of the attractor and, therefore, of the signal $\varphi(t)$.

Flow is directly known only in the simplest model cases. Sometimes it is possible to determine the existence of the generating flow and reconstruct its attractor, i.e., to get its image in a finite space. In principle, the latter can be done using the procedure of Grassberger and Procaccia (1983a, b). The a priori condition of the applicability of the procedure is the existence of the attractor, which generates the signal and which can be embedded in a finite space. In relation to the processes of the electrical activity of the cerebral cortex, this assumption is very questionable.

The method used by us formally resembles the approach (Grassberger and Procaccia 1983a, b). We present phenomenological considerations. The brain is a distributed system.

Its structurally detached elements, i.e., neurons, which are distributed entities, have the ability to delay. The direction of the evolution of the electrical processes in the neuron is determined by some background of the state of the neuron and its inputs (Lebedev 1990; Kogan 1979).

According to a fairly widespread point of view (Burns 1968; Bekhtereva 1988), neural associations are responsible for the processing of information in the brain and the nature of electrical processes at the macro level. In turn, they also possess the property of delay, and its time is estimated in the order mentioned previously. This phenomenon took place in the model proposed previously: To specify the initial conditions in the ring neuron structure, it was necessary to set time of passage of the excitation wave.

All of this leads to the idea of using $C[-h, 0]$ as the phase space, i.e., the continuous space on the interval $[-h, 0]$ functions, and considering the signal $\varphi(t)$ as the curve $\varphi(t + s) \in C[-h, 0](s \in [-h, 0])$ in it. The parameter h below indicates the most typical oscillatory period in the human EEG, i.e., $h \approx 100$ ms. The latter formal consideration is also suggested by model representations. Note that if we start from the model of oscillations of the activity of neural populations (Lebedev and Lutsky 1972), we would have to choose $h \approx 50$ ms. (The corresponding experimental data are shown below, but they are less meaningful.)

As the pattern (sample) of activity, we take the segment of an EEG before a given moment of time t_*, i.e., the function $\varphi(t_* + s)$ of $s \in [-h, 0]$, where h is duration (length) of the pattern.

The proposed approach allows us to characterize the geometric structure of the set of patterns by a number that we call the "pseudo-correlation dimension" (PCD). Explain the formalism for the calculation of PCD.

Represent the pattern $\varphi(t_* + s)(s \in [-h, 0])$ as the net function:

$$zt_* = (\varphi(t_* - (n-1)\Delta t), \varphi(t_* - (n-2)\Delta t), \ldots, \varphi(t_*)), \qquad (8.1.1)$$

which we call the "discrete pattern of length" n. In (8.1.1) $h = \Delta t(n-1)$. To speed up calculation, we determine ρ distances between patterns in pseudometric, thus thinning out net functions:

$$\rho(z_{t*}, z_{t**}) = \max_{0 \le i \le 1[(n-1)/\tau]} |\varphi(t_* - i\tau\Delta t) - \varphi(t_{**} - i\tau\Delta t)|, \qquad (8.1.2)$$

where $\tau(1 \le \tau \le n-1)$ is an integer, and $[*]$ is an integer part.

Consider the time series of the observations of the EEG of the signal of length N: $\varphi(t_1), \varphi(t_2), \ldots, \varphi(t_N)$, where $t_i = i\Delta t$ Then at our disposal is $N - n + 1$ of discrete patterns $z_{t_n}, \ldots, z_{t_N}$. Randomly form the set Z of $N_{ex}(1 \le N_{ex} \le N - n + 1)$ patterns, which we call the "reference."

Under the pseudo-correlation integral (PCI), we understand the expression:

$$c(l) = \sum_{Z_{t_i} \in Z} \left[\sum_{j=n}^{i-v} \theta(l - \rho(Z_{t_i}, Z_{t_j})) + \sum_{i+v}^{N} \theta(l - \rho(Z_{t_i}, Z_{t_j})) \right], \qquad (8.1.3)$$

where $\theta(*)$ is the function of Heaviside, and $v > 0$ is some number. It is easy to see that the expression in square brackets is the number of hits of patterns z_t for $t = t_n, \ldots, t_{i-v}, t_{i+v}, \ldots, t_N$ in l—the neighborhood of reference $z_{t_i} \in Z$. Patterns z_t for $t = t_{i-v+1}, t_{i+v-1}$ in the study of population of this neighborhood are not considered.

Input value (8.1.3) differs from the correlation integral (Grassberger and Procaccia 1983a, b) only by the fact that l—neighborhood is considered in the pseudo-metric. Note also that the correlation integral is usually introduced in the language of probabilities (see Loskutov and Mikhailov 1990).

In processing an EEG, we deal with a digital signal where the values $\varphi(t_i)$ are the integers within a certain range $[0, l_{max}]$, It is also natural to take as a whole the radius l of the studied neighbourhoods of reference patterns. Note the existence of such l_0 that at $l < l_0$ in l, i.e., where the neighbourhood of reference patterns does not obtain any other pattern, and therefore $C(l) = 0$ at $l < l_0$ but $C(l_0) \ne 0$ (In any case, this was observed in our experiments.)

Consider the function $\log_2 C(l)$ for $l \in [l_0, l_{max}]$ Choose some l_{CR} from the interval $l_{max} > l_{CR} \ge l_0 + 1$ (critical point) and approximate $\log_2 C(l)$ by linear spline (broken line), the argument of which is $\log_2 l$ On the interval $l \in l_0, l_{CR}$, it is a linear function of the specified argument:

$$\log_2 C(l) \approx r \log_2 l + b, \qquad (8.1.4)$$

applying the best mean-square approximation. Denote its mistake by $\Delta_1(l_{CR})$. For $l > l_{CR}$, as an approximation we choose a constant equal to average value $\log_2 C(l)$, and denote the corresponding mean-square error denote by $\Delta_2(l_{CR})$. Choose the critical point l_{CR} so that the value $\Delta = \max(\Delta_1(l_{CR}), \Delta_2(l_{CR}))$ is minimal.

We call the PCD of the signal $\varphi(t)$ the number r in formula (8.1.4). Despite the sign of approximate equality in (8.1.4), the method of calculation of the PCD is unique.

Here we make some comments. Assume that components of the vectors z_t are real (non-integral). Let us look at the set z_t as the implementation of vector random variable. Suppose that it is uniformly distributed on a smooth m-dimensional (compact) manifold. Divide the value of the correlation integral $C(l)$ by the number of support vectors. It is easy to see that with growth of the length of the sample, increasing the number of support vectors, and decreasing l_{CR}, we obtain in (8.1.4) $r \rightarrow m$. Thus, we have the above-described (adapted) method of calculating the dimension of the manifold, on which the random variable is distributed. In case of a violation of smoothness, the number r is not an integer.

Because below we compare the results with the works of other investigators, it is necessary to make some remarks. As already mentioned, PCD is different from the correlation integral (see Grassberger and Procaccia 1983a, b); in fact, only in that neighborhoods of vectors (8.1.1) is the pseudo-metric defined (8.1.2). Furthermore, according to the procedure of Grassberger and Procaccia, just as here, we build approximation (8.1.4) (in general at small l) and determine the number r, which depends on n the number of coordinates of vectors (8.1.4). The value r is often called the "correlation dimension" and the number n the "dimension of the embedding space." If, starting from some dimension n_0 of the embedding space, correlation dimension stops growing, then we say that there is "saturation." It is in this case, according to Grassberger and Procaccia (1983a, b), that the existence of flow can be asserted, the attractor of which can be embedded in a finite-dimensional space and is crucial for the signal $\varphi(t)$. We call the number n_0 the "true dimension of the embedding space" and the corresponding number r the "true correlation dimension of the signal" $\varphi(t)$.

Determination of the number r from the approximate equality (8.1.4) is an important problem. To calculate this, one can often proceed as follows (Başar et al. 1988). Choose some $l_0 < l_* < l_{**} < l_{max}$ (discard the extreme cases). For $l \in [l_*, l_{**}]$, numerically determine tangent of the angle of the slope of the curve as the function $\log_2 l$. Then the obtained values are averaged. Here we are not satisfied with the uncertainty in the choice of the numbers l_* and l_{**}.

Another method was proposed by Xu and Xu (1988). We build the graph of the tangent of the specified angle depending on the $\log_2 C(l)$. In most cases, it will have area of slow change, i.e., an almost horizontal plateau. The height of the midpoint of the plateau is taken for the correlation dimension. Here again there is uncertainty in the determination of the boundary points of the plateau. There are cases where the plateau is not clearly expressed.

Referring back to Ephremova et al. (1991), we choose and record the number $l_* > l_0$ of empirical considerations, and then by the method of least squares we build an approximation for all of the intervals $l \in [l_0, l_{**}]$, where $l_{max} \geq l_{**} \geq l_*$. Among the obtained numbers r, we choose the maximum number. Approbation of the described methods showed that the calculated value r is "at the disposal" of the researcher.

8.2 Description of Experimental Methods and Their Approbation

Observations were carried out on young people ages 18–21 years. More than 70 people took part in the experiments. In a sound-proofed and shielded chamber, subjects were sitting in a relaxed position with their eyes open in the waking state. Their EEG was recorded monopolarly (points O_2 and F_4) in the occipital and frontal derivations. The indifferent electrode was on the earlobe. The results were recorded in analogue form on magnetograph with a passband of 0–9 kHz.

Occipital deviation was chosen because the most expressed alpha rhythm is expressed in this area, which, according to some authors, plays an important role in the storage and reproduction of information (Pavlova and Romanenko 1988; Lebedev 1990; Bekhtereva 1988). In the future, potentials of the electroencephalogram would be transformed by 12-bit analog-to-digital converter. Input-signal magnitude was selected so that after processing its maximum amplitude was 9–10 binary digits. Based on the fact that the frequency of the signal conversion must exceed the frequency of the fastest part of the process by 5–10 times, we chose a sample frequency of 250 Hz. Thus, in formula (8.1.1), the discrete time step is $\Delta t = 4$ ms.

We agree with Lebedev (1990) that the characteristic time for neural associations is approximately 100 ms. Therefore, as the duration of patterns of the activity we chose sequentially 64, 80, 96 and 112 ms, and the lengths of n discrete patterns in (8.1.1) were, respectively, 17, 21, 25, and 29. Control points within patterns, in which the latter were compared and were taken in 16 ms, i.e., in formula (8.1.2) $\tau = 4$. This is consistent with the recommendations (Mayer-Kress and Layne 1987; Frochling et al. 1988). Obviously, the number of control points corresponding to the lengths of patterns were 5, 6, 7, and 8.

The value of PCD is statistical characteristics, so the number of observations must be sufficiently large. We chose the number of experimental points $N = 10,000$, which corresponds to a 40-s epoch of analysis. As a result, at our disposal are approximately 400 patterns non-overlapping in time (the total number of patterns is $N - n$, i.e., approximately N). This, we hope, is sufficient to obtain statistical characteristics. Also, the time of the order of a minute is taken empirically in psychology for a standard in the study of short-term memory (Golubeva 1980; Hoffman 1986), the relation of which with the correlation dimension we investigated. Based on the number of patterns non-overlapping in time, we estimated the required number of reference patterns (vectors). In our opinion, it can be taken to equal 500.

In determining the level of contribution of each specific reference pattern, the DCI is removed from the EEG itself, and the adjacent left and right segments of the EEG are 40 ms in length. Take into consideration the quantization step in formula (8.1.3) $v = 10$.

In general, the procedure of calculating the correlation dimension is largely empirical, and researchers have propose different rationales for the choice of parameters in the processing of their experimental data (see Başar et al. 1988, 1989; Xu and Xu 1988; Soong and Stuart 1989).

With respect to the epoch of analysis, the approach of some neuroscientists (Xu and Xu 1988; Soong and Stuart 1989) is extremely unique. It is believed that if a researcher is interested in fast transient changes, then he or she chooses a time within a few seconds. To get general characteristics of the process to identify integral, stable indicators (in particular, at a noisy signal), the epoch of analysis increases to several minutes.

In the first case, to obtain the necessary number of experimental points for statistical analysis, it is proposed to take the small step of quantization. In this regard, we note that because the analogue signal of an EEG is continuous, components of the discrete pattern at small Δt will be approximately the same. Vectors z_{t_i} will be located in the neighborhood of the straight line in R^n, which forms equal angles with the coordinate axes. The result of calculation of the correlation dimension can be predicted. It will be approximately equal to one (Xu and Xu 1988).

There are the objections in some papers (Pessa et al. 1988; Babloyantz 1990) with respect to the analysis of lengthy segments of EEG. The EEG signal is a process with unpredictable dynamics, and investigators wonder what will be the result of mixing intervals with different dimensionality. The answer is trivial: The method allows us to determine the average dimension. Another question is this: What does the latter provide for practice. This is already known fundamentally. However, most researchers, noting the nonstationary dynamicity of EEG process, take especially homogeneous intervals for the analysis of the EEG. If we consider that the variability is a reflection of the normal operation of the brain, then it is necessary to analyze the EEG without an artificial sort of operation. Even Walter pointed out that there is one quantitative characteristic, the value of which is sensitive to the degree of "intellectual brilliance." It is a measure of variation of the frequency spectra. "The most original people with traits of genius seem to have high mobility in this sense: to get for them a number of similar averages, it is necessary to take these averages over a very long period—at least a few minutes. "Dumb" brain gives indistinguishable from each other results at averaging of half a dozen ten-second analysis" (Walter 1966, p. 231).

If the study of lengthy EEG segments provides the opportunity to obtain characteristics allowing meaningful interpretation, then the analysis of the correlation dimension of EEG is valid. As we see it, as presented in the next paragraph, the results, along with data (see Dvorak et al. 1985; Başar et al. 1988; Soong and Stuart 1989), solve this problem positively.

We now discuss some practical details of the processing of the results of separate experiments.

Figure 8.1 presents examples of graphs of the dependence of $\log_2 C(l)$ on $\log_2 l$ for a 40-s epoch of analysis along with 500 reference vectors. The curves in the figure from left to right correspond to the lengths of patterns lasting 64, 80, 96, and 112 ms.

Figure 8.2 shows the dependences of angular coefficients $tg\alpha(l)$ tangential to these curves as functions of $\log_2 l$. The curves are slightly shifted to the left because

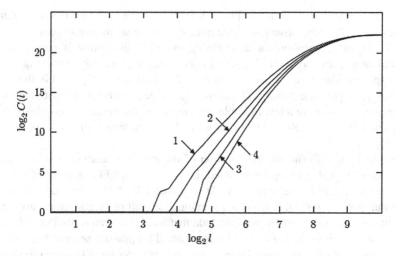

Fig. 8.1 Graphs of pseudo-correlation integrals. *Curves 1–4* correspond to the durations of patterns of 64, 80, 96, and 112 ms

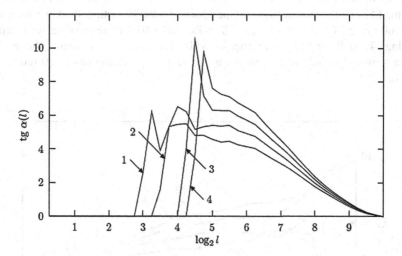

Fig. 8.2 Graph of angular coefficients tangential to the curves of pseudo-correlation integrals. *Curves 1–4* correspond to the durations of patterns lasting 64, 80, 96, and 112 ms

at numerical implementation we used a certain procedure of averaging over the window. Areas of horizontal stabilization of curves of (Başar et al. 1988, 1989) correspond to the level of the PCD. Obviously, we can only conditionally identify these areas.

According to Xu and Xu (1988), we should construct parametric curve $x = \log_2 C(l), y = tg\alpha(l)$ and take the height of the horizontal plateau for the value of

PCD. The method is illustrated in Fig. 8.3. Implementation of the EEG is the same. Certain difficulties are observed in the allocation of areas of stabilization.

Our use of the method for determining the PCD dimension is based on the obvious properties shown in Fig. 8.1 of curves that are naturally approximated by linear splines. The result of the proximity of the right curve (Fig. 8.1) is shown in Fig. 8.4. It is quite satisfactory. The correlation dimension coincides with the slope of the area of spline. In a test example considered by us, the algorithm yields PCD values of 4.50, 5.11, 5.50, and 6.06, respectively, for patterns lasting 64, 80, 96, and 112 ms.

To study the influence of the duration of the epoch of analysis, we conducted control studies with each epoch doubled, i.e. we took an EEG segment of 80 s. In this case, for the EEG, we obtained PCD values of 4.45, 5.18, 5.50, and 6.17 at the same durations of patterns mentioned previously. In all other control experiments, the difference was also small. We conclude that indeed we can restrict ourselves to the epoch analysis of 40 s of EEG. This is essential for practice because the study of long intervals of EEG leads to a large amount of computer time. However, it should be noted that for longer patterns lengthening of the epoch of analysis is required.

We also considered the problem of the influence of the number of reference patterns on the PCD value. For the control, we took 1000 references. In this case, for the EEG under consideration (at the epoch of analysis lasting 40 s), we obtained the following PCD values—4.25, 5.8, 5.79, and 6.07—corresponding to patterns lasting 64, 80, 96 and 112 ms, respectively. In other control experiments, the results were similar. We believe that by doing so we justify the choice of 500 reference vectors.

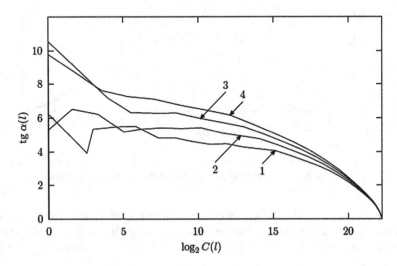

Fig. 8.3 Graph of angular coefficients of pseudo-correlation integrals as functions of them. *Curves 1–4* correspond to the patterns lasting 64, 80, 96, and 112 ms, respectively

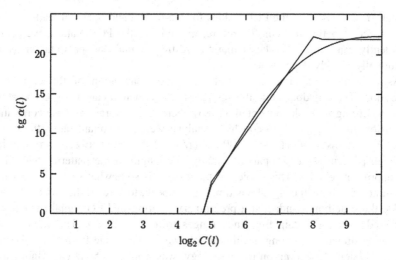

Fig. 8.4 The result of approximation by the linear spline of the pseudo-correlation integral for the pattern lasting 112 ms

We also give information on the correlation dimension for the 1st and 10th standard patterns: 4.53, 5.52, 5.79, and 6.7 corresponding, respectively, to PCD values of 4.04, 4.96, 5.23, and 5.17. These results clearly indicate heterogeneity of the EEG signal, which is consistent with the generally accepted point of view (see Devyatkov and Grindel 1973; Başar 1990).

Experiments have been conducted to study PCD saturation. For this, durations of patterns were increased to 192 ms; however, PCD saturation was not found.

8.3 Results and Discussion

Consider first, averaged by groups of examinees, the PCD values of monopolar EEG leads from the frontal F_4 and occipital O_2 zones. For patterns lasting 96 and 112 ms, data are given in Table 8.1. Based on the differences of PCD, we suggest that the complexity of the EEG in the frontal area is higher than in the occipital area.

Table 8.1 Averaged values of PCD EEG in areas F_4 and O_2 for patterns lasting 96 and 112 ms (*SD* standard deviation)

Area	PCD EEG		SD		Number of examinees
	$h = 96$ ms	$h = 112$ ms	$h = 96$ ms	$h = 112$ ms	
F_4	6.2	6.8	0.27	0.46	3478
O_2	5.5	6.1	0.45	0.56	78

Statistical analysis of the PCD values for the frontal and occipital leads showed no connection between them. Therefore, in studying the PCD values, we cannot confidently reject the hypothesis that the resting frontal and occipital areas are functionally weakly dependent.

As previously mentioned, we did not expect saturation of the correlation dimension. There is doubt that the saturation phenomenon can be observed at all. Also, according to Babloyantz et al. (1985), there is no saturation of the correlation dimension. In that paper, 30-s epochs of analysis were taken quantization the step of signal $\Delta t = 10$ ms, product $\tau\Delta t = 100$ ms. ($\tau\Delta t$ was the time between control points in which patterns were compared.) Finally, the length of the patterns themselves was approximately 1 s. This choice of parameters is somewhat surprising because for a time of 100 ms the signal has time, so to speak, to take all the values. In later works, these authors conducted a preliminary filtering of EEG signal, which was processed only for certain frequency ranges from 5 to 15 Hz. The value of the correlation dimension among healthy people was in the range from 5 to 7 Hz and among the sick, depending on the pathology, was from 2 to 5 Hz (see Babloyantz and Destexhe 1986; Babloyantz 1990).

Mayer-Kress and Layne (1987), when analyzing the alpha rhythm on 10-s intervals of EEG with quantization step $\Delta t = 2$ ms and value $\tau\Delta t \approx 10$–40 ms, also found no saturation of the correlation dimension at pattern lengths of approximately hundreds of milliseconds. However, these recordings were obtained bipolarly and therefore cannot be directly compared with the results presented here.

The next important fact we should highlight is the considerable individual PCD variability. Most studies have been focused on the ideology of attractors and not on the EEG itself. That is why they only ascertain the fact of the variability of the correlation dimension.

Regarding variability, for example, Pessa et al. (1988) stated that among 10 subjects, the value of the EEG correlation dimension was approximately 8.3 with only slight differences in the second-decimal digit. However, as the investigators noted, the number of experimental points was only 100. At such a length of experimental series, the correlation dimension is not statistically valid.

Summing up the preliminary results, we say that the EEG, when recorded in a patient in a quiet state, reflects the state of dynamic chaos. The PCD is averaged over the time (integral) feature. It depends on the individual and varies in the interval from 5 to 8 (for patterns lasting 112 ms).

Mayorov and Myshkin (1993a, b) audited the assumption that the individual features of the chaotic dynamics of EEG reflect specificity of brain activity and are objective characteristics, by which—in some way—one can judge the information capabilities of the brain. This approach stems from the hypothesis of frequency and phase coding in the brain, and the above-suggested neural network model is capable of storing information. From the model, it follows that the more "degrees of freedom" the system has, i.e., the greater the variety of oscillations, the higher its capacity to store information. In turn, the diversity of the system's oscillations is quite natural to characterize by the correlation dimension of its attractor. Thus, the assessment of such dynamic process as EEG by means of calculating the PCD seems to be adequate.

The value of short-term memory has been taken as a measure of the information capacity of the brain n. The amount of short-term memory to decimal numbers was determined by means of adaptive computer technique (Myshkin 1988). It was compared with the PCD calculated for patterns lasting 112 ms in the occipital lead of an EEG. A graph of experimental dependence for the group mean values (see Table 8.2) is shown in Fig. 8.5. The x-axis corresponds to the correlation dimension, whereas the vertical axis corresponds to the amount of short-term memory.

Table 8.2 Averaged group values of PCD EEG and the amount of short-term memory for 78 examinees (*SD* standard deviation)

Group number	PCD EEG		Amount of memory		No. of subjects monitored
	Averaged value	SD	Averaged value	SD	
1	6.77	0.12	5.57	0.4	10
2	6.52	0.06	5.86	0.6	10
3	6.36	0.05	5.38	0.5	10
4	6.21	0.06	5.69	0.6	10
5	6.06	0.05	5.29	0.4	10
6	5.88	0.06	5.28	0.5	10
7	5.72	0.06	5.16	0.5	10
8	4.92	0.06	4.99	0.5	8

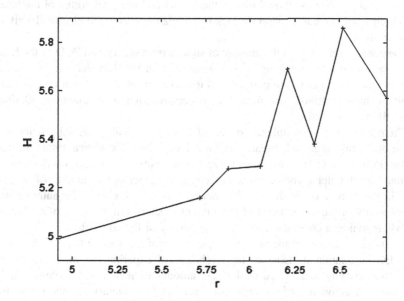

Fig. 8.5 Experimental dependence of short-term memory H of the pseudo-correlation dimension r

Even visually there is a tendency of dependence. A larger amount of short-term memory responds to larger PCD values. Statistical analysis showed that the coefficient of bivariate correlation for the entire sample of 78 subjects (without grouping results) is +0.407, i.e., $p < 0.01$. Note that we did not run the sample sorting because the selection of experimental data can sometimes unintentionally justify very strange results.

Refusal of sortings has its drawbacks. The PCD, as a random variable, is distributed approximately according to normal law. The number of subjects for whom the correlation dimension of EEG is large or small are rather small (i.e., rare). As a consequence, data for small and large values of the correlation dimension are underrepresented in the sample.

The sample satisfies the condition: Removing any randomly selected item from it does not change the trend of dependence, thus leaving it significant ($p < 0.01$).

Table 8.2 presents mean group values of the correlation dimension and the amount of short-term memory. They are obtained by regular ranking. Averaged quantities also have a positive and significant correlation of the bivariate coefficient of +0.79 at $p < 0.01$.

With decreasing the length of the patterns, the statistically significant association of PCD and short-term memory disappears. In our opinion, this underscores the correct choice of the pattern duration in the neighborhood of the most typical oscillatory period of the EEG.

Statistical analysis of the relationship between the amount of short-term memory and PCD in frontal derivations of the EEG did not lead to significant results.

Remember that the dimensions of EEG themselves in the frontal and occipital areas did not correlate. These facts do not contradict the data of neurophysiology and psychology. The functional role of the frontal and occipital cortex of the brain is different, and that can explain the lack of correlation (at rest) of electrical activity of these areas.

The connection between the amount of short-term memory and PCD of the EEG signal only in the occipital region is consistent with the data that the mnemonic process mainly occurs in the parietal and temporal areas of the brain (see Beritov 1969; Bekhtereva 1980). The frontal areas perform integrative functions (Beritov 1969; Blum et al. 1988).

The relationship of complexity of the EEG structure with potential possibilities of the brain has interested researchers for a long time (Bekhtereva 1980). In particular, Pasinkova (1988), to characterize the diversity of potential oscillations in the range of the alpha rhythm, took the average product of the number of oscillations in the spindle of rhythm on the spindle length (index aT). The author notes monotonicity of the dependence of the amount of short-term memory of aT but on the whole makes a conclusion about the positivity of the connection.

Let us discuss some of the possible applications of methods of the calculation of the EEG correlation dimension. In psychological experiments, improvement of mnemonic activity and, in particular, the amount of memorized material under conditions of activation of the organism, increase of voluntary attention, positive motivation, etc., has repeatedly been observed. These states of psychological activity at the physiological level have similar symptoms (Pavlova and Romanenko 1988). In the EEG high-frequency components appear, and thus evolves the process of desynchronization, i.e., in general it increases the diversity of oscillations. Similarly, the process of thinking is reflected by cortical rhythmics.

Common methods of analyzing brain bioelectrical activity, e.g., spectral analysis, assessment of the coherence level, can actually describe the complexity of oscillations only qualitatively, whereas the use of the correlation dimension can "measure complexity" by numbers. Xu and Xu (1988) discovered valid but opposite changes of correlation dimension at different stages of psychic activity. In general, the tendency of memory to deteriorate in a brain state characterized by a low EEG correlation dimension has been confirmed. This is consistent with our model representations.

In connection with the above-mentioned information, there can be objections to the method of our experiments itself. Indeed, the PCD was measured with the subject at rest and the amount of short-term memory measured with the subject in the active state. In planning the work, we assumed that the state of rest is a kind of "wait state," in which are manifested the potential possibilities of the brain are manifested. Moreover, it appears that in the active state, the complexity of processes and the strong interaction of all areas of the brain can make it difficult to identify the relationship between the correlation dimension and such brain characteristics as the amount of short-term memory.

References

Adey, V. R. (1967). Organization of brain structures in terms of communication and storage of information. In *Modern problems of electrophysiology of the central nervous system* (pp. 324–340). Moscow: Nauka.

Adey, W. R. (1974). Organization of the brain: The brain as a noisy computer. In *The main problems of electrophysiology of the brain* (pp. 350–366). Moscow: Nauka.

Adrian, E. D. (1934). Electrical activity of the nervous system. *Archives of Neurology and Psychiatry, 34*(6), 1125–1136.

Adrian, E. D., & Moruzzi, B. (1939). Impulses in the pyramidal tract. *Journal of Physiology, 97*(2), 153–159.

Andersen, P., & Eccles, J. (1962). Inhibitory phasing of neural discharge. *Nature, 4855*(196), 645–647.

Arduini, A., & Whitlock, D. G. (1953). Spike discharges in pyramidal system during recruitment waves. *Journal of Neurophysiology, 16*(4), 430–436.

Atsev, E. S., Arutyunova, V. S., & Narikashvili, V. S. (1968). Relationship between the slow electrical activity and discharge of individual neurons in the somatosensory cortex of the cat. In *Problems of dynamic localization of brain functions* (pp. 164–174). Moscow: Nauka.

Babloyantz, A. (1990). Chaotic dynamics in brain activity. In E. Başar (Ed.), *Chaos in brain function* (pp. 42–48). Berlin, Heidelberg: Springer.

Babloyantz, A., & Destexhe, A. (1986). Low dimensional chaos in an instance of epilepsy. *Proceedings of the American Academy of Sciences, 83*, 3513.

Babloyantz, A., Nicolis, C., & Salasar, M. (1985). Evidence of chaotic dynamics of brain activity during the sleep cycle. *Physics Letters (A), 111*, 152–156.

Başar, E. (1990). Chaotic dynamics and resonance phenomena in brain function: Progress, perspectives and thoughts. In E. Başar (Ed.), *Chaos in brain functions* (pp. 1–30). Berlin, Heidelberg: Springer.

Başar, E., Başar-Eroğlu, C., & Röschke, J. (1988). Do coherent patterns of the strange attractor EEG reflect deterministic sensory-cognitive states of the brain? *From chemical to biological organization*, 297–306.

Başar, E., Başar-Eroğlu, C., Röschke, J., & Schilt, I. (1989). Chaos and alpha-preparation in brain functions. In *Models of brain functions* (pp. 365–395). Cambridge: Cambridge University Press.

Bekhtereva, N. P. (1980). *Healthy and diseased human brain*. Moscow: Nauka.

Bekhtereva, N. P. (1988). Mechanisms of activity of human brain. In N. P. Bekhtereva (Ed.), *Neurophysiology of man*. Leningrad: Nauka.

Belsky, Y. L., Vedenyapin, A. B., Dmitriev, A. S., Zenkov, L. R., Starkov, S. O., & Vasiliev, P. P. (1993). Diagnosis of pathological states of the brain based on the analysis of the electroencephalogram by methods of nonlinear dynamics. *Radio Engineering and Electronics, 38*(9), 1625–1635.

Beritov, I. S. (1969). *Structure and functions of the cerebral cortex*. Moscow: Nauka.

Bernstein, N. A. (1990). *Physiology of movements and activity*. Moscow: Nauka.

Blum, F., Leiserson, A., & Hofstefer, L. (1988). *Brain, mind and behaviour*. Moscow: Mir.

Brookchart, J. M., & Zanchetti, A. (1956). The relation between electrocortical waves and responsiveness of corticospinal system. *Electroencephalography and Clinical Neurophysiology, 8*(3), 427–444.

Burns, B. N. (1968). *Uncertainty in the nervous system*. Moscow: Mir.

Devyatkov, N. D., & Grindel, O. M. (1973). Investigation of instability of temporal characteristics of human EEG by method of phase-frequency analysis. *Herald of the Academy of Medical Sciences of the USSR, 5*, 41–45.

Dvorak, I., Siska, J., Wackermann, I., et al. (1985). Evidence for interpretation of the EEG as a deterministic chaotic process with a low dimension. *Activitas Nervosa Superior, 28*(3), 228–231.

Elul, R. (1967). Amplitude histogram of the EEG as an indicator of the cooperative behaviour of neurons. *Electroencephalography and Clinical Neurophysiology, 23*(1), 87.

Elul, R., & Adey, W. R. (1966). Nonlinear relationship of spike and waves in cortical neurons. *The Physiologist, 8*, 98–104.

Ephremova, T. M., Kulikov, M. A., & Rezvova, I. R. (1991). Participation of nonlinear dynamic processes in the formation of high-frequency EEG of the rabbit. *ZHVND, 41*(5), 998–1006.

Frochling, H., Crutchfield, J. P., Farmer, J. D., Packard, N. H., & Shaw, R. (1988). On determining the dimension of chaotic flows. *Physica D, 3*, 605–617.

Frost, D., Jr., & Kellaway, G. A. (1965). Correlation of spontaneous EEG activity and unit discharges in isolated cerebral cortex of the cat. *Electroencephalography and Clinical Neurophysiology, 18*(5), 517.

Gavrilov, V. V. (1987). The ratio of EEG and impulse activity of neurons in the behaviour of the rabbit. In *EEG and neuronal activity in psychophysiological research* (pp. 33–34). Moscow: Nauka.

Georgiev, G. V. (1966). The ratio of slow (rhythm of voltage) and impulse activity in ECG of the rabbit. *ZHVND, 16*(1), 76.

Golovchinsky, V. B. (1965). The ratio of surfaces of EEG and discharges of single neurons of the first somatosensory cortex without and with intraanesthetic. *Journal of Physiology of the USSR named after I. M. Sechenov, 51*, 1159–1163.

Golubeva, E. A. (1980). *Individual characteristics of human memory*. Moscow: Pedagogy.

Grassberger, P., & Procaccia, I. (1983a). Characterization of strange attractors. *Physical Review Letters, 50*(5), 346–349.

Grassberger, P., & Procaccia, I. (1983b). Measuring the strageness of strage attractors. *Physica D: Nonlinear Phenomena, 9*(1–2), 189–208.

Hoffman, I. (1986). *Active memory*. Moscow: Progress Publishers.

Jacome, D. E., & Morilla-Pastor, D. (1990). Unreactive EEG: Pattern in locked in syndrome. *Electroencephalography and Clinical Neurophysiology, 1*(1), 31–36.

Kogan, A. B. (1979). *Functional organization of neural systems of the brain.* Leningrad: Nauka.

Lebedev, A. N. (1990). Human memory, its mechanisms and limits. In *Research of memory* (pp. 104–118). Moscow: Nauka.

Lebedev, A. N., & Lutsky, V. A. (1972). EEG rhythms are the result of the interaction of oscillating processes. *Biophysics, 17*(3), 556–558.

Livanov, M. N. (1972). *Spatial organization of the processes of the brain.* Moscow: Nauka.

Loskutov, A. Y., & Mikhailov, A. S. (1990). *Introduction to synergetics.* Moscow: Nauka.

Mayer-Kress, G., & Layne, S. P. (1987). Dimensionality of human electroencephalogram. In S. N. Koslow, A. J. Mandell, & M. F. Shlesinger (Eds.), *Perspectives in biological dynamics and theoretical medicine.* Austria: Urban & Schwarzeberg.

Mayorov, V. V., & Myshkin, I. Y. (1993a). Computation of one correlation characteristics of the electroencephalogram and its relation to the volume of short-term memory. *Radio Engineering and Electronics, 38*(5), 900–909.

Mayorov, V. V., & Myshkin, I. Y. (1993b). Correlation dimension of the EEG and its relation to the volume of short-term memory. *Psychological Journal, 14*(2), 62–75.

Myshkin, I. Y. (1988). Systematic approach to the problem of basic psychophysical law. *Psychological Journal, 9*(6), 83–91.

Pasinkova, A. (1988). Relationship of the spectral characteristics of the electroencephalogram with a volume of short-term memory. In *Psychology of cognitive processes.* Moscow: Nauka.

Pavlova, L. P., & Romanenko, A. F. (1988). *System approach to the psychophysiological study of human brain* (p. 209). Leningrad: Science.

Pessa, E., De Pascalis, V., & Marucci, F. S. (1988). The detection of strange attractors in brain dynamic through EEG date analysis. In *Proc. 4th Conf. Int. Organ. Psychophysiol.,* Prague.

Pribram, K. (1975). *Languages of the Brain.* Moscow: Mir.

Shulgina, G. I. (1978). *Bioelectrical activity of the brain and the conditioned reflex.* Moscow: Nauka.

Soong, A. C. K., & Stuart, C. I. J. M. (1989). Evidence of chaotic dynamics underlying the human alpha-rhythm electroencephalogram. *Biological Cybernetics, 62*(1), 55–62.

Trush, V. D., & Kornievsky, A. V. (1978). *Computers in neurophysiological research.* Moscow: Nauka.

Walter, G. (1966). *Living brain.* Moscow: Mir.

Xu, N., & Xu, J. (1988). The fractal dimension of EEG as a physical measure of conscious human brain activities. *The Bulletin of Mathematical Biology, 50*(5), 559–565.

Zhadin, M. N. (1984). *Biophysical mechanisms of formation of the electroencephalogram.* Moscow: Nauka.

Chapter 9
Estimates of Differences of Evoked Potentials

The evoked potential (EP) is a segment of the electroencephalogram that is registered immediately after the start of the stimulus. In our studies, we used flashes of light as the stimulus. We usually consider EP duration of 500–600 ms. The most common method of investigating EP is associated with their averaging (Ivanitsky et al. 1984). The stimulus is presented a sufficient number of times, and a library of EP recordings is collected. Then the EP recordings are averaged over the mass. Averaged EP recordings (AEPs) for various stimuli are compared both visually and with the use of statistical criteria such as Wilcoxon. Visually the form of the AEP is relatively stable. This allows classifying (albeit very arbitrarily) the time zones of the AEP. For the AEV, we find a time average. Time intervals, where the value of AEV exceeds the average, are called the "zones (waves) of positivity," and those that are correspondingly less than the average are called the "zones (waves) of negativity." For positive and negative waves, we use the notation P_t, N_t, where t is time of their start.

If we average randomly selected areas without prior stimulus presentation (background activity), then it is also possible to observe oscillations in the averaged EEG tracings, but their amplitude is less than those on the AEP recording. Note that the amplitude differences are usually not great, but the durations of the zones of the positivity and negativity have are significantly different. Below we consider the evaluation criteria of differences of EPs that are not related to their averaging (Mayorov and Myshkin 1995; Myshkin and Mayorov 1996).

9.1 The Phenomenon of EPa

Although the method of measuring EPs has been in use for >40 years, it continues to raise controversy about EPs components and whether they reflect the world of objects and the processes of intracerebral activity. In other words, is specific information, carried by a stimulus, and the processes of its processing coded in the form of EPs? There are mixed results and different points of view on this account.

From the standpoint of the cognitive approach, EPs are interpreted as a reflection of the processes of reception and the processing of information (Ivanitsky et al. 1984;

© Springer International Publishing Switzerland 2015
S. Kashchenko, *Models of Wave Memory*,
Lecture Notes in Morphogenesis, DOI 10.1007/978-3-319-19866-8_9

Lytarev and Shostak 1994). Numerous investigations have shown the benefit of the fact that EPs reflect different parameters of external stimuli such as intensity, modality, and complexity, specificity in the nerve structures. EPs also reflect the subjective importance of stimulation. The two most identified components of AEP are as follows:

1. There exist components attributable to the time interval of 100–200 ms. According to Naatanen et al. (1986), they correspond to preprocessing of the stimulus.
2. The exists wave P_{300}, which is associated with the estimation of informative efficiency of stimulus and decision-making (Kostandov 1988; Ivanitsky et al. 1984; Kanunnikov and Vetosheva 1988).

It should be noted that the data on time parameters of information processing based on the analysis of EPs are contradictory. Thus, according to one source (Kochubey et al. 1988), the processing of comparable information in visual stimuli is carried out within the first 80 s. According to other sources, during this interval the analysis of "physical characteristics" of stimuli occurs, and the synthesis of the received information and past experiences (Ivanitsky et al. 1984) begins after 100 ms. At the same time, physical symptoms—such as intensity, contrast, color, shape, and position in the field of view and others—appear in addition to the influence exerted by them on the parameters of the early components, thus modulating components in the range of 100–300 ms (N_{100}, P_{130}, N_{160}, P_{200}, P_{300}) [237]. Along with the traditional view that early waves (≤ 100 ms) reflect the physical characteristics of the stimulus, there is evidence of their importance in the activation of early mechanisms of selective attention (see Cowan 1988).

In contrast, from the cognitive-approach point of view, EPs reflect only the processes of the receipt and transfer of information, whereas the processing is associated with the activation of small number of neurons and is not reflected in the EPs (Schwartz 1976). At first glance, the similarity of responses with simultaneous abduction from different areas of scalp, as well as the uniformity of responses with simultaneous special and temporal averaging (Gnezditsky 1990), is evidence of the nonspecificity of EP. However, we offer a different interpretation. In fact, the above-mentioned results reveal the important fact of unity of the nature of the brain's response to different stimuli. The periodicity of oscillations of the amplitude of responses reflects periods of change in the activity of neural populations. Already in early works on EP research, it was found that the nature of the responses depends on the specifics of the EEG at the time of stimulation, for example, on the phase of the alpha wave (Dustman and Beck 1965). The configuration of EP, of course, also influences the level of the functional state of the brain reflected in the nature of its electrical activity. Taking into account these data, we believe that EPs reflect changes directly related to the characteristics of the stimulus as well as the level of excitability of the structures that generate the EPs.

Researchers (see Livanov 1965; Shilnikov et al. 2004) have shown that the involvement of neurons is a cyclical process of reaction. This is the way to understand the oscillating nature of EPs. Further development of this idea has led to a hypothesis about the mechanisms of phase-frequency coding in brain structures (Lebedev 1985, 1990). From standpoint of this hypothesis, the image that arises in response to the stimulus is activated cyclically. Variability of the structure of EPs can be explained as follows. The formation of the total pattern of electrical activity at each consecutive stimulus (with constant parameters) involves different groups of neurons. Therefore, even when the activity (process) is at one point, the actual structure of EPs can be different. The wave structure of EPs still has some specificity, which is consistent with the above-stated facts and was confirmed by the experimental data presented previously.

From the position of the hypothesis developed by us, such specificity is possible not for the whole EP but only for time zones where there is synchronization of neural populations. We tested this idea in an experiment where the independent variable was stimuli of different colors. The literature describes the changes of EP on the presentation of stimuli of different colors (Novikova et al. 1979; Grigorieva 1987). We conducted a qualitative analysis of the observed changes. More important and difficult is the problem of extracting and quantifying differences in the EPs, thus making it possible to identify them.

Each researcher assumes some basic idea. According to our model concepts, in the EP there must exist zones where classes of the corresponding segments of the EP differ for various stimuli. These zones have alternate zones where the differences are masked.

The problem of finding the difference in EPs can be solved as follows. We take two sets of EPs generated by two different stimuli. Let us turn to the discrete time. Call the pattern the "segment in discrete time." It is determined by two parameters: (1) the time of each launch, which will be measured from the time of stimulation; and (2) duration of each launch. We must show that two classes of patterns are obtained for different experimental conditions and that they are heterogeneous as "cut" from different classes. We choose the attribute for the patterns—the weighted average. Assign a critical level for the attribute value. If it is statistically valid that the value attribute for one class of patterns is more than the critical value and for the other class is less than the critical value, then we consider that the classes of patterns vary. Such approach gives one of the possible variants of the identification of individual EPs.

9.2   Description of Statistical and Research Methods

Consider the following statistical methods used in the analysis of the EP. By R^N denote N-dimensional vector space. Suppose there are two sufficiently representative vector samples $\vec{X}_i (i = 1, \ldots, n_1)$ and $\vec{Y}_j (j = 1, \ldots, n_2)$, belonging to the space R^N. We raise the question of their homogeneity, i.e., whether both samples are generated by one and the same random variable. To solve the problem, we use two methods.

9.2.1   The Method Based on the Existence of the Separating Plane (Delta-Tailed)

The method is based on a geometric idea: heterogeneous sample vectors (points in space) that one can try to separate from each other (at least partially) by the plane. The method presented in the form of the sequence of rules is easily implemented in software.

Denote by $M[X_i]$ and $M[Y_j]$ vectors, obtained by means of coordinate-wise averaging, respectively, of samples $\vec{X}_i (i = 1, \ldots, n_1)$ and $\vec{Y}_j (j = 1, \ldots, n_2)$, We introduce the vector $\vec{a}_1 = (\vec{n}_1/n) M[X_i] - (n_2/n) M[Y_j]$, where $n = n_1 + n_2$ is the number of elements in the pooled sample. We normalize the found vector: $\vec{a} = \vec{a}_1 / \sqrt{\vec{a}_1, \vec{a}_1}$. Here $(*, *)$ is a scalar product. Arrange the numbers $(\vec{a}, \vec{X}_i)(i = 1, \ldots, n_1)$ in ascending order and the numbers $(\vec{a}, \vec{Y}_j)(j = 1, \ldots, n_2)$ in descending order. Denote by i_0 the smaller number, for which $(\vec{a}, \vec{X}_i) > (\vec{a}, \vec{Y}_i)$. If such the number does not exist, then the samples are certainly not homogenous. Suppose there exists the specified number. Introduce the number $b_0 = 0.5[(\vec{a}, \vec{X}_{i0}) - (\vec{a}a, \vec{Y}_{i0})]$ and the linear function (weighted average) $\Pi(\vec{Z}) = (\vec{a}, \vec{Z}) - b_0$, where $\vec{Z} \in R^N$, Denote by θ_X the number of elements of the sample $\vec{X}_i (i = 1, \ldots, n_1)$, satisfying inequality $\Pi(\vec{X}_i) > 0$. Denote by θ_Y the number of elements $\vec{Y}_j (j = 1, \ldots, n_2)$, for which $\Pi(\vec{Y}_j) < 0$. Consider the quantity (statistics):

$$\delta_n = \sqrt{n}[\theta_X/n_1 + \theta_Y/n_2 - 1].$$

As the null hypothesis. we will take the assumption that samples X_i and Y_j are generated by the same random variable. In the case of validity of the null hypothesis for the representative samples $n_1 \gg 1, n_2 \gg 1$, inequality

$$-\frac{\alpha}{2}\left(\sqrt{\frac{n}{n_1}} + \sqrt{\frac{n}{n_2}}\right) \leq \delta_n \leq \frac{\alpha}{2}\left(\sqrt{\frac{n}{n_1}} + \sqrt{\frac{n}{n_2}}\right) \equiv \delta_{Cr} \qquad (9.2.1)$$

holds with probability $p \geq (\varphi(\alpha) - \varphi(-(-\alpha)))^2$, where the normal distribution function is $\varphi(\alpha)$.

The proof of this statement follows directly from the limit theorem of Moiv Laplace. If inequality (9.2.1) is violated, then the null hypothesis can be rejected with error probability, at most, at $1 - p$. Below assume that $\alpha = 2.2$. The result is $p = 0.945$.

Let as a result of calculations the null hypothesis be refuted. Hyperplane $\Pi(\vec{Z}) = 0$ divides the space R^N into two half spaces: R_X^N, where $\Pi(\vec{Z}) > 0$ and R_Y^N, where $\Pi(\vec{Z}) < 0$. Call half–spaces R_X^N and R_Y^N "eigenspaces," respectively, for samples $\vec{X}_i (i = 1, \ldots, n_1)$ and $\vec{Y}_j (j = 1, \ldots, n_2)$. It is easy to see that a randomly selected element of the pooled sample $(\vec{X}_i (i = 1, \ldots, n_1), \vec{Y}_j (j = 1, \ldots, n_2))$ more frequently ends up in its own half-space. Thus, the hyperplane $\Pi(Z) = 0$ provides a partial separation of samples. The described method of constructing a separating plane is not the best one. However, below we will use it in view of its simplicity.

If, as a result of calculation, the null hypothesis is refuted, then the meaningful value becomes the value $(\theta_X + \theta_Y)/n$, i.e., the percentage of correctly identified (in their own half-spaces) vectors $\vec{X}_i$ and $\vec{Y}_j$.

9.2.2 Method Based on the Relation of Closeness (Gamma Criterion)

This criterion was developed by Levin (1993). It is based on another geometric idea: analysis of the relation "to be the closest." Let θ_X be the number of elements in the sample $\vec{X}_i \in R^N (i = 1, \ldots, n_1)$, for which the nearest element in the pooled sample $(\vec{X}_i(i = 1, \ldots, n_1), \vec{Y}_j(j = 1, \ldots, n_2))$ is among $\vec{X}_i \in R^N$. When searching for the closest element, the given element itself is not considered. Similarly, let θ_Y be the number of elements in the sample $\vec{Y}_j$, for which the closest in the pooled sample is one of the elements $\vec{Y}_j$.

Introduce the quantity (statistics):

$$\gamma_n = \sqrt{n}(\theta_X/n_1 + \theta_Y/n_2 - 1),$$

where $n = n_1 + n_2$ is the number of elements in the pooled sample. If we take as a null hypothesis the assumption that the samples $\vec{X}_i \in R^N (i = 1, \ldots, n_1)$ and $\vec{Y}_j \in R^N (i = 1, \ldots, n_2)$ are homogeneous, then we can calculate the mathematical expectation of the quantity γ_n and estimate the standard deviation (at $n \geq 4$):

$$M[\gamma_n] = -\sqrt{n}/(n - 1),$$

$$\sigma^2 \leq \frac{n - 2}{(n - 1)^2} + \frac{1}{(n - 1)(n - 3)}\left[n\left(1 - \frac{2}{n_1} - \frac{2}{n_2}\right) + 6\right] \equiv \overline{\sigma}^2.$$

The critical value of the statistics will be chosen [89] using the inequality of Chebyshev

$$P[|\gamma_n - M[\gamma_n]| > t] \leq \frac{\sigma^2}{t^2},$$

where, as usual, $P[*]$ is the probability of the event. Let p be the level of significance. Find the value t from the equation

$$\frac{\overline{\sigma}^2}{t^2} = 1 - p$$

and we believe

$$\gamma_{Cr} = M[\gamma_n] + \frac{\overline{\sigma}}{\sqrt{1-p}} = -\frac{\sqrt{n}}{n-1} + \frac{\overline{\sigma}}{\sqrt{1-p}}.$$

We will reject the null hypothesis of homogeneity of samples if it appears that

$$\gamma_n \geq \gamma_{Cr}.$$

Below we write $t = 0.99$. Furthermore, in our software implementation of the gamma-criterion, the distance $\rho(\vec{X}, \vec{Y})$ between the vectors X and Y is calculated by formula:

$$\rho(\vec{X}, \vec{Y}) = \sum_{k=1}^{N} |x_k - y_k| + \sum_{k=1}^{N-1} |(x_{k+1} - x_k) - (y_{k+1} - y_k)|.$$

Discuss the method of experiments (conducted in collaboration with Myshkin). The subjects were 14 people aged 17–21 years. Investigations were carried out in a shielded, darkened room. The subjects were seated in a comfortable armchair. Initially at rest, with their eyes open, monopolar electroencephalogram (EEG) during 2 min. The active electrode was placed at point O_z according to the International 10–20 System. The indifferent electrode was applied to the earlobe. After recording the background (EEG), registration of EPs was performed. The subjects were presented with four series of light pulses, respectively, in white, red, green, and yellow. To obtain different color stimuli, we used standard color filters (photic stimulator "Orion"). For equalization of stimuli according to brightness, neutral filters of different density were used (equalization was performed by taking into account the subjective assessment of brightness).

EPs were recorded with subjects in a state of passive wakefulness. The subjects asked to fix their eyes to the center of the stimulus field. The flash occurred a distance of 1.5 m from the subject's eyes. The stimuli were presented at an interval of 2 s. The break between the series of stimuli was 3–5 min. Potentials were amplified by electroencephalograph (8EEG.III (RFT)) with an upper bandwidth of 200 Hz and a time constant of 0.3 s. AEPs were converted by 12-bit analog-to-digital converter with a sampling rate of 250 Hz. The epoch analysis of EPs lasted 512 ms. For further analysis, we recorded at least 57 individual responses to each of the types of stimuli. For qualitative analysis, we used the averaging procedure of the EP. Under statistical analysis, the entire waveform of a single response, i.e., the digital series describing the configuration EP with a sampling interval of 4 ms, was used.

Thus, in discrete time, each EP is a 128-dimensional vector. Recall that the pattern is the EP segment in discrete time, which is distinguished by its start and duration. Patterns are also vectors. The dimension of the patterns we chose was equal to 16. In continuous time, each of them is the essence of the EP segment of a duration of 64 ms, which is slightly more than half of the α-rhythm that is most pronounced in the quiescent state for most individuals.

For two classes of EPs, corresponding to two different stimuli, the moments of the start of the patterns were identically fixed. Appropriate samples of patterns were compared according to the delta and gamma criteria. All possible moments of the start of patterns were enumerated.

For better understanding of methods, we immediately formulate an important result. For each subject for any pair of stimuli, the zones of the start of the patterns, where the statistical criteria indicated the heterogeneity (difference) of the respective classes of patterns, were discovered. For all subjects, zones of difference were alternated with intervals where statistical criteria revealed no differences (zone masking). The zones, allocated by the delta and gamma criteria areas, were somewhat different. Recall that the criteria are based on various geometric concepts.

9.3 The Results of Research and Their Discussion

Visual comparison of the AEPs showed that responses to identical stimuli are different from subject to subject, but there exists a significant similarity to stimuli of different colors within the individual. In Fig. 9.1, graphs of centered AEPs in response to stimuli of four colors in two subjects (A) and (B) are shown. Note that in the papers of neuroscientists, the y-axis is usually downward. In case of (A), the answers are quite similar to each other. In case (B), there are visual differences. Despite the widespread use of AEPs, the researchers note that one loses the unique information contained in the original response (Tatko 1988). Meshchersky (1986) explicitly states that EPs should be considered an averaged artifact.

As an illustration, consider Fig. 9.2. It represents an AEP, summed (averaged) in respect to all of the subjects to stimuli of all four colors. Summation of the responses for all of the subjects eliminates individual differences observed in the perception of stimuli of different colors. However, the general nature of the response is the same, which is reflected in cyclical oscillations of the amplitude of the EPs. Therefore, in our view, the assessment of differences in the AEP will not be effective. We believe that the coding of sensory information is purely individual and the differences are to be found in the original, not the averaged, responses. The experimental results below are compared using statistical methods (delta and gamma criteria), thus allowing to judge if the conditions, in which they are received, have changed.

Figure 9.3 shows the results of processing of the responses to stimuli of green and red colors by means of the criterion of separating plane (delta criterion). The numbers on the left side of the figure are the serial numbers of the subjects. The steps in the graphs are the areas of statistically significant differences of post-stimulus patterns. Below is the time scale indicating the start of the patterns. Attention is drawn to significant individual differences in the location of the zones of difference on the time axis for different subjects.

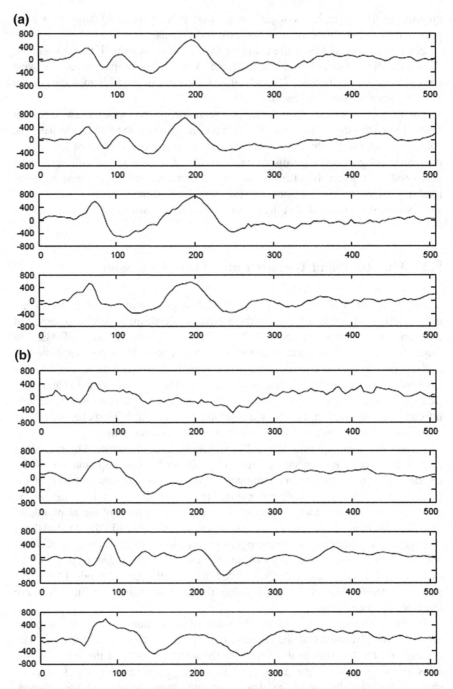

Fig. 9.1 Graphs of averaged EPs for subjects *A* and *B* to stimuli of different colors. Each graph shows the bottom-up AEPs for green, red, white, and yellow colors. In case (**a**), the responses are similar to each other, and in case (**b**) there are visual differences

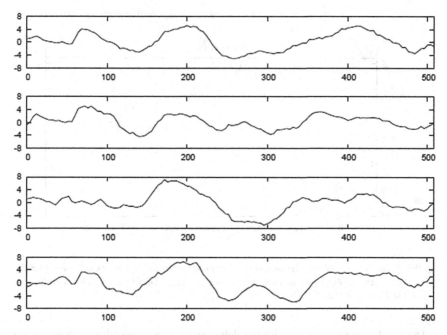

Fig. 9.2 Graphs of averaged EPs to stimuli of different colors additionally averaged over all subjects. On the graph are shown the bottom-up AEPs for green, red, white, and yellow colors

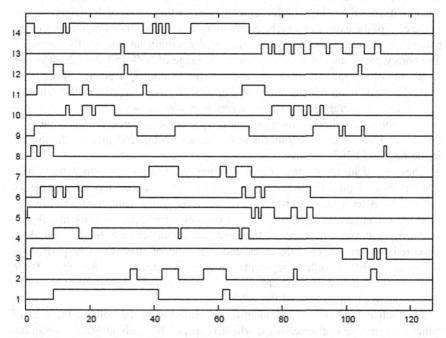

Fig. 9.3 Graph of the zones of differences of EP to stimuli of green and red colors by delta criterion for each subject. The numbers on the *left* are the serial numbers of the subjects; the steps correspond to zones of statistically significant differences; and at the *bottom* is shown the time in msec

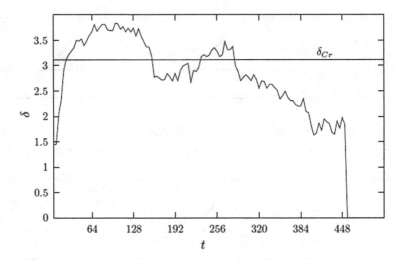

Fig. 9.4 Graph of the values averaged over all subjects with delta statistics depending on the position of the start of 60 ms poststimulus patterns for red and green colors. The *horizontal line* δ_{Cr} is the critical value of statistics ($p = 0.945$)

The results, in accordance with the delta criterion, averaged for 14 subjects are given in Fig. 9.4. The *x*-axis is the moments of the start of patterns, and the *y*-axis is the averaged values of delta statistics according to the set. A line, parallel to the *x*-axis, cuts the critical level of the value of statistics (see 9.2.1) of the delta criterion, in the case of the excess of which, the hypothesis of homogeneity of samples of the patterns is rejected. There are two statistically significant zones of differences. Temporary parameters of the zones are in the range of ≅30–150 and 200–265 ms.

Results of the analysis of these same patterns to stimuli of green and red color using the gamma criterion are shown in Fig. 9.5. The *y*-axis represents the values of gamma criteria averaged statistics in accordance with the set of subjects. Analysis of the graph yields slightly different results by zones of differences. One zone is in the range of ≅45–153 ms; another is in the range of ≅233–253 ms; and the third is in the range of ≅305–317 ms.

The use of delta and gamma criteria for comparison of poststimulus patterns with other pairs of light stimuli for all the subjects also showed repetitive zones of differences. After averaging, pattern were observed in the same time intervals.

Control comparison of EPs on any fixed color for different pairs of subjects revealed significant differences in the delta and gamma criteria. In all cases, there were reliable periods of differences and the zone of masking. Comparison of background EEG recording segments showed no significant differences. These results (1) indicate the working capacity of criteria; and (2) suggest the existence of specific areas of differences of the EPs.

For each individual, we can identify the stimulus on the pattern, the start of which is in the area of differences, which is special for each subject. Used in the delta criteria sign Π (if $\Pi > 0$, then there is one color, and if $\Pi < 0$, then there is the

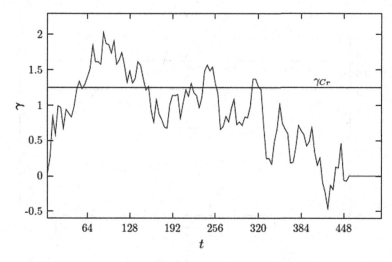

Fig. 9.5 Graph of the values averaged over all subjects with the gamma statistics depending on the position of the start of 60-ms poststimulus patterns for red and green colors. The *horizontal line* γ_{Cr} is the critical value of statistics ($p = 0.99$)

other color) allows us to properly specify the stimulus in 63 % of cases. This percentage is guaranteed by the accepted level of reliability. In our experiments, the identification of the stimulus on the pattern was correctly implemented at an average of 71 %; for some individuals, this level reached 90 %.

The obtained results allow us to consider that in response to a certain stimulus, specific (reflecting stimulus) patterns of activity occur. Proofs of specificity lies in the fact that (1) there are differences of patterns to stimuli of different colors; (2) the poststimulus patterns to identical stimuli for different subjects were varied; and (3) the poststimulus patterns differed from the background EEG. Zones of differences were individual in nature, but most often the differences between the background and the poststimulus EEG patterns were observed in the range of 160–240 ms.

When comparing the data among subjects (Fig. 9.3), we observed a significant time scatter of zones of differences in patters to stimuli of green and red colors. Researchers give quite contradictory information about the reflection in the components of the EP of the stages of signal processing by the brain. Thus, according to data (Renault et al. 1982), the identification, selection, and comparison of visual stimuli (by the criteria of EP) falls between 182 and 286 ms. Assessment of the comparability of stimuli occurs already at 22 ms with visual stimulation (Tatko 1988) and at 40 ms with sound stimulation (Kochubey et al. 1988). Such divergent results, in our opinion, are related to the specifics of the individual temporal characteristics of the EP. Our observations indicate that the temporal characteristics of the EP are purely individual.

Another result obtained in our research is that the differences in EPs occur periodically in phases. Another confirmation is shown in the graph in Fig. 9.6. It is

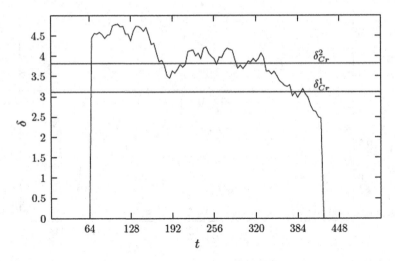

Fig. 9.6 Graph of the values of delta statistics obtained after adjusting the phase and averaged over all subjects depending on the position of the start of 64 ms of patterns for red and green colors. The *horizontal line* δ_{Cr}^1 indicates the critical value of statistics for significance level ($p = 0.945$), and the line δ_{Cr}^2 indicates the significance level ($p = 0.99$)

created by modification of the method using the criterion of the separating plane. In this case, when analyzing poststimulus patterns, we applied a technique that we call "phase adjustment." Its essence is the following. Choose A stimulus (I.E., color). Find the corresponding averaged EP. Cut from it A window of duration ranging form 64 to 400 ms. In discrete time, we obtain a 100-dimensional vector. In each original EP, find the 100-dimensional pattern, which is the least different pattern in its root-mean-square from the vector of the window. We call the pattern the "adapted EP," which indicates the area of the EP that is closest to the area selected by the window part of the averaged EP. Simultaneously, the adapted EP is a phase shift of the original EP. For each of the stimuli, find its own set of adapted EPs, to which we apply the procedure used in processing of the original EPs.

The graph given in Fig. 9.6 shows that at a level of significance $p = 0.945$, there is one large zone of differences in the range $\cong 60$–370 ms. For the level of significance $p = 0.99$, there are two such zones: the first is located at the range $\cong 60$–180 ms, and the second in the range $\cong 200$–330 ms. Note that without adjusting the phase at a significance level, $p = 0.99$ the pronounced zones of differences were absent. We can give the following explanation for this. The latent periods of different single EPs will not be the same because the excitability of reacting elements will be unique for each time point. The use of phase adjustment allows us to level the spread of latent periods of responses to individual stimuli that, as a result, are reflected in the appearance of statistically reliable zones of differences in the late components of the EP.

What is the cause of differences in the manifestations of the phase of the wave structure of the EP? These differences were reported in Tatko (1988) and explained only by a hypothesis of cyclic recurrence of processes of the reception and processing of information, the essence of which, unfortunately, is not presented in this work. We explain this phenomenon in line with the above-suggested model of the neural network. As noted previously, the external stimulus causes synchronization activity of the nerve cells. In this case, two neuronal populations interact: (1) the former neuron's own periodic activity and (2) the activity associated with the reaction to the stimulus. The interaction of these processes provides a cyclic activation, which is reflected in the oscillation of the EP. This pattern reflects the specific activity of the stimulus. The degree of synchronization of the elements as one progressed through cycles declines. The process exists in the system for a while and then gradually fades away.

Thus, experiments have shown the periodic presence of zones of differences in the compared classes of EPs. In contrast, the data (Gnezditsky 1990) testify in favor of nonspecificity of the EP averaged over space and time. From the standpoint of the hypothesis (Livanov 1965) on the chain of neural processes, the EP can indeed be regarded as a variable (and therefore not specific) reaction of the brain characterized by the periodic activation of different neuronal populations in response to external stimulation and their excitability associated with periodic changes.

However, there are data on the specificity of waveforms of response in different zones (Lehmann and Skrandies 1979). Thus, in the EP, there are common nonspecific components, which are reflected in the oscillations of potentials, and there are its own specific differences observed in both the EP recording from different zones of the brain and the EP recording in one zone to stimuli of different modalities. Consequently, there are at least two reasons for the specificity of the EP: (1) the specificity of the stimulus; and (2) the specific response of neurons in different brain zones.

The above-mentioned data suggest that changes occur in both early and late components of the EP. The process is periodic as is evidenced by cyclically emerging zones of differences in the EP.

References

Cowan, N. (1988). Evolving conceptions of memory storage, selective attention, and their mutual constraints within human information-processing system. *Psychological Bulletin, 104*(2), 163.

Dustman, R. E., & Beck, E. C. (1965). Phase of alpha brain wave, reaction time and visually evoked potentials. *Electroencephalography and Clinical Neurophysiology, 18*(5), 433–440.

Gnezditsky, V. V. (1990). Isolation of the evoked potential to single stimulus—method of spatial synchronous averaging. *Human Physiology, 16*(3), 120–126.

Grigorieva L.P. (1987). On some regularities of systemic organization of visual perception in normal and impaired vision. In *Neurophysiological determinants of processes of human information processing* (pp. 220–238). Moscow: Academy of Sciences of the USSR, Institute of Psychology.

Ivanitsky, A. M., Strelets, V. B., & Korsakov, I. A. (1984). *Information processes of the brain and psychic activity*. Moscow: Nauka.

Kanunnikov I.E., & Vetosheva V.I (1988) Modern views on the importance of psychophysiological significance of P_{300}. *Human Physiology, 14*(2), 314.

Kochubey B. I., Snezhevsky P. B., & Rutman E. M. (1988). Very early changes of the potential associated with voluntary attention: The chance discovery of fact. In *Psychophysiology of Cognitive Processes* (pp. 66–71). Moscow: Academy of Sciences of the USSR, Institute of Psychology.

Kostandov, E. A. (1988). *Conscious and unconscious forms of higher nervous activity in Mechanisms of the human brain activity*. Leningrad: Nauka.

Lebedev A. N. (1985). Encoding information in memory by coherent waves of neural activity. In *Psychophysiological patterns of perception and memory* (pp. 6–33), Moscow: Nauka.

Lebedev, A. N. (1990). Human memory, its mechanisms and limits. *Research of memory* (pp. 104–118). Moscow: Nauka.

Lehmann, D., & Skrandies, W. (1979). Multichannel mapping of spatial distributions of scalp potential fieds evoked by checkerboard reversal to different retinal areas. In D. Lehmann & E. Callaway (Eds.), *Human Evoked Potentials*. New York: Plenum Press.

Levin A. Y. (1993). On a consistent multidimensional nonparametric criterion of homogeneity. *Russian Mathematical Surveys, 48*, 6(294), 155–156.

Livanov, M. N. (1965). *Neuro kinetics in problems of modern neurophysiology* (pp. 37–62). Leningrad: Nauka.

Lytarev, S. A., & Shostak, V. I. (1994). Classification of images according to mapping of visual evoked potentials. *Human Physiology, 20*(1), 13–21.

Mayorov V. V., & Myshkin I. Y. (1995). *Criterion of quantitative assessment of differences of induced brain potentials*. Moscow: Russian Academy of Sciences, Institute of Scientific Information on Social Sciences, 50 504, 15.6: 21.

Meshchersky, R. M. (1986). Psychophysiological isomorphism: Reality or illusion? *Problems of Psychology, 4*, 82–90.

Myshkin, I. Y., & Mayorov, V. V. (1996). An assessment of differences of evoked cortical potentials on visual stimuli in humans. *ZHVND, 46*(3), 488–495.

Naatanen, R., Alho, K., & Soames, M. (1986). *Brain mechanisms of selective attention in cognitive psychology* (pp. 85–92). Moscow: Nauka.

Novikova, L. A., Grigorieva, L. P., Tolstova, V. A., & Fursova, V. A. (1979). The use of evoked potentials for the study of visual perception. *Human physiology, 5*(3), 527–534.

Renault, B., Lesevre, N., Remond, A., & Ragot, R. (1982). Onset and offset of brain events as indices of mental chronometry. *Science, 215*(4538), 1413–1415.

Schwartz, M. (1976). Averaged evoked responses and the encoding of perception. *Psychophysiology, 13*(6), 546–553.

Shilnikov, L. P., Shilnikov, A. L., Turaev, D. V., & Chua, L. (2004). *Methods of qualitative theory in nonlinear dynamics, Part I*. Moscow-Izhevsk: Institute of Computing Research.

Tatko V. L. (1988). Chronology of stages of visual information processing: Evoked potentials of the human brain. In *Physiological Aspects of the Activity of Human Operator* (Vol. 35, pp. 49–61). Moscow: Proceedings of All-Russian Research Institute of Technical Aesthetics. Series Ergonomics.

Appendix A
Differential-Difference Equations with Delay

Consider differential equation

$$\dot{u} = f(u(t), u(t - h_0), \ldots, u(t - h_m)). \qquad (A.0.1)$$

Here $u \in R^n$ and holds inequality $0 < h_0 < h_1 < h_2 < \cdots < h_m = h$. Equation (A.0.1) is called the "differential-difference equation with delay." Equations of this type are common in applications; in particular, we used them in modeling neurons. These equations are the object of serious mathematical research.

For Eq. (A.0.1), write the initial Cauchy problem (see, e.g., Èl'sgol'ts 1971, Myshkis 1972). Denote by $C_n[-h, 0]$ the continuous space on the interval $[-h, 0]$ functions with values R^n in Let t_0 be the initial moment of time, and $u_0(s) \in C_n[-h, 0]$ The Cauchy problem for Eq. (A.0.1) is to find the solution for $t > t_0$ its solution $u(t)$ such that $u(t_0 + s) \equiv u_0(s)$ for $s \in [-h, 0]$.

For simplicity, we assume that in Eq. (A.0.1) that the function f is continuously differentiable in all its arguments. Let $t \in (t_0, t_0 + h_0)$. Then the values of arguments are $t - h_0, \ldots, t - h_m < t_0$, and the function $u(t)$ is known—$u(t - h_j) = u_0 (t - h_j)$—for $j = 0, \ldots, m$. On this interval, Eq. (A.0.1) goes into the ordinary differential equation

$$\dot{u} = f(u(t), u_0(t - h_0), \ldots, u_0(t - h_m))$$

with initial condition for the solution $u(t_0) = u_0(0)$. By virtue of the assumptions made about some interval, the ordinary differential equation is uniquely solvable. It may be that over time solutions less than h_0, the solution goes to infinity. Otherwise, in the case favorable to us, the obtained solution is determined at the point $t = h_0$, which allows holding the next step. In this case, we shift the initial moment to the point $t_0 + h_0$, and as the new initial function we take $u(t_0 + h_0 + s)$. Continuing this process, we obtain the solution defined on the whole temporal semiaxis $t > t_0$ unless, of course, at some stage the solution will not go to infinity. Thus, the Cauchy problem for any initial condition is solvable either on the entire semiaxis $t > t_0$, or on some finite time interval.

© Springer International Publishing Switzerland 2015
S. Kashchenko, *Models of Wave Memory*,
Lecture Notes in Morphogenesis, DOI 10.1007/978-3-319-19866-8

Note that at the point $t = t_0$, the left derivative of the solution $u(t)$ may not exist. It is easy to see that at the point $t = t_0 + h$ (unless, of course, the solution does not go to infinity), the derivative of the solution exists and is continuous. In general, if the solution can be extended to the point $t = t_0 + kh$ (here $k \geq 1$ is an integer), then at this point exists and is continuous in the kth derivative of the solution. Thus, the smoothness of the solution of the differential-difference equation with delay increases with time.

The above-stated simple algorithm for constructing a solution for the differential-difference equation is called "the method of steps."

Example 1 (see Èl'sgol'ts 1971): Let us solve the equation

$$\dot{u}(t) = 6u(t - 1)$$

for the initial value $t_0 = 1$ and the initial function $u_0(s) = 1 + s$ defined on the interval $-1 \leq s \leq 0$.
Considering the interval $t \in [1, 2]$, obtain the equation

$$\dot{u}(t) = 6(t - 1), \ u(1) = u_0(0) = 1.$$

Its solution takes the form:

$$u(t) = 3(t - 1)^2 + 1.$$

Let us now consider the interval $t \in [2, 3]$. We have:

$$\dot{u}(t) = 6\left(3(t - 2)^2 + 1\right) \equiv 18(t - 2)^2 + 6, \quad u(2) = 4.$$

We obtain the solution:

$$u(t) = 6(t - 2)^3 + 6t - 8.$$

Thus, we obtain:

$$u(t) = \begin{cases} 3(t - 1)^2 + 1, & t \in [1, 2]; \\ 6(t - 2)^3 + 6t - 8, & t \in [2, 3]. \end{cases}$$

It is clear that the process of constructing a solution can be continued indefinitely.

The phenomenon of nonextendability of solutions of differential equations with delay is quite common.

Example 2 Consider the equation

$$\dot{u} = u^2 + u(t-1),$$

for the initial value $t_0 = 0$ and the initial function $u_0(s) = 1$ defined on the interval $-1 \le s \le 0$.

For $t \in [0, 1]$, the equation takes the form

$$\dot{u} = u^2 + 1, \quad u(0) = 1.$$

Its solution is $u(t) = tg(t + \pi/4) \to \infty$ at $t \to \pi/4 < 1$. Thus, the solution for a finite time goes to infinity.

In the next section, application of the method of steps will be demonstrated by more interesting example of specification of the neuron's oscillation period.

References

Èl'sgol'ts, L. E., & Norkin S. B. (1971). *Introduction to the theory of differential equations with deviating argument*. Moscow: Nauka.

Myshkis, A. D. (1972). *Linear differential equations with delayed argument*. Moscow: Nauka.

Appendix B
Amendment to the Period of the Solution of the Neuron

In Theorem 1.6.1, we give an estimate of zero approximation for the period of the solution of Eq. (1.5.7) of the neuron. For the convenience of the reader, we write this equation once more:

$$\dot{u} = \lambda[-1 - f_{Na}(u) + f_K(u(t-1))]u. \tag{B.0.1}$$

As in Theorem 1.6.1, it is assumed that positive sufficiently smooth functions $f_{Na}(u)$ and $f_K(u)$ tend monotonically to zero as $u \to \infty$ faster than $O(u^{-1})$, i.e., hold the following relations:

$$f_{Na}(u) \le Cu^{-(1+\varepsilon)}, \tag{B.0.2}$$

$$f_K(u) \le Cu^{-(1+\varepsilon)}, \tag{B.0.3}$$

at some $\varepsilon > 0$. We also recall the notation:

$$\alpha = f_K(0) - f_{Na}(0) - 1, \tag{B.0.4}$$

$$\alpha_1 = f_K(0) - 1, \tag{B.0.5}$$

$$\alpha_2 = f_{Na}(0) + 1. \tag{B.0.6}$$

According to Theorem 1.6.1, assume that these numbers are positive. We use the method of steps to clarify the period of the solution of Eq. (B.0.1) (see Mayorov et al. 2003).

Introduce the class S_λ of initial functions $\varphi(s)$, which are continuous on the interval $s \in [-1, 0]$ and satisfy the condition:

$$\lambda^{-n} \exp(2\lambda\alpha s) \le \varphi(s) \le \lambda^{-n} \exp(\lambda\alpha s/2), \tag{B.0.7}$$

where $n \ge 1$ is a fixed integer. The initial conditions of the periodic solutions belong to the class S_λ (we will not introduce a special notation for the periodic solution and its initial condition).

© Springer International Publishing Switzerland 2015
S. Kashchenko, *Models of Wave Memory*,
Lecture Notes in Morphogenesis, DOI 10.1007/978-3-319-19866-8

Theorem B.0.1 *Under the above-mentioned conditions for the period of **T** of the solution $u(t)$ of Eq. (B.0.1), $\lambda \to \infty$ is a valid asymptotic representation*

$$T = T_0 + \Delta T + o(\lambda^{-n}), \tag{B.0.8}$$

where, in accordance with Theorem 1.6.1, $T_0 = \alpha_1 + 2 + \alpha_2/\alpha$, and

$$\Delta T = \frac{1}{\lambda} \int\limits_0^\infty \left[\frac{f_K(u) - \alpha_1}{\alpha_1 - f_{Na}(u)} + \frac{1}{\alpha} \times \frac{\alpha - f_K(u)}{1 + f_{Na}(u)} \right] \frac{du}{u} < 0. \tag{B.0.9}$$

Proof Divide the interval $t \in [0, T]$ into a number of subintervals. A graph of the periodic solution of Eq. (B.0.1) and the corresponding partition are shown in Fig. B.1. The very form of the solution corresponds to Theorem 1.6.1. We compute the values t_1, t_2, t_3 and T.

1. Let $t \in [0, t_1]$. Calculations show that $t_1 = O(n \ln \lambda/\lambda)$. Given that $u(s) = \varphi(s)$ for $s \in [-1, 0]$, Eq. (B.0.1) takes the form:

$$\dot{u} = \lambda[-1 - f_{Na}(u) + f_K(\varphi(t-1))]u. \tag{B.0.10}$$

Here $t - 1 \in [-1, 0]$. Using the definition of the initial class of functions S_λ, we obtain:

$$f_K(\varphi(t-1)) = f_K(0) + O(\varphi(t-1)) = f_K(0) + o(\lambda^{-n}). \tag{B.0.11}$$

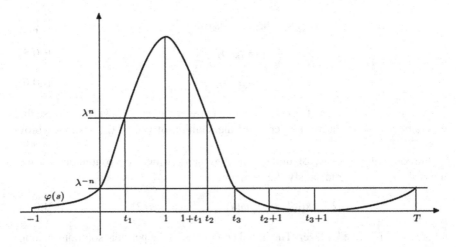

Fig. B.1 Schematic graph of the periodic solution of the equation of the neuron and the partition of the period into subintervals

Moreover, in view of (B.0.7), the term $o(\lambda^{-n})$ is exponentially small in λ. Neglecting this term, we obtain equation

$$\dot{u} = \lambda[\alpha_1 - f_{\text{Na}}(u)]u. \tag{B.0.12}$$

Here we have used the notation (B.0.5). Due to the fact that $u(0) = \lambda^{-n}$ and $u(t_1) = \lambda^n$, from Eq. (B.0.12) it immediately follows that

$$\frac{1}{\lambda} \int_{\lambda^{-n}}^{\lambda^n} \left[\frac{1}{\alpha_1 - f_{\text{Na}}(u)}\right] \frac{du}{u} = t_1.$$

In the integral, we derive the singular terms in the standard way. We obtain

$$t_1 = \frac{1}{\lambda} \left[\frac{n \ln \lambda}{\alpha} + \int_{\lambda^{-n}}^{1} \left[\frac{1}{\alpha_1 - f_{\text{Na}}(u)} - \frac{1}{\alpha}\right] \frac{du}{u}\right]$$

$$+ \frac{1}{\lambda} \left[\frac{n \ln \lambda}{\alpha_1} + \int_{1}^{\lambda^n} \left[\frac{1}{\alpha_1 - f_{\text{Na}}(u)} - \frac{1}{\alpha_1}\right] \frac{du}{u}\right].$$

Both integrals on the right side converge as $\lambda \to \infty$. For the first of them, by virtue of notation (B.0.4) and (B.0.5) for α and α_1, respectively, we have asymptotic equality

$$\int_{\lambda^{-n}}^{1} \left[\frac{1}{\alpha_1 - f_{\text{Na}}(u)} - \frac{1}{\alpha}\right] \frac{du}{u} = \frac{1}{\alpha} \int_{\lambda^{-n}}^{1} \left[\frac{f_{\text{Na}}(u) - f_{\text{Na}}(0)}{\alpha_1 - f_{\text{Na}}(u)}\right] \frac{du}{u}$$

$$= \frac{1}{\alpha} \int_{0}^{1} \left[\frac{f_{\text{Na}}(u) - f_{\text{Na}}(0)}{\alpha_1 - f_{\text{Na}}(u)}\right] \frac{du}{u} + O(\lambda^{-n}).$$

For the second integral we obtain

$$\int_{1}^{\lambda^n} \left[\frac{1}{\alpha_1 - f_{\text{Na}}(u)} - \frac{1}{\alpha_1}\right] \frac{du}{u} = \frac{1}{\alpha_1} \int_{1}^{\infty} \left[\frac{f_{\text{Na}}(u)}{\alpha_1 - f_{\text{Na}}(u)}\right] \frac{du}{u} + o(\lambda^{-n}).$$

Here we use the notation (B.0.2) for $f_{\text{Na}}(u)$. For further discussion, the expression for t_1 can be conveniently rewritten as:

$$t_1 = \frac{n \ln \lambda}{\lambda} \left(\frac{1}{\alpha} + \frac{1}{\alpha_1} \right) + \frac{1}{\lambda} \int\limits_0^1 \left[\frac{1}{\alpha_1 - f_{Na}(u)} - \frac{1}{\alpha} \right] \frac{du}{u}$$

$$+ \frac{1}{\lambda} \int\limits_1^\infty \left[\frac{1}{\alpha_1 - f_{Na}(u)} - \frac{1}{\alpha_1} \right] \frac{du}{u} + O(\lambda^{-n}). \tag{B.0.13}$$

In the resulting expression no simplification is held because it, oddly enough, complicates the subsequent computations.

2. Let $t \in [t_1, 1]$. On this interval $u(t) > \lambda^n$. The original equation (B.0.1) still has the form (B.0.10) $(t - 1 \in [-1, 0], u(t - 1) = \varphi(t - 1))$ and valid is the formula (B.0.11) (as $t \to 1$ the term $o(\lambda^{-n})$ is replaced by $O(\lambda^{-n})$). Additionally, from the estimate (B.0.2) for $f_{Na}(u)$, we have $f_{Na}(u) = o(\lambda^{-n})$.

Neglecting in Eq. (B.0.10) terms $O(\lambda^{-n})$, we obtain the simple equation

$$\dot{u} = \lambda[f_K(0) - 1]u, \quad u(t_1) = \lambda^n.$$

If we consider Eq. (B.0.5) for α_1, then for its solution at $t = 1$ we have

$$u(1) = \lambda^n \exp[\lambda \alpha_1 (1 - t_1)]. \tag{B.0.14}$$

3. Consider the interval $t \in 1, 1 + t_1$. On it, by virtue of estimate, (B.0.2) and formula (B.0.14) $f_{Na}(u) < \exp(-\lambda \alpha_1)$. This term in Eq. (B.0.1) can be neglected, and the equation takes the form

$$\dot{u} = \lambda[-1 + f_K(u(t - 1))]u.$$

From the last equation with the initial condition (B.0.14), we obtain

$$u(1 + t_1) = \lambda^n \exp \left[-\lambda t_1 + \lambda \int\limits_1^{1+t_1} f_K(u(s - 1)) ds + \lambda \alpha_1 (1 - t_1) \right]. \tag{B.0.15}$$

We transform the integral in the formula:

$$I = \lambda \int\limits_1^{1+t_1} f_K(u(s - 1)) ds$$

$$= \lambda \int\limits_0^{t_1} f_K(u(s)) ds = \lambda \int\limits_0^{t_1} \frac{f_K(u(s))}{\lambda[-1 - f_{Na}(u(s)) + f_K(u(s - 1))]u} \dot{u}(s) ds. \tag{B.0.16}$$

In the denominator of the fraction $f_K(u(s - 1)) = f_K(\varphi(s - 1)) = f_K(0) + o(\lambda^{-n})$. Term $o(\lambda^{-n})$ will be neglected (leaving a mark of exact equality). Because $u(0) = \lambda^{-n}$, and $u(t_1) = \lambda^n$, we obtain:

$$I = \int_{\lambda^{-n}}^{\lambda^n} \left[\frac{f_K(u)}{\alpha_1 - f_{Na}(u)} \right] \frac{du}{u}.$$

Proceeding as in the first step, we single out the singular component, $I = I_1 + I_2$, where

$$I_1 = \frac{n f_K(0)}{\alpha} \ln \lambda + \int_0^1 \left[\frac{f_K(u)}{\alpha_1 - f_{Na}(u)} - \frac{f_K(0)}{\alpha} \right] \frac{du}{u} + O(\lambda^{-n}). \tag{B.0.17}$$

$$I_2 = \int_1^\infty \left[\frac{f_K(u)}{\alpha_1 - f_{Na}(u)} \right] \frac{du}{u} + o(\lambda^{-n}). \tag{B.0.18}$$

Formula (B.0.16) takes the form:

$$u(1 + t_1) = \lambda^n \exp[-\lambda t_1 + I_1 + I_2 + \lambda \alpha_1(1 - t_1)]. \tag{B.0.19}$$

4. Let now $t \in [1 + t_1, t_2]$ (value $u(1 + t_1)$ be given by formula (B.0.19), and $u(t_2) = \lambda^n$). On this interval, the values $u(t)$ and $u(t - 1)$ are exponentially large in λ, and therefore the values of the function $f_{Na}(u(t))$ and $f_K(u(t - 1))$, by virtue of estimates (B.0.2) and (B.0.3), are exponentially small. Neglect these small values in (B.0.1) and obtain a quite simple equation:

$$\dot{u} = -\lambda u,$$

from which, by virtue of initial condition (B.0.19), it follows that

$$u(t_2) = \lambda^n \exp[-\lambda(t_2 - 1) + I_1 + I_2 + \lambda \alpha_1(1 - t_1)] = \lambda^n.$$

Moment of time t_2 is easily found:

$$t_2 = 1 + \lambda^{-1}(I_1 + I_2) + \alpha_1(1 - t_1). \tag{B.0.20}$$

5. On the interval $t \in [t_2, t_3]$ (recall that in $u(t_3) = \lambda^n$), function $f_K(u(t - 1))$ is exponentially small in λ. Neglecting this term in (B.0.1), we arrive at equation

$$\dot{u} = -\lambda[1 + f_{Na}(u)]u.$$

Proceeding as in the analysis of Eq. (B.0.12), we obtain the integral representation:

$$t_3 - t_2 = \frac{1}{\lambda} \int_{\lambda^{-n}}^{\lambda^n} \left[\frac{1}{1 + f_{Na}(u)} \right] \frac{du}{u},$$

where we use $u(t_2) = \lambda^n, u(t_3) = \lambda^{-n}$. As in the first step, choose the integral singular component: $t_3 - t_2 = J_1 + J_2$, where

$$J_1 = \frac{n \ln \lambda}{\lambda \alpha_2} + \frac{1}{\lambda} \int_0^1 \left[\frac{1}{1 + f_{Na}(u)} - \frac{1}{\alpha_2} \right] \frac{du}{u} + O\left(\lambda^{-(n+1)} \right). \qquad (B.0.21)$$

$$J_2 = \frac{n \ln \lambda}{\lambda} + \frac{1}{\lambda} \int_1^\infty \left[\frac{1}{1 + f_{Na}(u)} - 1 \right] \frac{du}{u} + o\left(\lambda^{-(n+1)} \right). \qquad (B.0.22)$$

Thus, moment of time t_3 is found, and by (B.0.20)

$$t_3 = 1 + \lambda^{-1}(I_1 + I_2) + \alpha_1(1 - t_1) + J_1 + J_2. \qquad (B.0.23)$$

6. Consider interval of time $t \in [t_3, t_2 + 1]$. Values $f_{Na}(u)$ and $f_{Na}(0)$ differ by no more than the amount $O(\lambda^{-n})$. Therefore, we replace in (B.0.1) $f_{Na}(u)$ by $f_{Na}(0)$. Furthermore, we note that $u(t - 1) > \lambda^n$, and therefore the values of the function $f_K(u(t - 1))$ are exponentially small. Equation (B.0.1) becomes very simple:

$$\dot{u} = -\lambda \alpha_2 u,$$

where we used formula (B.0.6) for α_2. Because $u(t_3) = \lambda^{-n}$, we obtain

$$u(t_2 + 1) = \lambda^{-n} \exp[-\lambda \alpha_2(t_2 + 1 - t_3)]. \qquad (B.0.24)$$

7. For the interval of time $t \in [t_2 + 1, t_3 + 1]$, the value $u(t)$ is exponentially small in λ and holds equality $f_{Na}(u) = f_{Na}(0) + o(\lambda^{-n})$. Discarding (B.0.1) in small terms, we obtain the following equation:

$$\dot{u} = \lambda[-\alpha_2 + f_K(u(t - 1))]u.$$

(Here we use formula (B.0.6).) The latter equation is analogous to the equation analyzed in the third step. We obtain

$$u(t_3 + 1) = \exp\left[-\lambda \alpha_2(t_3 - t_2) + \lambda \int_{t_2 + 1}^{t_3 + 1} f_K(u(s - 1))ds \right] u(t_2 + 1).$$

By virtue of (B.0.24)

$$u(t_3 + 1) = \lambda^{-n} \exp\left[-\lambda \alpha_2 + \lambda \int_{t_2 + 1}^{t_3 + 1} f_K(u(s - 1))ds \right]. \qquad (B.0.25)$$

Proceeding as stated previously, we obtain the asymptotic representation for the integral:

$$\lambda \int_{t_2+1}^{t_3+1} f_K(u(s-1))\mathrm{d}s = K_1 + K_2,$$

where

$$K_1 = \frac{nf_K(0)}{\alpha_2}\ln \lambda + \int_0^1 \left[\frac{f_K(u)}{1+f_{Na}(u)} - \frac{f_K(0)}{\alpha_2}\right]\frac{\mathrm{d}u}{u} + O(\lambda^{-n}), \qquad (B.0.26)$$

$$K_2 = \int_1^\infty \frac{f_K(u)}{1+f_{Na}(u)}\frac{\mathrm{d}u}{u} + o(\lambda^{-n}), \qquad (B.0.27)$$

As a result, (B.0.25) takes the form:

$$u(t_3+1) = \lambda^{-n}\exp[-\lambda\alpha_2 + K_1 + K_2]. \qquad (B.0.28)$$

8. On the interval of time $t \in [t_3+1, T]$ (T is the period, i.e., $u(T) = \lambda^{-n}$) values of functions $f_{Na}(u)$ and $f_K(u)$ are asymptotically close to $f_{Na}(0)$ and $f_K(0)$ (the difference is not more than $o(\lambda^{-n})$). Equation (B.0.1) takes the form:

$$\dot{u} = \lambda\alpha u,$$

from which it follows that

$$u(T) = \exp[\lambda\alpha(T-(t_3+1))]u(t_3+1) = \lambda^{-n}.$$

From the latter equality and (B.0.28) we obtain

$$\lambda\alpha T - \lambda\alpha(t_3+1) - \lambda\alpha_2 + K_1 + K_2 = 0.$$

Find the period T using formula (B.0.23) for t_3:

$$T = 2 + \alpha_1 + \frac{\alpha_2}{\alpha} + \frac{1}{\alpha}(I_1 + I_2) - \alpha_1 t_1 + J_1 + J_2 - \frac{1}{\lambda\alpha}(K_1 + K_2).$$

The term, independent of λ, is the value T_0 from Theorem B.0.1. Substituting into the latter formula the values $I_1, I_2, t_1, J_1, J_2, K_1, K_2$ in accordance with formulas (B.0.13), (B.0.17), (B.0.18), (B.0.21), (B.0.22), (B.0.26), and (B.0.27) and collecting similar terms in $\ln \lambda/\lambda$, we see that their sum is zero. After collecting similar terms, the sums of integrals are reduced to the form (B.0.9) for ΔT, i.e., the main part of the correction to the period. Here untransformed integrands are used, to which we paid attention in formula (B.0.13).

We show that the obtained correction to the period of oscillations of the membrane potential of the neuron is negative. Give the integrand (B.0.9) for ΔT to the common denominator. The numerator of the fraction A can be written as

$$A = f_K(u)[\alpha - \alpha_1 + f_{Na}(u)] + \alpha f_{Na}(u)[f_K(u) - \alpha_1 - 1].$$

We have $\alpha = -1 - f_{Na}(0) + f_K(0) = \alpha_1 - f_{Na}(0)$. Hence $f_{Na}(0) = \alpha_1 - \alpha$. Furthermore, from (B.0.5), it follows that $f_K(0) = \alpha_1 + 1$. Using the monotone decrease of functions $f_{Na}(u)$ and $f_K(u)$, we estimate the value A:

$$A < f_K(u)[\alpha - \alpha_1 + f_{Na}(0)] + \alpha f_{Na}(u)[f_K(0) - \alpha_1 - 1] = 0.$$

Thus, the theorem is proved.

Reference

Mayorov, V. V., Myachin, M. L., & Paramonov, I. A. (2003). On the amendment to the period of solution of equation with delay, describing the dynamics of a neuron. In *Proceedings of All-Russian Scientific Conference dedicated to the 200th anniversary of Yaroslavl State University named after P.G. Demidov*, (pp. 45–46). Yaroslavl.

Appendix C
Model of the Saltatory Conduction of the Nerve Pulse

In this section, we discuss the model of the conduction of excitation along the myelinated nerve fiber (see Anufrienko et al. 2004, 2007). This model describes the spread and interaction of excitation waves in a chain of nodes of Ranvier. The system of equations is analyzed analytically; we also show the results of numerical modeling.

C.1 The Notion of Saltatory Conduction

In classical neurophysiology, the propagation of nerve impulses was often compared to the burning of gunpowder. However, although the individual properties of the nerve fiber [the law "all or nothing" (i.e., refractoriness)] are really reproduced by the power model, a much better idea of the passage of the pulse is evident when an iron wire is placed in nitric acid.

In one of the experiments, the iron wire was covered with glass tubes located at some distance from each other. When the wire was immersed in concentrated nitric or sulfuric acid, a wave of chemical reaction propagated along the wire from one open area to another, thus "jumping" over the areas covered with tubes. Such a character of propagation of chemical wave along the iron wire is called "saltatory" conduction ("saltare" means "to jump").

There are myelinated and unmyelinated nerve fibers. Myelinated fibers are coated with a lipid layer called the "myelin sheath." This sheath is not solid: At a distance of 1.5–2 mm from each other along the entire length of the axon are the so-called "nodes of Ranvier." In the area of the nodes, there is no myelin sheath, and the length of each node is equal to 0.5–2.5 μm (Schade and Ford 1976). A schematic illustration of the nerve fiber is shown in Fig. C.1.

The myelin sheath has high electrical resistance; that is why the pulses propagate abruptly from node to node along the axon. The role of the myelin sheath and its gaps (the nodes of Ranvier) in the electrical excitation of nerve fiber has been demonstrated experimentally. Experiments showed that the myelin sheath is a good insulator and that a response occurs only in the nodes of Ranvier under the action of excitatory current. If through the node we pass an outward current that exceeds the threshold

© Springer International Publishing Switzerland 2015
S. Kashchenko, *Models of Wave Memory*,
Lecture Notes in Morphogenesis, DOI 10.1007/978-3-319-19866-8

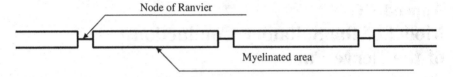

Fig. C.1 Schematic illustration of the nerve fiber

value (approximately 4×10^{-10} A), then the potential of the membrane of the node changes almost according to the law of "all or nothing." As a result the action current appears; it is approximately five times higher than the threshold current. It flows outside through the neighboring node and excites it. Multiple repetition of this process ensures the propagation of the nerve impulse along the fiber (Tasaki 1971).

Saltatory conduct contributes to a significant increase of the speed of impulse conduction because for its conduction, only depolarization in the area of the nodes is required. The speed of the conduction along myelinated fibers is 20–25 times higher than that along unmyelinated fibers of the same diameter (Schade and Ford 1976).

C.2 Model of Threshold Neuron

Recall that the neuron as detector generates pulses only as a result of a sufficiently strong action. For the description of the neuron as detector in the right side of Eq. (1.5.1), we should add the term responsible for the leakage current: $I_1 = g_1(u_1 - u)$, where $u_1 > 0$ is the equilibrium potential for the ions involved in the leakage, and g_1 is the coefficient of conductivity. As a result, for the membrane potential $u(t) > 0$, we obtain similar to Eq. (1.5.7):

$$\dot{u} = \lambda[-1 - f_{Na}(u) + f_K(u(t-1))]u + \varepsilon. \tag{C.2.1}$$

Here parameter $\lambda \gg 1$ reflects a high rate of flow of electric processes, and parameter $0 < \varepsilon \ll 1$ takes into account leakage of currents passing through the membrane of the threshold neuron. Positive sufficiently smooth functions $f_{Na}(u)$ and $f_K(u)$ decrease monotonically to zero as $u \to \infty$ faster than $O(u^{-1})$. They describe the state of the sodium and potassium channels. We introduce the parameters:

$$\alpha = 1 + f_{Na}(0) - f_K(0) > 0, \tag{C.2.2}$$

$$\alpha_1 = f_K(0) - 1 > 1, \tag{C.2.3}$$

$$\alpha_2 = f_{Na}(0) + 1 > \alpha_1, \tag{C.2.4}$$

Parameters α_1 and α_2 are identical to the parameters previously introduced (B.0.5) and (B.0.6), and the number α has the opposite sign compared to the number α familiar to us from formula (B.0.4).

Consider that

$$f_K(0) - f_{Na}(1) - 1 > 0. \tag{C.2.5}$$

This number is important and is related to a threshold. Associate the spike of the neuron with moment of time t_s, such that $u(t_s) = 1$ and $u(t) < 1$ at $t_s - 1 < t < t_s$. Inequality (C.2.5) ensures that at this moment generation of the spike begins. Here Eq. (C.2.1) has a positive state of equilibrium:

$$u = u_* \approx \frac{\varepsilon}{\lambda \alpha} \ll 1.$$

Consider the question of its sustainability. Linearizing equation (C.2.1) in the neighborhood u_*, we obtain to within $o(1)$:

$$\dot{u} = -\lambda\alpha(u - u_*) + \frac{\varepsilon}{\alpha}\left[-f'_{Na}(u_*)(u - u_*) + f'_K(u_*)(u(t-1) - u_*)\right].$$

We write out the characteristic quasipolynomial:

$$\mu = -\lambda\alpha + \frac{\varepsilon}{\alpha}\left[-f'_{Na}(u_*) + f'_K(u_*)\exp(-\mu)\right].$$

Single out the real part:

$$\mathrm{Re}\mu = -\lambda\alpha + \frac{\varepsilon}{\alpha}\left[-f'_{Na}(u_*) + f'_K(u_*)\exp(-\mathrm{Re}\mu)\cos(\mathrm{Im}\mu)\right].$$

Taking into account that $\lambda \gg 1$, and $\varepsilon \ll 1$, we determine that all the roots of the characteristic quasipolynomial are located on the complex plane to the left of the imaginary axis. Consequently, the equilibrium state $u = u_*$ is exponentially stable.

Analyze Eq. (C.2.1) as $\lambda \to \infty$. To model the action on the neuron, we will use a special set of the initial conditions:

$$u(s) = \varphi(s), \quad s \in [-1, 0].$$

The class of initial functions consists of continuous on the interval $s \in [-1, 0]$ functions $\varphi(s)$, thus satisfying the conditions:

$$\varphi(0) = 1, \quad 0 \le \varphi(s) \le \max(\exp(\lambda\alpha s/2), \lambda^{-1})$$

Denote this class of functions by S. Associate conventionally the start and the end of the spike with moments of time when $u(t)$ crosses the unit value, respectively, with positive and negative speed.

Using step-by-step asymptotic integration method, we obtain formulas describing the dynamics of the membrane potential of the threshold neuron:

$$u(t) = \begin{cases} \exp[\lambda \alpha_1 (t + o(1))], & t \in [\delta, 1 - \delta], \\ \exp[\lambda (\alpha_1 - (t-1) + o(1))], & t \in [1 + \delta, 1 + \alpha_1 - \delta], \\ (\varepsilon + o(1))/\lambda \alpha_2, & t \in [1 + \alpha_1 + \delta, 2 + \alpha_1 - \delta], \\ (\varepsilon + o(1))/\lambda \alpha_2, & t > 2 + \alpha_1 + \delta. \end{cases} \quad \text{(C.2.6)}$$

Here $0 < \delta \ll 1$ is an arbitrary small fixed number. From these formulas it follows that at $t \geq \alpha_1 + 2 + \delta$, the approximate equality $u \approx u_*$ holds as long as the neuron does not generate a new spike, which can only be caused by an external signal.

C.3   Point Model of Saltatory Conduction of Excitation

The myelinated area of a nerve fiber represents a distributed system described in this case by a parabolic differential equation in partial derivatives. At the same time, the most important features of the dynamics of the spread of excitation can be obtained on a simplified model, in which the myelinated nerve fiber area is a point object.

C.3.1   Description of the Model

Consider the area of nerve fiber containing $N + 1$ nodes of Ranvier. Assign to nodes the numbers 0 to N and denote by $u_i(t)$ their membrane potentials. Denote by $v_i(t)$ $(i = 1, \ldots, N)$. the potential myelinated area located between the nodes with numbers $i - 1$ and i, $v_i(t)(i = 1, \ldots, N)$. Membrane potentials of the nodes of Ranvier and myelinated areas will be measured from the level of maximum hyperpolarization, so $u_i(t) \geq 0$ and $v_i(t) \geq 0$. The current, flowing through the myelinated area, is given by formula (Tasaki 1971):

$$I = C \frac{dv_i}{dt} + \frac{v_i}{R_0}.$$

Here R_0 is the resistance of the membrane coated with myelin, and C is the effective capacity. The resistance of the myelin sheath is strong. In addition, the membrane changes very quickly during the passage of the pulse, so the ohmic component can be neglected (Tatko 1988). According to Kirchhoff's law, the current flowing through myelinated area with the number i is equal to:

$$I = \frac{u_{i-1} - v_i}{R} + \frac{u_i - v_i}{R},$$

where R is the resistance of passage of the myelinated node area. Finally, we obtain the following equation:

$$\frac{dv_i}{dt} = \frac{1}{RC}(u_{i-1} - 2v_i + u_i).$$

Summing the balance of currents passing through the nodes of Ranvier, we obtain the following for its potential:

$$Lu = \exp(-\lambda\sigma)(v_i - 2u_i + v_{i+1}), \tag{C.3.1}$$

where

$$Lu = \dot{u}_i - (\lambda[-1 - f_{Na}(u_i) + f_K(u_i(t-1))]u_i + \varepsilon.$$

The meaning of the parameter $\sigma(0 < \sigma < \alpha_1)$ is as follows: The multiplier $\exp(-\lambda\sigma)$ inhibits weak signals. In other words, the node can generate a spike only in the case when $v_i(t) + v_{i+1}(t)$ is much greater than $u_i(t)$.

Thus, to describe the process of the pulse propagation along the axon we obtain the following system of differential equations:

$$\dot{u}_0 = \lambda[-1 - f_{Na}(u_0) + f_K(u_0(t-1))]u_0 + \varepsilon + \exp(-\lambda\sigma)(v_1 - u_0), \tag{C.3.2}$$

$$\dot{u}_i = \lambda[-1 - f_{Na}(u_i) + f_K(u_i(t-1))]u_i + \varepsilon \\ + \exp(-\lambda\sigma)(v_i - 2u_i + v_{i+1}), \quad i = 1, \dots, N-1; \tag{C.3.3}$$

$$\dot{u}_N = \lambda[-1 - f_{Na}(u_N) + f_K(u_N(t-1))]u_N + \varepsilon + \exp(-\lambda\sigma)(v_N - u_N), \tag{C.3.4}$$

$$\dot{v}_i = \lambda(u_{i-1} - 2v_i + u_i), \quad i = 1, \dots, N. \tag{C.3.5}$$

Here the parameter $\lambda \gg 1$ reflects the high rate of occurrence of electrical processes; and parameter $0 < \varepsilon \ll 1$ takes into account the leakage of currents passing through the membranes of nodes. Positive sufficiently smooth functions $f_{Na}(u_i)$ and $f_K(u_i)$ decrease monotonically to zero as $u \to \infty$ faster than Ou_i^{-1}. Each of the nodes acts as the neuron as detector, i.e., we consider as valid formulas (C.2.2) to (C.2.4) for α, α_1 and α_2. In addition, we assume that inequality (C.2.5) holds as does inequality

$$0 < \sigma < \alpha_1. \tag{C.3.6}$$

Inequality (C.3.6) ensures the inhibition of weak signals on the border of nodes and the myelinated area.

System (C.3.2) to (C.3.5) describes the sequence of the nodes of Ranvier, which are interconnected through myelinated areas. Note that the system of equations has the equilibrium state

$$u_i = v_i = u_* \approx \frac{\varepsilon}{\lambda \alpha}.$$

C.3.2   Investigation of Stability of Equilibrium Position

We show that the equilibrium state

$$u_i = v_i = u_* \approx \frac{\varepsilon}{\lambda \alpha}$$

is exponentially stable. Linearized in the neighbourhood of u_* system (C.3.2) to (C.3.5) has the form:

$$\dot{u}_0 = \frac{\varepsilon}{\alpha}\left[-f'_{Na}(u_*)(u_0 - u_*) + f'_K(u_*)(u_0(t-1) - u_*)\right]$$
$$- (\lambda \alpha + \exp(-\lambda \sigma))(u_0 - u_*) + \exp(-\lambda \sigma)(v_1 - u_*); \qquad (C.3.7)$$

$$\dot{u}_i = \frac{\varepsilon}{\alpha}\left[-f'_{Na}(u_*)(u_i - u_*) + f'_K(u_*)(u_i(t-1) - u_*)\right]$$
$$- (\lambda \alpha + 2\exp(-\lambda \sigma))(u_i - u_*) + \exp(-\lambda \sigma)((v_i - u_*);$$
$$+ (v_{i+1} - u_*)); \quad i = 1, \ldots, N-1; \qquad (C.3.8)$$

$$\dot{u}_N = \frac{\varepsilon}{\alpha}\left[-f'_{Na}(u_*)(u_N - u_*) + f'_K(u_*)(u_N(t-1) - u_*)\right]$$
$$- (\lambda \alpha + \exp(-\lambda \sigma))(u_N - u_*) + \exp(-\lambda \sigma)(v_N - u_*); \qquad (C.3.9)$$

$$\dot{v}_i = \lambda((u_{i-1} - u_*) - 2(v_i - u_*) + (u_i - u_*)), \quad i = 1, \ldots, N. \qquad (C.3.10)$$

Denote by $y(t) \in R^{2N+1}$ the vector function

$$y(t) = \begin{pmatrix} u_0(t) - u_* \\ v_1(t) - u_* \\ u_1(t) - u_* \\ \vdots \\ v_N(t) - u_* \\ u_N(t) - u_* \end{pmatrix}.$$

System (C.3.7) to (C.3.10) is linear, but it is conveniently rewritten as:

$$\dot{y} = F(y). \qquad (C.3.11)$$

Here

$$F_1(y) = \frac{\varepsilon}{\alpha}\left[-f'_{Na}(u_*)y_1 + f'_K(u_*)y_1(t-1)\right] - (\lambda\alpha + \exp(-\lambda\sigma))y_1$$
$$+ \exp(-\lambda\sigma)y_2;$$

$$F_{2i+1}(y) = \frac{\varepsilon}{\alpha}\left[-f'_{Na}(u_*)y_{2i+1} + f'_K(u_*)y_{2i+1}(t-1)\right]$$
$$- (\lambda\alpha + 2\exp(-\lambda\sigma))y_{2i+1}$$
$$+ \exp(-\lambda\sigma)(y_{2i} + y_{2i+2}), \quad i = 1,\ldots,N-1;$$

$$F_{2N+1}(y) = \frac{\varepsilon}{\alpha}\left[-f'_{Na}(u_*)y_{2N+1} + f'_K(u_*)y_{2N+1}(t-1)\right]$$
$$- (\lambda\alpha + \exp(-\lambda\sigma))y_{2N+1} + \exp(-\lambda\sigma)y_{2N};$$

$$F_{2i}(y) = \lambda(y_{2i-1} - 2y_{2i} + y_{2i+1}), \quad i = 1,\ldots,N.$$

As usual, we will look for a solution of Eq. (C.3.11) in the form:

$$y(t) = \exp(\mu t)h,$$

where $h \in R^{2N+1}$ is the constant vector. We obtain:

$$\mu h = \begin{pmatrix}
a_0 & \exp(-\lambda\sigma) & 0 & 0 & \ldots & 0 & 0 & 0 \\
\lambda & -2\lambda & \lambda & 0 & \ldots & 0 & 0 & 0 \\
0 & \exp(-\lambda\sigma) & a_1 & \exp(-\lambda\sigma) & \ldots & 0 & 0 & 0 \\
0 & 0 & \lambda & -2\lambda & \ldots & 0 & 0 & 0 \\
\ldots & \ldots & \ldots & \ldots & \ldots & \ldots & \ldots & \ldots \\
0 & 0 & 0 & 0 & \ldots & a_{N-1} & \exp(-\lambda\sigma) & 0 \\
0 & 0 & 0 & 0 & \ldots & \lambda & -2\lambda & \lambda \\
0 & 0 & 0 & 0 & \ldots & 0 & \exp(-\lambda\sigma) & a_N
\end{pmatrix} h.$$

$$(C.3.12)$$

Here

$$a_0 = a_N = -(\lambda\alpha + \exp(-\lambda\sigma)) + \frac{\varepsilon}{\alpha}\left(-f'_{Na}(u_*) + f'_K(u_*)\exp(-\mu)\right);$$

$$a_i = -(\lambda\alpha + 2\exp(-\lambda\sigma)) + \frac{\varepsilon}{\alpha}\left(-f'_{Na}(u_*) + f'_K(u_*)\exp(-\mu)\right); \quad i = 1,\ldots,N-1.$$

We show that for all roots of Eq. (C.3.12) the following condition is satisfied:

$$Re\mu < 0.$$

At $Re \geq 0$, discarding the terms $o(1)$, we obtain:

$$\mu h = \lambda \begin{pmatrix} -\alpha & 0 & 0 & 0 & \dots & 0 & 0 & 0 \\ 1 & -2 & 1 & 0 & \dots & 0 & 0 & 0 \\ 0 & 0 & -\alpha & 0 & \dots & 0 & 0 & 0 \\ 0 & 0 & 1 & -2 & \dots & 0 & 0 & 0 \\ \dots & \dots & \dots & \dots & \dots & \dots & \dots & \dots \\ 0 & 0 & 0 & 0 & \dots & -\alpha & 0 & 0 \\ 0 & 0 & 0 & 0 & \dots & 1 & -2 & 1 \\ 0 & 0 & 0 & 0 & \dots & 0 & 0 & -\alpha \end{pmatrix} h.$$

The characteristic equation has the form:

$$P(\mu) = -\lambda^{2N+1} \left(\alpha + \frac{\mu}{\lambda}\right)^{N+1} \left(2 + \frac{\mu}{\lambda}\right)^{N} = 0.$$

At $Re\mu > 0$, we have $P(\mu) = 0$. Therefore, as $\lambda \to \infty$, all of the roots of Eq. (C.3.12) are located on the complex plane to the left of the imaginary axis. This means that the equilibrium position

$$u_i = v_i = u_* \approx \frac{\varepsilon}{\lambda\alpha}$$

of system (C.3.2) to (C.3.5) is exponentially stable.

C.3.3 Investigation of the System of Equations Describing the Point Model of Saltatory Conduction of Excitation

Analyze system (C.3.2) to (C.3.5) as $\lambda \to \infty$. We stimulate excitation of the zero node by setting for it the initial conditions from class S, i.e. $u_0(s) = \varphi_0(s) \in S$. For the rest of nodes, we assume that

$$u_i(s) = u_*, \quad s \in [-1, 0].$$

For all myelinated areas we write

$$v_i(0) = u_*.$$

By virtue of (C.2.5) and from the initial conditions, it follows that at $t = 0$ the spike of the zero node starts, i.e., formulas (C.2.6) are valid. The membrane potential of other nodes is close to u_*, because for their activation we need strong enough action from the outside.

The membrane potential of the zero node, in the absence of action on it by the first myelinated area, is defined by formulas (C.2.6). These formulas, which give the membrane potential of the first myelinated area, are as follows:

$$v_1(t) = \begin{cases} \exp[\lambda\alpha_1(t+o(1))], & t \in [\delta,\ 1-\delta], \\ \exp[\lambda(\alpha_1 - (t-1) + o(1))], & t \in [1+\delta, t_* - \delta], \\ \exp[\lambda\alpha_1(t-\tau+p(1))], & t \in [t_* + \delta, 1+\tau - \delta], \\ \exp[\lambda(\alpha_1 - (t-\tau-1) + o(1))], & t \in [1+\tau+\delta,\ 1+\alpha_1+\tau-\delta], \\ (\varepsilon+o(1))/\lambda\alpha_2, & t \in [1+\alpha_1+\tau+\delta, 2+\alpha_1-\delta], \\ (\varepsilon(\alpha+\alpha_2)+o(1))/2\lambda\alpha\alpha_2, & t \in [2+\alpha_1+\delta, 2+\alpha_1+\tau+\delta], \\ (\varepsilon+o(1))/\lambda\alpha, & t > 2+\alpha_1+\tau+\delta. \end{cases}$$

$$(C.3.13)$$

Here $\tau = \sigma/\alpha_1$ [$\tau < 1$ by virtue of formula (C.3.6)] and

$$t_* = 1 + \alpha_1\tau/(\alpha_1+1). \qquad (C.3.14)$$

The results of investigation show that there is no action on the zero node at $t > 0$, and on the other nodes—at $t > i\tau$ (i is the number of the node). Therefore, the myelinated areas cannot influence the node of Ranvier when the spike is generated, and for some time after the spike occurs, until the node reaches the equilibrium state ($u_i(t) = u_* + o(1)$). This period of time is called the "refractory period," and its duration is:

$$T_R = \alpha_1 + 2 + o(1).$$

During this period, Eqs. (C.3.2) to (C.3.4) can be integrated independently of the others. This phenomenon has a simple biological sense. It is known [137] that the myelinated fiber cannot conduct a high-frequency pulse signal because the nodes have to "recover."

From formulas (C.2.6) and (C.3.13), it follows that

$$v_1(t) \approx \begin{cases} u_0(t), & 0 < t < t_*, \\ u_1(t), & t_* < t, \end{cases}$$

where t_* is given by formula (C.3.14). Thus, the membrane potential of myelinated area has two points of a maximum

$$t_{\max}^1 = 1 + o(1),$$
$$t_{\max}^2 = 1 + \tau + o(1)$$

and in between them a point of the local minimum

$$t_{\min} = t_* + o(1).$$

This is consistent with the biological data (Tasaki 1971) (compare Figs. C.2 and C.4).

Analysis of system (C.3.2) to (C.3.5) shows that the spike of the ith node of Ranvier starts at moment of time $t = i\tau + o(1)$, and the membrane potential of the node is defined by formula:

$$u_i(t) \approx u_0(t - i\tau), \quad t > i\tau, \quad i = 1, \ldots, N.$$

We obtain the membrane potential of the ith myelinated area by a time shift of the potential of the first area

$$v_i(t) \approx v_1(t - (i - 1)\tau), \quad t > (i - 1)\tau.$$

Thus, at the excitation of the zero node of Ranvier along the chain of the nodes, a wave of pulses (spikes) in the direction of increasing numbers of nodes will be propagated. If the node is excited in the middle of the chain, then the wave propagates in two directions. Excitation of the last node generates a wave of pulses, thus propagating in the direction of decreasing numbers of nodes. With the simultaneous excitation of two extreme nodes, there occur two waves travelling towards each other, which mutually cancel each other out at their collision.

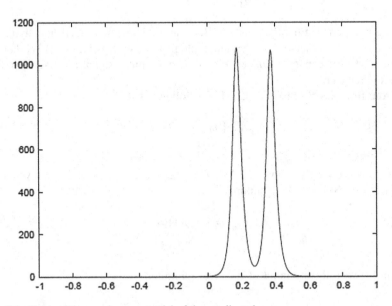

Fig. C.2 Graph of the membrane potential of the myelinated area

The obtained results are fully consistent with biological data. We investigated system (C.3.2) to (C.3.5) numerically determine the following parameter values:

$$\lambda = 7, \quad \varepsilon = 0.05, \quad \sigma = 1.011,$$
$$f_{Na}(u) = R_1 \exp(-u^2), \quad f_K(u) = R_2 \exp(-u^2),$$
$$R_1 = 1.2, \quad R_2 = 2.1.$$

The results of calculation are shown in Figs. C.2 and C.3. Figure C.2 shows the graph of the potential of myelinated area located between the nodes assigned numbers 3 and 4. It has two points of maximum, which coincide with the maxima of corresponding nodes, and in between there is the point of minimum. Figure C.3 shows the membrane potentials of nodes at $N = 6$ from zero to six, respectively. Spikes of nodes consist of areas of rapid growth and almost-as-rapid decrease of membrane potential. The duration of the descending area is slightly longer than that of the ascending area ($\alpha_1 = 1.1$). The height of spikes is close to the calculated $(\exp(\lambda\alpha_1))$.

Figures C.4 and C.5 show oscillograms recorded at different areas of the nerve fiber. Figure C.4 shows the record at the area of a nerve fiber of 1-mm length without the nodes of Ranvier. In this case, the current running through the myelin sheath has two peaks separated by interval of 0.1 msec, which corresponds to the time during which the pulse passes from one node to another. The peaks on the oscillogram occur when these nodes are excited. The record shown in Fig. C.5 corresponds to the membrane potential of the node of Ranvier. Tasaki (1971)

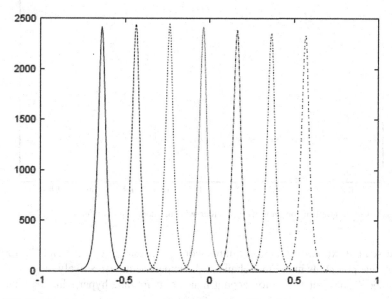

Fig. C.3 Graph of the membrane potentials of the nodes of Ranvier

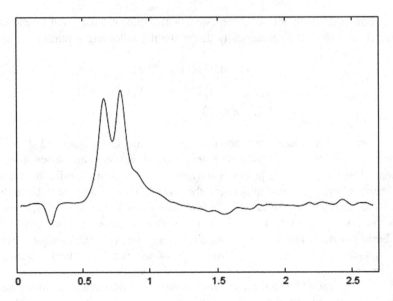

Fig. C.4 Oscillogram of the area of nerve fiber without the nodes of Ranvier

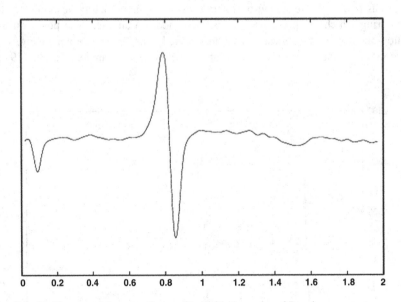

Fig. C.5 Oscillogram of the area of nerve fiber with the nodes of Ranvier

discussed the experimental curves in his work. Note that graphs C.3 and C.5 for the membrane potentials in the model and in the experiment, respectively, do not agree. The model does not take into account that the membrane hyperpolarizes after the pulse.

References

Anufrienko, S. E., & Mats, A. S. 2007. Model of saltatory conduction of excitation along the branching nerve fiber. *Modelling and Analysis of Information Systems, 14*(2), 17–19.

Anufrienko, S. E., Mayorov, V. V., Myshkin, I. Y., & Gromov, S. A. (2004). Investigation of the system of equations with delay, modelling saltatory conduction of excitation. *Modelling and Analysis of Information Systems, 11*(1), 3–7.

Schade, J., & Ford, D. (1976). *Fundamentals of Neurology.* Moscow: Mir.

Tasaki, I. (1971). *Nervous excitement.* Moscow: Mir.

Appendix D
Networks of Neurocellular Automata

Proposed by Shabarshina (1994, 1996, 2000); see also Mayorov et al. (2006), the model of neurocellular automaton axiomatizes as described in Sects. 1.6 and 2.3 discussing the dynamics of neuron without external action and with external action. Special features include (1) the spike is transformed into a momentary pulse; (2) during the refractory period the membrane potential does not change; and (3) and an exponentially decaying oscillation threshold is introduced (Kryukov and Borisyuk 1986). Neurocellular automaton is a type of integrative threshold element (Rabinovich et al. 2006).

The history of cellular automata begins with the work of John von Neumann (1971) about the construction of self-replicating machines. Biological preconditions when developing the model of cellular automata, as we know them, were first used by Wiener and Rosenbluth (1961). Cellular automata are widely used to model physical and chemical processes (systems of "reaction-diffusion type, flow of fluids, processes of growth and reproduction). This appendix includes neural networks that are based on neurons, each of which is a cellular automaton periodically generating pulses. Such networks can be used to generate sequences with the predetermined distribution of pulses; however, below, as an example, we consider a simpler problem, that of the self-organization of networks into subsets of synchronously operating automata.

D.1 Neurocellular Automata as Formal Neurons

Each of the neurocellular automata (NCAs) discussed below is described by the state vector $u^{(t)}, p^{(t)}$ and the parameters T_R, q. Here $u^{(t)} \geq u^0 > 0$ is the membrane potential, and t is current moment of time. If for all $s \in [t, t+\tau], (\tau > 0)$ holds inequality $u^{(t)} < p^{(t)}$, then the values of the membrane potential at moments of time t and $t+\tau$ are related by $u^{(t+\tau)} = u^{(t)} \exp(\Theta\tau)$, where $\Theta > 0$. Value $p^0 \geq p^{(t)} > 0$ is the generation threshold. We assume that $p^{(t+\tau)} = p^{(t)} \exp(-p\Theta\tau)$, where $p > 0$. If, at some moment of time, value $u^{(t)}$ is equal to $p^{(t)}$, then the NCA generate an instantaneous pulse (spike) and sends it to all of the related automata. After the

© Springer International Publishing Switzerland 2015

S. Kashchenko, *Models of Wave Memory*,
Lecture Notes in Morphogenesis, DOI 10.1007/978-3-319-19866-8

generation of pulse, the automaton for the time T_R goes into the refractory state (state of absolute immunity of the NCA to the action of other automata). In this state, the membrane potential has the value u^0. At the time of pulse generation, the threshold is set equal to p^0 (in the future, as already mentioned, it decreases exponentially). Under the action of the actualized spike, a transmitter is released at the synapse, which is located on the automaton's receiver as long as the automaton receiver is not in a refractory state. Let T_1 be the lifetime of the neurotransmitter. Consider that $T_1 < T_R$. However, if the automaton's receiver has generated the pulse under the action of the neurotransmitter, then the synapses of the neurotransmitter adjacent to it breaks down. In the presence of a neurotransmitter at one synapse, the dynamics of the membrane potential are subject to the following law:

$$u^{(t+\tau)} = u^{(t)} \exp(\Theta\tau(1+q)).$$

The value q in the model is called the "synaptic weight," and it characterizes the efficiency of the neurotransmitter's action. In the case then the time interval $[t, t+\tau]$ of the neurotransmitter is present at n synapses, the formula for the membrane potential takes the form:

$$u^{(t+\tau)} = u^{(t)} \exp\left[\Theta\tau\left(1 + \sum_{i=1}^{n} q_i\right)\right],$$

where q_i are the synaptic weights of the individual synapses. We call the synapse for $q > 0$ "excitatory" and that for $q < 0$: inhibitory."

In the isolated state, each NCA generates spikes periodically in time T_2, which is determined from the condition

$$u^0 \exp(\Theta(T_2 - T_R)) = p^0 \exp(-p\Theta T_2). \tag{D.1.1}$$

The formally determined dynamics of the membrane potential of the NCA corresponds to the development of the potential of a biological neuron. Note that here we are speaking of pacemaker neurons, which are oscillators. For biological neuron detectors generating a pulse only under the influence of an external action, the axiomatics should be slightly modified.

D.2   The Simplest Homogeneous Fully Connected Network: The Ring of Three Automata

Consider the system of three automata united in a ring where each element must have access to two remaining elements. Assume that all of the synaptic weights are identical and equal to q, i.e., the network is homogeneous. Denote by $t_1^n, n = 1, 2, 3, \ldots$, the consecutive moments of the start of spikes of the first NCA. Let t_2^n, t_3^n

be the corresponding sequences of the start of spikes of the second and third automata. Assume that

$$t_2^1 > t_1^1, \quad t_2^1 < t_1^1 + T_R, \quad t_3^1 > t_2^1, \quad t_3^1 < t_2^1 + T_1, \quad t_3^1 > t_1^1 + T_R, \quad t_3^1 < t_2^1 + T_R$$

Lemma D.2.1 *For intervals of time between spikes of adjacent automata the following limit relationships hold:*

$$\lim_{n \to \infty} t_1^n - t_3^n = \lim_{n \to \infty} t_2^n - t_1^n = \lim_{n \to \infty} t_3^n - t_2^n = \xi_*,$$

where

$$\xi_* = T_2(1+p)/(3(1+p)+q). \tag{D.2.1}$$

Explain the meaning of this statement. It is about stable oscillatory modes in the network of the NCA. The central point to find them is the choice of the initial conditions. Define the initial state of each NCA of the system as follows. Assume that at a fixed moment of time t_1^1, the spike of the first automaton started. Let spike of the second NCA occurs at some moment t_2^1, where $t_2^1 < t_1^1 + T_R$. This means that the first automaton, until moment of time t_2^1, is in a refractory state.

The third NCA, at the initial moment of time t_1^1, is in a refractory state and remains there for some time. By time t_2^1, it goes into the active state. Under the action of the spike of the second automaton on the synapses, the formed neurotransmitter lying on the third element. Let the spike of the third automaton occur at moment of time t_3^1, where $t_3^1 < t_2^1 + T_1$, but $t_3^1 > t_1^1 + T_R$. First, it means that (1) the neurotransmitter, which appeared under the action of the spike of the second automaton, had no time to decay until the moment t_3^1, and (2) by this time, the first automaton has come out of the refractory state.

The described choice of initial conditions allows us to use the following algorithm to investigate the dynamics of the NCA system. For any $t \in [t_1^1 + T_R, t_3^1]$, we can calculate the value of the membrane potential and the threshold value for the first NCA. In synapses belonging to the first NCA, the neurotransmitter is absent at this time (it will appear only at the time t_3^1). We can calculate t_2^1 the moment of start of the new spike of the first NCA. Similar arguments allow us to set the start time of new spikes t_2^2 and t_3^2, respectively, of the second and third NCAs. Thus, it is possible to analyze further the dynamics of all NCAs in the system.

The functioning of the system on the interval $[t_1, t_2]$ is called the "first clock cycle" of the passage of the excitation wave around the ring. The second clock cycle is opened by the spike of the first NCA at moment of time t_1^2. For the second clock cycle, we calculate the relevant moments of the start of spikes t_2^2, t_3^2, thus introducing temporary mismatches $\xi_2^1 = t_2^1 - t_1^1, \xi_3^1 = t_3^1 - t_2^1$ between the starts of the spikes of the first and second and the second and third NCAs. Denote by

$\xi_1^1 = t_1^2 - t_3^1$ a temporary delay of moment t_1^2 of the spike of the first NCA with respect to the start of spike of the third NCA. Let ξ_2^2 and ξ_3^2 be the mismatches of the moments of the start of the spikes of the first and second and the second and third NCAs, respectively, at the second stage of wave passage. Values ξ_1^1, ξ_2^2, and ξ_3^2 are expressed through ξ_2^1 and ξ_3^1. There arises the mapping describing the mismatches of the start of the spikes at consecutive clock cycles of the wave passage. It has a unique globally stable fixed point.

Proof of Lemma D.2.1 Formula for finding the value of the membrane potential of the first NCA at moment of time $t \geq t_3^1$ until a spike occurs has the following form:

$$u_1^{(t)} = u^0 \exp\big(\Theta(t - T_R) + \Theta q(t - t_3^1)\big).$$

Recall that the value u at the moment of the start of the spike is equal to the threshold value p^{t_1}, i.e.,

$$u^0 \exp\big[\Theta(t_1^2 - T_R) + \Theta q(t_1^2 - t_3^1)\big] = p^0 \exp[-\Theta p(t_1^2 - t_1^1)].$$

From this equation, taking into account (D.1.1), we can determine the moment t_1^2 as follows:

$$t_1^2 = (T_2(1+p) + qt_3^1)/(1+p+q). \tag{D.2.2}$$

Using the above-introduced above mismatches ξ_k^i for moments of the start of the spikes, we can rewrite the formula (D.2.2) in the form:

$$\xi_1^1(1+p+q) = (1+p)\big(T_2 - \xi_2^1 - \xi_3^1\big). \tag{D.2.3}$$

Similar calculations for the second and third NCAs allow us to obtain the relations:

$$\xi_2^2(1+p+q) = (1+p)\big(T_2 - \xi_1^1 - \xi_3^1\big). \tag{D.2.4}$$

$$\xi_3^2(1+p+q) = (1+p)\big(T_2 - \xi_2^2 - \xi_1^1\big). \tag{D.2.5}$$

Formulas (D.2.3) to (D.2.5) describe iterative process, thus reflecting the temporal relationships between the starts of spikes at consecutive clock cycles of the passage of the wave around the ring. This process converges. (We verify that it is the method of Seidel's solution of system of linear equations with a positive definite matrix.) For the NCA network, this means that over time the intervals between spikes of adjacent NCAs are stabilized. The limit point of the process is $\xi_1 = \xi_2 = \xi_3 = \xi_*$, where the value ξ_* is given by formula (D.2.1).

D.3 Self-organization of the Fully Connected Network into the Ring of Three Sets of Synchronously Operating Automata

Consider the fully connected homogeneous network NCA, i.e., we assume that each automaton has access to all other automata of network and that all synaptic weights are the same. Represent the network of automata as a union of three nonempty disjoint sets X_1, X_2, X_3, the number of elements in which is m_1, m_2, m_3 respectively. We say that automata of the set function synchronously if they generate pulses simultaneously. We are interested in the conditions under which consistently (and at the cycle) the synchronous generation of pulses by the automata of the first, second, third (and again the first, etc.) sets occurs. Introduce the following values:

$$\xi_1 = T_2(p+1)/\left[m_3\left(q + (p+1)\sum_{i=1}^{3} m_i^{-1}\right)\right],$$

$$\xi_2 = T_2(p+1)/\left[m_1\left(q + (p+1)\sum_{i=1}^{3} m_i^{-1}\right)\right], \qquad (D.3.1)$$

$$\xi_3 = T_2(p+1)/\left[m_2\left(q + (p+1)\sum_{i=1}^{3} m_i^{-1}\right)\right],$$

Theorem D.3.1 *Let* m_1, m_2, m_3 *be such that*

$$\xi_i < T_1, \quad T_R < \sum_{j=1}^{3} \xi_j - \xi_i < T_2, \quad i = 1, 2, 3.$$

There exists a stable mode of operation of the network, at which the automata belonging to each of the sets operate synchronously and the generation of pulses by the automata of the sets X_1, X_2, X_3 is carried out consistently and cyclically. At the same, time mismatches between the spikes of automata belonging to the third and first, first and second, and second and third sets are quantities ξ_1, ξ_2, ξ_3.

One of the highlights of proof of the theorem is the choice of the initial conditions. Define a priori for an NCA the moments of the start of the spikes so that spikes of the elements of set X_2 follow the spikes of the elements from set X_1. We assume that until the spikes of the NCAs from set X_1 end, NCAs from the set X_3 are in a refractory state. Let at the start of the spikes of the NCAs from the set X_2, the elements from set X_3 become active and the NCAs of set X_1 still be refractive. However, we assume that by the moments of the start of NCAs' spikes from set X_3, automata from set X_1 have come out of the state of refractoriness.

Denote by $0 \le t_{1,1} \le t_{1,2} \le \cdots \le t_{1,m_1}$ the moments of the start of the spikes of the automata from set X_1. Furthermore, let the spikes of automata from set X_2 start at the moments $t_{2,1} \le t_{2,2} \le \cdots \le t_{2,m_2}$, where $t_{2,1} > t_{1,m_1}$. Consider that NCAs from set X_3 generated pulses at the moments $t_{3,1} \le t_{3,2} \le \cdots \le t_{3,m_3}$, respectively, where $t_{3,1} > t_{2,m_2}, t_{3,1} > t_{1,m_1} + T_R$.

The described choice of initial conditions allows us to analyze the dynamics of NCAs of the system at the second clock cycle of the passage of the wave, which is opened at the moment $t'_{1,1}$ by spike of the first NCA of set X_1. We count the moments of the start of spikes at a new clock cycle beginning with $t'_{1,1}$ and denote them by $t'_{i,j}$, $i = 1, 2, 3$; $j = 1, \ldots, m_i$. Introduce a notation for temporal mismatches at the initial clock cycle of the wave. Write $\xi_{k,j} = t_{k,j} - t_{k,j-1} (k = 1, 2, 3)$, which are mismatches between spikes of the jth and $(j-1)$th NCAs within the sets. Let $\xi_{k,1} = t_{k,1} - t_{k-1,m_{k-1}} (k = 2, 3)$ be the mismatches between the start of the spikes of the first NCA of the kth set and the latter NCA of the $(k-1)$th set. Denote by $\xi'_{i,j}$ the analogous values at the second clock cycle of the passage of the wave. The spike of the first element of set X_1 (beginning of the second cycle of the passage of the wave) relative to spike of the latter element of the set X_3 is delayed by the value $\xi'_{1,1} = t'_{1,1} - t_{3,m_3}$.

Lemma D.3.1 *We have the formula:*

$$\xi'_{i,j} = (1+p)\xi_{i,j}/(1+p+(m_{i-1}+j-1)q), \qquad (D.3.2)$$

where $i = 1, 2, 3$, $j = 2, \ldots, m_i$, $m_0 = m_3$.

Proof At the second clock cycle of the passage of the wave, neurocellular automata of set X_1 are under the influence of spikes of the automata of the set X_3. Consider that $t_{3,m_3} < t'_{1,1} \le t'_{1,2} \le \cdots \le \cdots \le t'_{1,m_1}, t_{3,1} + T_1 > t'_{1,m_1}$. The start of the passage of the wave of spikes at the next clock cycle is opened by the spike of the first NCA of the first set. Calculate this moment as $t'_{1,1}$. The first automaton is only under the influence of spikes of NCA from X_3; that is why the dynamics of its development in the time interval $t_{3,m_3} < t \le t'_{1,1}$ is described by the following formula:

$$u^{(t)} = u^0 \exp\left[\Theta(t - T_R) + \Theta q \left(\sum_{i=1}^{m_3-1} i(t_{3,i+1} - t_{3,i}) + m_3(t - t_{3,m_3}) \right) \right].$$

Because the value of the membrane potential at the moment of start of spike generation is equal to the threshold, then for $t'_{1,1}$ we obtain:

$$t'_{1,1} = t_{3,m_3} + \left((T_2 - t_{3,m_3})(1+p) - q \sum_{k=1}^{m_3-1} k(t_{3,k+1} - t_{3,k}) \right) / (1 + qm_3 + p).$$

$$(D.3.3)$$

To find moment of time $t'_{1,2}$ of the start of the spike of the second NCA from set X_1 we must take into account that for the second automaton to the action of pulses coming from the set X_3 is added the pulse of the first NCA from set X_1. The result is as follows:

$$t'_{1,2} = t'_{1,1} + \left((T_2 - t'_{1,1} + t_{1,2})(1+p) - q \sum_{k=1}^{m_3-1} k(t_{3,k+1} - t_{3,k}) \right) / (1 + qm_3 + p).$$

(D.3.4)

This expression is transformed considering (D.3.3). As a result, is the following is valid for formula (D.3.2).

$$\xi'_{1,2} = t'_{1,2} - t'_{1,1}$$

Analogously, we obtain the relation (D.3.2) for $\xi'_{1,k}, k = 3, \ldots, m_1$. It remains to be noted that the same arguments can be made for automata of sets X_2 and X_3.

Corollary D.3.1 *Spikes of NCAs within sets are synchronized with time (with increasing number of clock cycle) and within the limit $\xi_{i,j} = 0$ for $j \geq 2$.*

This statement follows directly from Lemma D.3.1. Thus, we have proved the statement about synchronization from Theorem D.3.1.

Lemma D.3.2 *Let $\xi_{i,j} = 0$ for $j \geq 2$. Then the values $\xi'_{i,1}, \xi'_{k,1}$, where $i = 1, 2, 3; k = 2, 3$ (temporary mismatches between the starts of spikes of NCAs of neighbouring sets) satisfy the linear relations:*

$$(1 + p + qm_3)\xi'_{1,1} + (1+p)(\xi_{2,1} + \xi_{3,1}) = T_2(p+1), \qquad (D.3.5)$$

$$(1+p)\xi'_{1,1} + (1 + p + qm_2)\xi'_{2,1} + (1+p)\xi_{3,1} = T_2(p+1), \qquad (D.3.6)$$

$$(1+p)\xi'_{1,1} + (1+p)\xi'_{2,1} + (1 + m_1 q + p)\xi_{3,1} = T_2(p+1). \qquad (D.3.7)$$

Proof Calculations similar to those in the proof of Lemma D.3.1 show:

$$t'_{2,1} = t'_{1,m_1} + \left((T_2 - (t'_{1,1} + t'_{1,m_1} + t_{2,1})(1+p) \right.$$

$$\left. - q \sum_{k=1}^{m_1-1} k(t_{1,k+1} - t_{1,k}) \right) / (1 + qm_1 + p). \qquad (D.3.8)$$

$$t'_{2,j} = t'_{2,j-1} + (t_{2,j} - t_{2,j-1})(p+1)/(1 + (m_1 + j - 1)q + p), \qquad (D.3.9)$$

$$t'_{3,1} = t'_{2,m_2} + \left((T_2 - (t'_{1,1} + t'_{2,m_2} - t_{3,1})\right)(1+p)$$

$$-q \sum_{k=1}^{m_2-1} k(t_{2,k+1} - t_{2,k}))/(1 + qm_2 + p). \tag{D.3.10}$$

$$t'_{3,l} = t'_{3,l-1} + (t_{3,l} - t_{3,l-1})(p+1)/(1 + (m_2 + l - 1)q + p),$$
$$j = 2, \ldots, m_2, \quad l = 2, \ldots, m_3. \tag{D.3.11}$$

Formulas (D.3.8) to (D.3.11) acquire clearer form if we use the notation for temporary mismatches between spikes. As a result, we obtain:

$$\xi'_{1,1} = \left[\begin{array}{c} \left(T_2 - \sum_{r=2}^{m_1} \xi_{1,r} - \sum_{s=1}^{m_2} \xi_{2,s} - \sum_{t=1}^{m_3} \xi_{3,t}\right)(p+1) \\ -q \sum_{i=2}^{m_3} (i-1)\xi_{3,i} \end{array}\right] (1 + m_3 q + p)^{-1}, \tag{D.3.12}$$

$$\xi'_{2,1} = \left[\begin{array}{c} \left(T_2 - \sum_{r=2}^{m_2} \xi_{2,r} - \sum_{s=1}^{m_3} \xi_{3,s} - \sum_{t=1}^{m_1} \xi_{1,t}\right)(p+1) \\ -q \sum_{i=2}^{m_1} (i-1)\xi'_{i,1} \end{array}\right] (1 + m_1 q + p)^{-1}, \tag{D.3.13}$$

$$\xi'_{3,1} = \left[\begin{array}{c} \left(T_2 - \sum_{r=2}^{m_3} \xi_{3,r} - \sum_{s=1}^{m_1} \xi'_{1,s} - \sum_{t=1}^{m_2} \xi'_{2,t}\right)(p+1) \\ -q \sum_{i=2}^{m_2} (i-1)\xi'_{2,i} \end{array}\right] (1 + m_2 q + p)^{-1}, \tag{D.3.14}$$

$$\xi'_{i,j} = (1+p)\xi_{i,j}/(1 + (m_3 + j - 1)q + p), \quad j = 2, \ldots, m_1, \tag{D.3.15}$$

$$\xi'_{2,j} = (1+p)\xi_{2,j}/(1 + (m_1 + j - 1)q + p), \quad j = 2, \ldots, m_2, \tag{D.3.16}$$

$$\xi'_{3,j} = (1+p)\xi_{3,j}/(1 + (m_2 + j - 1)q + p), \quad j = 2, \ldots, m_3. \tag{D.3.17}$$

By virtue of D.3.1, we can write $\xi'_{k,j} = \xi_{k,j} = 0$ for $j \geq 2$. As a result, we obtain the formulas (D.3.5) to (D.3.7), and this completes the proof of Lemma.

Proof of Theorem D.3.1 Formulas (D.3.5) to (D.3.7) describe iterative process. By the values $\xi_{2,1}, \xi_{3,1}$ are consistently calculated $\xi'_{1,1}, \xi'_{2,1}, \xi'_{3,1}$. After replacing $\xi_{i,1}$ to $\xi'_{i,1}$ ($i = 2, 3$) we can continue calculations. The process converges because iterative relations represent the Seidel method for solving the system of linear equations with a positively definite and symmetric matrix. Components of its limit point (ξ_1, ξ_2, ξ_3) are given by formulas (D.3.1). Thus, the theorem is proved.

Thus, it is shown that the homogeneous fully connected network can be organized into a ring of sets of simultaneously operating automata. Note that the problems of synchronization of the elements of complex systems are highly relevant (see, e.g., Kleptsyn et al. 1984). We give clear interpretation of the obtained results. Let the automata be located on the plane. We consider the generation of pulses by the automata of some set as the activation of a black-and-white image. The network can store an infinitely long sequence of three visual images. This can be regarded as a model of information storage in a dynamic form.

D.4 Organization of Fully Connected Weakly Inhomogeneous Network of Neurocellular Automata into the Ring Structure from Sets of Synchronously Operating Automata

Consider a fully connected weakly heterogeneous network of NCAs, i.e., assume that each automaton has access to all other automata of the network and that the synaptic weights only slightly differ from each other. We represent the network of automata as a union of three nonempty pairwise disjoint sets X_1, X_2, X_3, the number of elements in which is m_1, m_2, m_3 respectively. As noted previously, we are interested in conditions under which the synchronous generation of pulses by automata of the first, second, third, again the first and etc., sets occurs consistently and at a cycle.

Assign numbers to the automata of the network with two indices: the first is the number of the set that owns the automaton; the second is the automaton's serial number within the set.

Denote the synaptic weight of impact of the kth element of the ith set on the jth NCA of the lth module by $q_{l,j}^{i,k} > 0$, $i, l = 1, 2, 3$, $k = 1, 2, \ldots, m_i$, $j = 1, 2, \ldots, m_i$.

The functioning of the system in the interval between the adjacent-in-time pulses of the first NCA of the first set is called the "clock cycle" of the passage of the excitation wave. Denote by $t_{i,j}^k$ $(i = 1, 2, 3; j = 1, 2, m_i)$ the moments of the generation of pulses by automata of the net at the kth clock cycle of the passage of the wave.

We study oscillatory modes, at which at any clock cycle of the passage of wave distribution of moments of pulse generation by the network's automata is subject to the following conditions.

1. Pulses of elements within sets follow in ascending numerical order of elements, and the first NCA of the ith set generates a nonzero output signal after the latter m_{i-1}th NCA of the $i - 1$th set. We write out the corresponding inequalities:

$$t_{i,j+1}^k \geq t_{i,j}^k; \quad t_{i,1}^k > t_{i-1,m_i-1}^k; \quad t_{1,1}^k > t_{3,m_3}^{k-1};$$
$$(i = 1, 2, 3; \; j = 1, \ldots, m_i - 1). \tag{D.4.1}$$

2. Automata from set X_{i-1} have a direct impact on automata from set X_i. We understand that under direct impact, the NCA spikes from set X_i occurred before they would destroy on their own neurotransmitter, which was released on the synapses under the impact of pulses coming from set X_{i-1}, i.e.,

$$t^k_{1,m_1} < t^{k-1}_{3,1} + T_1; \quad t^k_{2,m_2} < t^k_{1,1} + T_1; \quad t^k_{3,m_3} < t^k_{2,1} + T_1. \tag{D.4.2}$$

3. Each automata from set X_i comes out of the refractory state by the time pulses are generated by the elements of the X_{i-1}th set, but they have no time to generate the pulse spontaneously:

$$t^{k-1}_{1,m_1} + T_R < t^k_{3,j} < t^{k-1}_{1,1} + T_2, \tag{D.4.3}$$

$$t^k_{2,m_2} + T_R < t^k_{1,j} < t^k_{2,1} + T_2, \tag{D.4.4}$$

$$t^k_{3,m_3} + T_R < t^k_{2,j} < t^k_{3,1} + T_2, \tag{D.4.5}$$

Assume that inequalities (D.4.1) to (D.4.5) are held at any cycle of the passage of the wave over the network. This supposition allows us to use the following algorithm of the investigation of the dynamics of the NCA system. For any $t \in [t^{k-1}_{1,1} + T_R, t^{k-1}_{3,1}]$, we can calculate the value of the membrane potential and the value of the threshold for the first NCA of the first set. Beginning from moment of time $t^{k-1}_{3,1}$, this NCA is under the action of the group of pulses of the excitatory NCAs of set X_3. Taking into consideration that the value of the membrane potential at the moment of spike generation is close to the threshold value, we can calculate $t^k_{1,1}$—the moment of spike generation of the new pulse of the first NCA of set X_1. Similar arguments allow us to state the moments of the new pulses $t^k_{1,j}$ and $t^k_{2,l}, t^k_{3,l}$ ($l = 1, \ldots, m_i$, $i = 1, 2, 3$) of the rest automata of the sets X_1, X_2, X_3. Having analyzed in such the way the further dynamics of all the NCAs of the network, we obtain the following system of equations:

$$(t^k_{1,1} - \overline{t^{k-1}_3}) \left(1 + p + \sum_{s=1}^{m_3} q^{3,s}_{1,1} \right)$$

$$= (1 + p) \left(T_2 + t^{k-1}_{1,1} - \overline{t^{k-1}_3} \right) \sum_{s=1}^{m_3} q^{3,s}_{1,1} \left(t^{k-1}_{3,s} - \overline{t^{k-1}_3} \right), \tag{D.4.6}$$

$$(t_{1,i}^k - \overline{t_3^{k-1}})\left(1+p+\sum_{s=1}^{m_3} q_{1,1}^{3,s} + \sum_{r=1}^{i-1} q_{1,i}^{1,r}\right)$$

$$= (1+p)\left(T_2 + t_{1,i}^{k-1} - \overline{t_3^{k-1}}\right)\sum_{s=1}^{m_3} q_{1,i}^{3,s}\left(t_{3,s}^{k-1} - \overline{t_3^{k-1}}\right) + \sum_{r=1}^{i-1} q_{1,i}^{1,r}\left(t_{1,r}^{k-1} - \overline{t_3^{k-1}}\right),$$

$$\text{(D.4.7)}$$

$$(t_{j,1}^k - \overline{t_{j-1}^k})\left(1+p+\sum_{s=1}^{m_{j-1}} q_{j,1}^{j-1,s}\right)$$

$$= (1+p)\left(T_2 + t_{j,1}^{k-1} - \overline{t_{j-1}^k}\right)\sum_{s=1}^{m_{j-1}} q_{j,1}^{j-1,s}\left(t_{j-1,s}^k - \overline{t_{j-1}^k}\right), \qquad \text{(D.4.8)}$$

$$(t_{j,i}^k - \overline{t_{j-1}^k})\left(1+p+\sum_{s=1}^{m_{j-1}} q_{j,i}^{j-1,s} + \sum_{r=1}^{i-1} q_{j,i}^{j,r}\right)$$

$$= (1+p)\left(T_2 + t_{j,i}^{k-1} - \overline{t_{j-1}^k}\right)\sum_{s=1}^{m_{j-1}} q_{j,i}^{j-1,s}\left(t_{j-1,s}^k - \overline{t_{j-1}^k}\right) + \sum_{r=1}^{i-1} q_{j,i}^{j,r}\left(t_{j,r}^k - \overline{t_{j-1}^k}\right),$$

$$\text{(D.4.9)}$$

where $i = 2, \ldots, m_j$, $j = 1, 2, 3$.

Here by $\overline{t_i^k}$ we denote the averaged moment of time of pulse generation by automata of set X_i, i.e.,

$$\overline{t_i^k} = \sum_{s=1}^{m_i} t_{i,s}^k / m_i.$$

It is convenient to investigate the convergence of the iterative process (D.4.6) to (D.4.9) using the notation for time periods between the moments of pulse generation.

Denote by $\xi_{1,j}^{k-1}$ the delay of spike of the jth NCA of the first set with respect to $\overline{t_3^{k-2}}$ averaged time of pulse generation by the automata of set X_3. Denote by $\xi_{s,j}^{k-1}$ the temporal mismatches between the $k-1$th moment of spike of the jth NCA of the sth set and the $\overline{t_{s-1}^{k-1}}$ averaged time of pulse generation by the automata of the $s-1$th set ($s = 2, 3$). By way of $\mu[\xi_i^{k-1}]$, we denote the average value $\xi_{i,j}^{k-1}$ for a given set ($i = 1, 2, 3$), i.e., a temporal mismatch between the averaged moments of pulse generation by the automata of sets that neighbor by virtue of number. Through $\xi_{i,j}^k$ and $\mu[\xi_i^k]$, denote the corresponding values at the next cycle of the passage of the excitation wave over the network.

Turning to the notation through temporary mismatches, formulas (D.4.6) to (D.4.9) can be written as:

$$\xi_{1,1}^k\left(1+p+\sum_{s=1}^{m_3}q_{1,1}^{3,s}\right)=(1+p)\left(T_2-\sum_{s=1}^{3}\mu\left[\xi_s^{k-1}\right]+\xi_{1,1}^{k-1}\right)$$

$$+\sum_{s=1}^{m_3}q_{1,1}^{3,s}(\zeta_{3,s}^{k-1}-\mu\left[\xi_3^{k-1}\right]),\qquad\text{(D.4.10)}$$

$$\xi_{1,j}^k\left(1+p+\sum_{s=1}^{m_3}q_{1,j}^{3,s}+\sum_{l=1}^{j-1}q_{1,j}^{1,l}\right)=(1+p)\left(T_2-\sum_{s=1}^{3}\mu\left[\xi_s^{k-1}\right]+\xi_{1,j}^{k-1}\right)$$

$$+\sum_{s=1}^{m_3}q_{1,1}^{3,s}\left(\zeta_{3,s}^{k-1}-\mu\left[\xi_3^{k-1}\right]\right)+\sum_{l=1}^{j-1}q_{1,j}^{1,l}\xi_{1,l}^{k},$$

$$\text{(D.4.11)}$$

$$\xi_{2,1}^k\left(1+p+\sum_{s=1}^{m_1}q_{2,1}^{1,s}\right)=(1+p)\left(T_2-\sum_{s=2}^{3}\mu\left[\xi_s^{k-1}\right]-\mu\left[\xi_1^{k}\right]+\xi_{2,1}^{k-1}\right)$$

$$+\sum_{s=1}^{m_1}q_{2,1}^{1,s}(\xi_{1,s}^{k}-\mu\left[\xi_1^{k}\right]),\qquad\text{(D.4.12)}$$

$$\xi_{2,j}^k\left(1+p+\sum_{s=1}^{m_1}q_{2,j}^{1,s}+\sum_{l=1}^{j-1}q_{2,j}^{2,l}\right)=(1+p)\left(T_2-\sum_{s=2}^{3}\mu\left[\xi_s^{k-1}\right]-\mu\left[\xi_1^{k}\right]+\xi_{2,j}^{k-1}\right)$$

$$+\sum_{s=1}^{m_1}q_{2,j}^{1,s}\left(\xi_{1,s}^{k}-\mu\left[\xi_1^{k}\right]\right)+\sum_{l=1}^{j-1}q_{2,j}^{2,l}\xi_{2,l}^{k},$$

$$\text{(D.4.13)}$$

$$\xi_{3,1}^k\left(1+p+\sum_{s=1}^{m_2}q_{3,1}^{2,s}\right)=(1+p)\left(T_2-\mu\left[\xi_3^{k-1}\right]-\sum_{s=1}^{2}\mu\left[\xi_s^{k}\right]+\xi_{3,1}^{k-1}\right)$$

$$+\sum_{s=1}^{m_2}q_{3,1}^{2,s}(\xi_{1,s}^{k}-\mu\left[\xi_2^{k}\right]),\qquad\text{(D.4.14)}$$

$$\xi_{3,j}^k\left(1+p+\sum_{s=1}^{m_2}q_{3,j}^{2,s}+\sum_{l=1}^{j-1}q_{3,j}^{3,l}\right)=(1+p)\left(T_2-\mu\left[\xi_3^{k-1}\right]-\sum_{s=1}^{2}\mu\left[\xi_s^{k}\right]+\xi_{3,j}^{k-1}\right)$$

$$+\sum_{s=1}^{m_2}q_{3,j}^{2,s}\left(\xi_{2,s}^{k}-\mu\left[\xi_2^{k}\right]\right)$$

$$+\sum_{l=1}^{j-1}q_{3,j}^{3,l}\xi_{3,l}^{k},\quad j=2,\ldots,m_i,\quad i=1,2,3.$$

$$\text{(D.4.15)}$$

Let $q_{i,j}^{k,l} = q^{k,l} + \delta_{i,j}^{k,l}$. We consider the question of convergence (D.4.10) to (D.4.15), which is provided: $q_{i,j}^k = q^{k,l}$.

It is obvious that from (D.4.10) to (D.4.15) follows:

$$\xi_{i,j}^k - \xi_{i,j-1}^k = (1+p)(\xi_{i,j}^k - \xi_{i,j-1}^k)/1 + p + \sum_{s=1}^m q_i^s, \quad i = 1,2,3; \ j = 1,\ldots,m_i.$$

Thus, with an increasing number of iterations, the difference $\xi_{i,j}^k - \xi_{i,j-1}^k \to 0$. This means that the values $\xi_{i,j}^k$ with respect to time are close to $\mu[\xi_i^k]$. Then the question of convergence of the mapping (D.4.10) to (D.4.15) is reduced to verification of the convergence of the mapping for the average mismatches:

$$\left(1+p+\sum_{s=1}^{m_3} q^{3,s}\right)\mu[\xi_1^k] + (1+p)(\mu[\xi_2^{k-1}] + \mu[\xi_3^{k-1}]) = (1+p)T_2, \quad \text{(D.4.16)}$$

$$\left(1+p+\sum_{s=1}^{m_1} q^{1,s}\right)\mu[\xi_2^k] + (1+p)(\mu[\xi_1^k] + \mu[\xi_3^{k-1}]) = (1+p)T_2, \quad \text{(D.4.17)}$$

$$\left(1+p+\sum_{s=1}^{m_2} q^{2,s}\right)\mu[\xi_3^k] + (1+p)(\mu[\xi_1^k] + \mu[\xi_2^k]) = (1+p)T_2. \quad \text{(D.4.18)}$$

Thus converges the iterative process (D.4.16) to (D.4.18) reflecting timing relationships between the averaged moments of pulse generation in consecutive cycles of the passage of the wave over the network. These relationships are a method of Seidel's solution of linear equations with a positive definite and symmetric matrix. The limit point of the process ξ^0 has the coordinates

$$\xi_i^0 = T_2(1+p)/\sum_{s=1}^{m_i} q^{i,s}\left[1 + (1+p)\sum_{j=1}^3 \left(\sum_{s=1}^{m_j} q^{j,s}\right)^{-1}\right], \quad (i = 1,2,3). $$

$$\text{(D.4.19)}$$

Inequalities (D.4.1) to (D.4.5) for the limit point have a simple form:

$$0 < \xi_i^0 < T_1, \quad \text{(D.4.20)}$$

$$T_R < \sum_{j=1}^3 \xi_j^0 - \xi_i^0 < T_2, \quad (i = 1, 2, 3). \quad \text{(D.4.21)}$$

From the general theory of difference equations [8] it follows that convergence of the process is not violated at sufficiently small changes in parameters, i.e., there exists such δ that at $\left|\delta_{i,j}^{k,l}\right| < \delta$ iterative process (D.4.10) to (D.4.15) has the limit point η^0 with coordinates slightly different from the coordinates of point ζ^0. Limit point η^0 satisfies inequalities (D.4.20) and (D.4.21).

Let the initial conditions of the iterative process be chosen in a sufficiently small neighborhood of point $\eta_i^0 (i = 1, 2, 3)$. By standard arguments, we can show that all iterations $\zeta_{i,j}^k (i = 1, 2, 3);\ j = 1, 2, \ldots, m_i)$ also satisfy (D.4.20) and (D.4.21). This proves the stability of the limit mode. The obtained result can be formulated as a theorem.

Theorem D.4.1 *Let the numbers $q^{i,j}$ be such that the values $\zeta_i^0 (i = 1, 2, 3);\ j = 1, 2, \ldots, m_i)$, as defined by formulas (D.4.16), satisfy inequalities (D.4.20) and (D.4.21). Then for any $\varepsilon > 0$ there exists such $\delta > 0$ that in the network with synaptic weights satisfying inequality $\left|q_{k,l}^{i,j} - q^{i,j}\right| < \delta i$, $(k = 1, 2, 3);\ j = 1, 2, \ldots, m_i);\ l = 1, 2, \ldots, m_k)$, the limit mode time intervals between spikes of automata in each set are equal to zero, i.e., the automata operate synchronously, and temporary mismatches η_i^0 between pulses of elements of the $i - 1$th and the ith sets satisfy inequality $\left|\eta_i^0 - \zeta_i^0\right| < \varepsilon$.*

Thus, it is shown that the inhomogeneous fully connected network can also be organized into a ring of sets of synchronously operating automata.

D.5 Why Are There Only Three Sets of Synchronously Operating Automata?

Recall and emphasize that the spike coming to the synapse of neuron-receiver, which is in a refractory state, does not cause the release of the neurotransmitter. [This assumption has biological conditions (Nicholls et al. 2003)]. The number of neurons in a fully weak heterogeneous network is large enough. The network can be partitioned into some number N of pairwise disjoint sets $X_1, \ldots, X_n$ so that oscillations of the NCA within each of them will be synchronous. At the same time, the NCA included in allocated sets $X_1, \ldots, X_n$, will generate pulses cyclically: $X_1, X_2, \ldots, X_n, X_1, X_2, \ldots$ given that the oscillatory mode is stable. Naturally, there are some limitations on the number of NCA of the network, the number of subsets, and the synaptic weights. These restrictions are similar to those of Theorem D.4.1.

References

Kleptsyn, A. F., Kozyakin, V.S., Krasnosel'skii, M.A., & Kuznetsov, N. A. (1984). On the effect of small desynchronization on stability of complex systems. II. *Automation and Remote Control, 3*, 42–47.

Kryukov, V. I., Borisyuk, G. N., et al. (1986). *Metastable or unstable states in the brain*. Pushchino.

Mayorov, V. V., Shabarshina, G. V., & Konovalov, E. V. (2006). Self-organization in a fully homogeneous network of neural cellular automata of excitatory type. In *Proceedings of the VIII All-Russian Conference "Neuroinformatics-2006"*, Part I (pp. 67–72). Moscow: Moscow Engineering Physics Institute.

Nicholls, J.G., Martin, A.R., Wallace, B.J., & Fuchs, P.A. (2003). *From neuron to brain*. Moscow: Editorial URSS.

Rabinovich, M. I., Varona, P., Selverston, A. I., & Abarbanel, H. D. I. (2006). Dynamical principles in neuroscience. *Reviews of Modern Physics*, *78*, October–December, 2–53.

Shabarshina, G. V. (1994). Conduction of excitation along the ring structure of neural cellular automata. In *Modelling and Analysis of Information Systems*. Yaroslavl, *2*, 116–121.

Shabarshina, G. V. (1999). Self-organization in homogeneous fully connected network of neural cellular automata of excitatory type. *Automation and Remote Control*, *2*, 112–119.

Shabarshina, G. V. (2000). Self-organization in slightly inhomogeneous fully connected network. In *Mathematical Modelling and Analysis of Information Systems*. Yaroslavl, *7*(1), 44–49.

Von Neumann, J. (1971). *Theory of self-reproducing automata*. Moscow: Mir.

Wiener, N., & Rosenbluth, A. (1961). Conduction of impulses in the heart muscle. *Cybernetic Collection*, *3*, 7–56 (Moscow: IL).

Appendix E
Networks of W-Neurons in the Problems of Storage of Cyclic Sequences of Binary Patterns

In our research with Shabarshina (1997, 1999), the construction called the "W-neuron network" which can memorize and reproduce a sequence of binary vectors (patterns), is proposed. The following is a review of studies on W-neurons. Note that W-neurons are a direct generalization of Wiener neurons (1961) and represent a kind of integrative threshold elements (Rabinovich et al. 2006).

E.1 Axiomatic Description of the W-Neuron

At any discrete moment of time t, the W-neuron is in one of three states. To describe the state of the W-neuron, introduce the value $s(t)$. The value $s(t) = 0$ corresponds to the wait state (rest); $s(t) = 1$ corresponds to the state of excitation; and, if $s(t) = -1$, the neuron is in a refractory state (i.e., resistant to external influence). The state of excitation lasts for one clock cycle. Let at time t the W-neuron be excited, i.e., $s(t) = 1$. Then, in the next $r_0 \geq 1$ of clock cycles, the W-neuron is in a refractory state: $s(t + i) = -1 (i = 1, \ldots, r_0)$. At moment of time $t + r_0 + 1$, it necessarily goes into a wait state, i.e., $s(t + r_0 + 1) = 0$.

At each moment of time t, the W-neuron produces an output signal $x(t) = \delta_0(s(t) - 1)$ where $\delta_0(v)$ is the pulse function: $\delta_0(0) = 1$ and $\delta_0(v) = 1$ at $v \neq 0$. Thus, the output signal is different from zero and is equal to one only in the state of excitation. The W-neuron has synaptic inputs of two types: summing and inhibitory. If at moment of time t the W-neuron is in wait state and one of absolutely inhibitory synapses receives signal, then at the next moment of time the element is in a state of refractoriness: $s(t + 1) = -1$, and remains therein for r_0 of clock cycles: $s(t + i) = -1$ for $i = 1, \ldots, r_0$. To describe the influence of summing synaptic inputs (let N be their number), we define for the W-neuron the value $u(t)$ and call it the "membrane potential." Let $x_i(t)$ be a signal (equal either to zero or one) arriving at a moment of time t on the ith synapse. Then we write

© Springer International Publishing Switzerland 2015
S. Kashchenko, *Models of Wave Memory*,
Lecture Notes in Morphogenesis, DOI 10.1007/978-3-319-19866-8

$$u(t) = q\delta_0(s(t-1))u(t-1) + \left[\sum_{i=1}^{N} w_i x_i(t)\right]\delta_0(s(t)), \qquad (E.1.1)$$

where the parameter $0 \leq q < 1$ and $\delta(*)$ is the pulse function. We call the numbers in (E.1.1) the "synaptic weights." Introduce the number u_0 and call it the "threshold value of the membrane potential." If at moment of time t the value $u(t) \geq u_0$, then at the next moment of time $s(t+1) = 1$ where the W-neuron goes into the excited state and produces an output signal $x(t+1) = 1$. It is convenient to write: $s(t+1) = \theta(u(t) - u_0)$, where $\theta(v)$ is the Heaviside function.

Here we explain the features of the summation of input signals. Let at moment of time $t = -1$ the value be $s(-1) = -1$ indicating that the W-neuron is in refractory state and at moment of time $t = 0$ the value be $s(0) = 0$ indicating that the W-neuron goes into the state of wait. Starting from zero moment of time, the summing synapses receive signals in a series: at moment of time $t = k(k = 1,...,N-1)$ comes a single signal on the kth synapse. Let up to moment of time $t = N - 1$ the W-neuron not pass to the refractory state. Then from (E.1.1) it follows that the value of the membrane potential

$$u(N-1) = \sum_{k=1}^{N} q^{n-k} w_k.$$

Thus, synaptic signals are summed, but the contribution of each of them in membrane potential decreases with time.

Note that at $q = 0$ in a wait state, the W-neuron responds to the synaptic effect, i.e., it produces an output signal in exactly the same way as the McCulloch–Pitts neuron. We call attention to one more fact. If in (E.1.1) we drop multiplier $\delta_0(s(t-1))$, the membrane potential of the W-neuron will not "be dropped" after the transition to a state of excitement. To the value of the membrane potential, the neuron will contribute its value at the moment of time preceding the state of excitement. It is very important to note that this is based on biological data. The value of the membrane potential at the moment of time preceding the state of excitation can significantly influence the time of the next excitation. In particular, the W-neuron can frequently generate single outputs (i.e., generate bursts).

The network of W-neurons operates in an oscillatory mode and also has an equilibrium state in which all W-neurons are in a wait state. Below we consider networks consisting of neural modules (associations). Each of them consists of N elements, which are connected by absolutely inhibitory synapses. As the result of such interaction, after excitation of any W-neuron, all elements of the module at the next clock cycle pass into the state of refractoriness. Connections between W-neurons of different modules are made by summing synapses.

We describe oscillatory modes that are of interest to us. At the initial moment of time, part of the W-neurons of one of the modules goes into an excited state and therefore generates single-output signals. At the next moment of time, output unit signals generate the W-neurons of another module. Nonzero output signals consistently generate all the other modules. We call this process the "clock cycle of the passage of the excitement wave." The next clock cycle of the passage of the wave starts with the generation of output unit signals on the part of neurons of the source module. This set does not necessarily coincide with the set of W-neurons, which become excited at the first clock cycle. At the second clock cycle, the wave of excitation passes through all of the modules in the same order as done at the first clock cycle. However, passage of W-neurons to the excited state, generally speaking, occurs at other than at the first clock cycle. In each module, sets of W-neurons excited at different clock cycles can intersect. After a number of clock cycles of the passage of the excitation wave, the W-neurons, having generated these signals at the first clock cycle, generate single-output signals in the source module. Later, the process is periodically repeated.

Below it will be shown that the summed weights of synapses can be chosen so that at any clock cycle in each module, predefined W-neurons will be in the excited state. The possibility of necessary training a single-layer perceptron is based on a simple property of it.

E.2 Possibility of Training the Single-Layer Perceptron

Consider a single-layer perceptron, i.e., a set of McCulloch–Pitts neurons, with common inputs. Let $X(t) = (x_1(t), \ldots, x_N(t))$ be the vector of input signals where $x_i(t)$ can take either zero or a single value. Such vectors are said to be "binary." Their set is denoted by B^N. Let $w_{k,1}, \ldots, w_{k,N}$ be the vector of synaptic weights of the kth McCulloch–Pitts neuron. The output signal of the kth neuron is

$$y_k(t+1) = \theta\left(\sum_{i=1}^{N} w_{k,i} x_i - b_k \right),$$

where the numbers b_k are called "threshold values." The perceptron generates output vector $Y(t+1) = (y_1(t+1), \ldots, y_l(t+1))$ where l is the number of neurons.

We introduce the standard notation matrix W, in which the rows are the vectors of synaptic weights (it is called the "synaptic") and threshold vector $b = (b_1, \ldots, b_l)$. Furthermore, for the vector $Z = (z_1, \ldots, z_l)$ we write $\theta(Z) = (\theta(z_1), \ldots, \theta(z_l))$. Then the output vector is $Y(t+1) = \theta(WX(t) - b)$.

Consider arbitrary set $X_1, \ldots, X_\xi \in B^N (\xi \leq N)$ of linearly independent (common position) input vectors. Let $Y_1, \ldots, Y_\xi \in B^l$ be an arbitrary set of output vectors. The following feature of the perceptron is known. We can choose the synaptic matrix W and threshold vector b so that the perceptron will answer to input vectors X_i by output vectors Y_i, i.e., $Y_i = \theta(WX_i - b)$ and the zero-output vector corresponds to the zero-input vector. At the same time, the components of the vector b will be positive.

Thus, we can preplan output vectors for given input vectors (to train the perceptron). The choice of synaptic weights and threshold values is realized constructively in accordance with certain rules, which are called the "rules of training a perceptron" (Dunin-Barkowsi 1978, Frolov and Muravyov 1987).

Consider here the algorithm of training a McCulloch–Pitts neuron based on the orthogonalization of input vectors. Let $X_1, \ldots, X_m, \bar{X}_1, \ldots, \bar{X}_l (m + l \leq N)$ be linearly independent. The McCulloch–Pitts neuron must generate a single-output signal for vectors $X_1, \ldots, X_m$ and a zero-output signal for vectors $\bar{X}_1, \ldots, \bar{X}_l$. From a geometric point of view, this means that the hyperplane $(w, X) - b = 0$ separates given sets of vectors.

Form the sequence of vectors

$$X_1 - X_m, \ldots, X_{m-1} - X_m, \bar{X}_1, \ldots, \bar{X}_l, X_m,$$

(they are linearly independent). Using the algorithm of Schmidt, we build an orthogonal sequence according to the given sequence. Let w be the latter vector obtained in the process of orthogonalization. By construction $(w, X_i - X_m) = 0$ for $i = 1, 2, \ldots, m - 1$, $(w, \bar{X}_i) = 0$ for $i = 1, 2, \ldots, l$. An important property follows from the algorithm of Schmidt:

$$(w, X_m) = \|w\|^2 > 0.$$

Let $b \in (0, \|w\|^2)$. Hyperplane $(w, X) - b = 0$ separates vectors $X_1, \ldots, X_m$ and $\bar{X}_1, \ldots \bar{X}_l$. Actually, $(w, X_i) - b = \|w\|^2 - b > 0, (w, X_i) - b = -b < 0$. Therefore, w is the sought synaptic vector, and b is the threshold value of the McCulloch–Pitts neuron, which interests us.

Note that the requirement of linear independence is rather strict. In fact, it is enough that the vector X_m does not to belong to the linear span of the vectors

$$X_1 - X_m, \ldots, X_{m-1} - X_m, \overline{X_1}, \ldots, \bar{X}_l.$$

E.3 Planning of Oscillatory Modes of the Predetermined Structure in the Networks of W-Neurons

Consider the net consisting of p modules and each of them containing N number of W-neurons. Remember that within the modules all of the W-neurons are connected with absolutely inhibitory synapses. Arbitrarily assign numbers to the modules. Also assign numbers to the W-neurons within each module. Let $S_k(t), U_k(t)$ and $X_k(t)(k = 1, \ldots, p)$ be, correspondingly, the vectors of states, membrane potentials, and vectors of output signals of the kth module at moment of time t. Denote by W_{kj}

the matrix with the vector in the current ith $(i = 1,\ldots,N)$ row consisting of the synaptic weights of action of W-neurons of the jth module on the ith W-neuron of the kth module. In accordance with formula (E.1.1) for the membrane potentials, we obtain:

$$U_k(t) = q \operatorname{diag}(\delta_0(S_k(t-1)))U_k(t-1) + \operatorname{diag}(\delta_0(S_k(t))) \sum_{j \neq k} W_{k,j}, X_j(t), \quad (E.3.1)$$

where the pulse $\delta_0(*)$ function is taken over by the coordinates, and diag(V) is a diagonal matrix with the vector V on the diagonal. The vector of the output signals of the kth module at moment of time $t + 1$ is

$$X_k(t+1) = \theta(U_k(t) - U_*), \quad (E.3.2)$$

where U_* is the vector consisting of the threshold values of membrane potentials. Below we assume that the threshold values for all W-neurons are equally fixed and that $u_0 > 0$.

We want to show that that the summed weights of synapses can be chosen so that at any clock cycle of the passage of the excitation wave through the network in each module, nonzero output signals generate predefined W-neurons.

Consider the sets of binary vectors $X_1^i, X_2^i, \ldots, X_p^i \in B^N (i = 1, \ldots, m; m \leq N)$. We assume that the vectors of each of the sets X_k^i (k is fixed) are linearly independent (common position).

Let the duration of the refractory state of W-neurons be $r_0 = p - 2$.

Statement E.3.1 *Synaptic matrices $W_{k,k-1}(k = 2, \ldots, p)$ and $W_{1,p}$ can be chosen so that there is a periodic mode of functioning of the network of W-neurons in which output signals are generated at consecutive times:*

$$X_1^1, X_2^1, \ldots, X_p^1, X_1^2, \ldots, X_p^2, \ldots, X_1^m, \ldots, X_p^m, X_1^1 \ldots \quad (E.3.3)$$

At each moment of time, a non zero output vector is formed only by one module. Vectors X_k^j ($j = 1,\ldots,m$) are generated by the kth module. This mode exists independently on the choice of synaptic weights $W_{k,j}(j \neq k - 1)$ for $k = 1, \ldots, p$ and $W_{1,j}(j \neq p)$.

We write the proof of this statement in two stages. First we show how to choose the synaptic matrices $W_{k,k-1}(k = 2, \ldots, p)$ and $W_{1,p}$; second we explain how initialization of the oscillatory mode is performed. We temporarily replace W-neurons with McCulloch–Pitts neurons. Choose the synaptic matrices $W_{k,k-1}(k = 2,\ldots,p)$ according to the rules of training the single-layer perceptron so that the neurons of the kth module $(k = 2,\ldots,p)$ would respond to the input vectors $X_{k-1}^i (i = 1, \ldots, m)$ (formed by the $k - 1$th module) by the output vector $X_k^i (i = 1, \ldots, m)$, that is,

$$X_k^i = \theta\left(W_{k,k-1}X_{k-1}^i - U_*\right). \tag{E.3.4}$$

In turn, the matrix $W_{1,p}$ is chosen so that for the input vectors $X_p^i (i = 1, \ldots, m)$, as the vectors of output signals of the first module, would serve, respectively, the vectors $X_1^{i+1} (i = 1, \ldots, m - 1)$, and for the vector X_p^m—vector X_1^1. By general position, we can make the appropriate choice of synaptic matrices. We return to the network of W-neurons.

Endow each W-neuron with an outside entrance. If at moment of time t the neuron is in a wait state $s(t) = (0)$, then the external signal $x_{ext}(t) = 1$ (i.e., it is absolutely inhibitory) transfers it to the next moment into the state of refractoriness, which will last for r_0 clock cycles. In turn, the signal $x_{ext}(t) = -1$ (excitatory) transfers W-neuron at the next moment by one cycle into the excitatory state (i.e., it generates a single output signal). The external zero signal has no effect on the neuron.

Let at zero moment of time all W-neurons be in a wait state. Using external inputs consistently at moments of time $t = k(k = 1, \ldots, p)$, we transmit absolutely inhibiting signals to the neurons of the kth module. At moment of time $t = p$, the neurons of the first module have already come out of the refractory state. At this moment, the neuron transmits the signal X_1^1 through external inputs to its elements. As a result, $X_1(p + 1) = X_1^1$, i.e., at moment of time $t = p + 1$, vector X_1^1 will be the output vector $X_1(p + 1) = X_1^1$, for the first module.

At moment of time $t = p + 1$. The W-neurons of the second module are in a wait state, and the modules with numbers $3, \ldots, p$ are in a refractory state. According to (E.3.1), the vector of the membrane potentials of the second module is $U_2(p + 1) = W_{2,1}X_1^1$. From (E.3.2) it follows that the vector of output signals of the second module is $X_2(p + 2) = \theta\left(W_{2,1}X_1^1 - U_*\right)$. According to (E.3.4), we obtain $X_2(p + 2) = X_2^1$.

Note that at $t = p + 2$ all W-neurons of the first module are in a refractory state. For some neurons, it is the natural state after excitation; others were exposed to the inhibitory action of the first neuron. At moment of time $t = p + 2$, only the neurons of the third module are in a wait state. Repeating the arguments presented previously, we see that the vector of the output signals is $X_3(p + 3) = X_3^1$.

It is clear that this construction can be continued. It is important that at any given time that one module generates the nonzero vector of output signals, W-neurons of the next module by number (after the Nth follows the first module) are in a wait state. The elements of the other modules are in a refractory state. At moment of time $t = p + 1 + mp$, the output vector of the first module is $X_1(p + 1 + mp) = X_1^1$. From this moment onward the process is periodically repeated.

We make a number of important remarks.

Remark 1 To initialize the oscillatory mode, one can use any module. Transfer sequentially, starting with the kth module, W-neurons to the refractory state. Establish at time $t = p + 1$ the output vector of the kth module $X_k(p + 1) = X_k^j (j = 1, \ldots, m)$. As a result, the network will generate the same periodic sequence

(E.3.3) of output vectors (with the shifted numeration). The process of establishing of the initial value of output vector is said to be making "germ."

Remark 2 The matrix of synaptic weights $W_{1,p}$ of the action of W-neurons of the pth module on the elements of the first module can be chosen so that the output vector of the first module for the input signal $X_p^i (j = 1, \ldots, m)$, coming from the pth module, would be vector X_1^j. In consecutive moments of time $t = k(k = 1, \ldots, p)$, transfer W-neurons of modules with numbers k to the refractory state. At moment of time $t = p + 1$, establish the output vector of the first module $X_1(p + 1) = X_1^j (j = 1, \ldots, m)$ (make the germ X_1^j). Then the network will period- ically reproduce "single-round" sequence of output vectors $X_1^j, \ldots, X_p^i, X_1^j, \ldots$. At any time, only one module generates a nonzero output vector, and vector $X_k^j (k = 1, \ldots, p)$ is formed by the kth module. Directing the matrix $W_{1,p}$ we can combine two or more "single-round" sequences into one.

Remark 3 Consider any way to bypass modules that does not have with the initial way of common-oriented elements (for example, in descending order of numbers). We will not change the previously selected matrices of synaptic weights. Quite similar to the previous, one can control the synaptic matrices in the direction of bypass so that the network could generate additionally one or several new predefined periodic sequences, i.e., output vectors. The initialization method of the corre- sponding oscillatory modes is identical to the one described previously. It is enough to transfer modules in an order corresponding to the direction of bypass to the refractory state and to make the germ of sequence into the first circumvention module.

E.4 The Use of the Ability of W-Neurons to Summarize Input Signals by Time

Previously a supposition was made about the linear independence of the output vectors of each module. For example, during the storage of word-sequences of characters of some alphabet, the common position is violated. In the word can be the same characters. At the reproduction of "iteration," one can use different ver- sions of the same character. W-neuron networks provide other opportunities associated with peculiarities of the summation of input signals in time.

We want to show the existence of a periodic mode of the W-neuron network in which the network reproduces the above-described sequence (E.3.3). However, some of the vectors X_j^i (j is fixed) can be equal to each other.

For simplicity, we restrict ourselves to a particular case. Let the duration of the refractory period be $r_0 = p - 3$. Then we can choose matrices $W_{k,k-1}, W_{k,k-2}$ of the synaptic weights (to train) so that (E.3.3) would serve as sequence of output signals.

For the proof, once again replace W-neurons of the kth module by McCulloch–Pitts neurons and consider a single-layer perceptron, the synaptic inputs of which we split into two sets. For one set of inputs, the synaptic is the matrix $W_{k,k-1}$, and for another is the matrix $qW_{k,k-2}$. The input vectors are denoted, respectively, by X_{k-1} and X_{k-2}. The perceptron output vector is $X_k = \theta(W_{k,k-1}X_{k-1} + qW_{k,k-2}X_{k-2} - U_*)$. Here, all the coordinates of vector U_* are equal to some $u_0 > 0$.

Below vectors X_k^j are the elements of sequence (E.3.3). We make a number of assumptions regarding the possibility of training a monolayer perceptron. They can be formulated as follows:

Proposition 1 *For $k = 3, \ldots, p$, the problem of perceptron training is solvable:*

$$\theta\left(W_{k,k-2}X_{k-2}^j - U_*\right) = 0 \qquad (j = 1, \ldots, m).$$
$$\theta\left(W_{k,k-1}X_{k-1}^j + qW_{k,k-2}X_{k-2}^j - U_*\right) = X_k^j \qquad (j = 1, \ldots, m).$$

Proposition 2 *We can train the perceptron with number $k = 1$ in the following way:*

$$\theta\left(W_{1,p-1}X_{p-1}^j - U_*\right) = 0 \quad (j = 1, \ldots, m),$$
$$\theta\left(W_{1,p}X_p^j + qW_{1,p-1}X_{p-1}^j - U_*\right) = X_1^{j+1} \quad (j = 1, \ldots, m-1),$$
$$\theta\left(W_{1,p}X_p^m + qW_{1,p-1}X_{p-1}^m - U_*\right) = X_1^1.$$

Proposition 3 *For the perceptron with number $k = 2$, the problem of training is solvable:*

$$\theta\left(W_{2,p}X_p^j - U_*\right) = 0 \quad (j = 1, \ldots, m),$$
$$\theta\left(W_{2,1}X_1^{j+1} + qW_{2,p}X_p^j - U_*\right) = X_2^{j+1} \quad (j = 1, \ldots, m-1),$$
$$\theta\left(W_{2,1}X_1^1 + qW_{2,p}X_p^m - U_*\right) = X_2^1.$$

These assumptions are likely to be fulfilled if the vectors $X_j^i (i = 1, \ldots, m)$ at fixed $j = 1, \ldots, p$ are linearly independent. In this case, we can assume that $W_{k,k-2} = 0$ for $k = 3, \ldots, p$ and also that $W_{1,p-1} = 0$ and $W_{2,p} = 0$. However, recall that we are interested in the situation where some of the vectors X_j^i (j is fixed) are equal to each other.

Let $Z \subset B^N$ be some set of linearly independent binary vectors (an alphabet). We assume that $X_j^i \in Z$. To fulfill Proposition 1, it is sufficient to require that at fixed $j = 2, \ldots, p$ there would be no identical pairs among the pairs of vectors $X_j^i, X_{j-1}^i (i = 1, \ldots, m)$. For proof, we note the following. First, it is easy to show that the vectors $\left(X_j^i, X_{j-1}^i\right) \in R^{2N} (i = 1, \ldots, m)$ at fixed $j = 2, \ldots, p$ are linearly independent. Second, let I_{j-1} be the set of such indices i that the vectors $X_{j-1}^i (i \in I_{j-1})$ account for all of set Z and that there are no identical vectors among them. Then the vectors $\left(X_j^i, X_{j-1}^i\right) \in R^{2N} (i = 1, \ldots, m)$ and $\left(X_{j-1}^i, 0\right) \in R^{2N}$ $(i \in I_{j-1})$ at fixed $j = 2, \ldots, p$ are linearly independent. It remains to refer to the possibility of training the perceptron.

Completely analogously sufficient conditions are formed under which Propositions 2 and 3 are fulfilled.

Let Propositions 1 through 3 be fulfilled and the synaptic matrices $W_{k,k-1}$ and $W_{k,k-2}$ be defined. We return to the network of W-neurons. The statement, analogous to that formulated in the preceding paragraph, is valid. There exists a periodic oscillatory mode, in which modules generate a sequence of output signals (E.3.3). At any moment of time, only one of the modules forms nonzero signals, and vectors X_k^i are output for the kth module.

Oscillatory-mode initialization can be performed almost in the same way as it is described in the preceding paragraph. Suppose that all of the neurons are in a wait state at zero time. In consecutive moments of time $t = k(k = 1, \ldots, p)$, using external inputs, transfer W-neurons of modules with the numbers k to the refractory state. At moment of time $t = p$ and acting through external inputs, the elements of the first module act so that at $t = p+1$ its output vector would be vector $X_1(p+1) = X_1^1$. Also, acting through the external inputs at moment of time $t = p+1$, the outputs of W-neurons of the second module transfer into the state $X_2(p+2) = X_2^1$. Thus, the germ of the sequence in this case are the first two vectors of sequence.

After initialization, the network will function in the planned oscillatory mode. Let us illustrate it. At moment of time $t = p+1$, neurons of the third module have just come out of the refractory state. They receive signals from the first module $\left(X_1(p+1) = X_1^1\right)$. The vector of the membrane potentials of the third module is $U_3(p+2) = W_{3,1}X_1^1$. For the output vector of the third module, in accordance with the choice of the matrix $W_{3,1}$ (Proposition 3), we obtain the value $X_3(p+2) = \theta\left(W_{3,1}X_1^1 - U_*\right) = 0$. At the next moment of time $t = p+2$, neurons of the third module receive signals $\left(X_2(p+2) = X_2^1\right)$ from W-neurons of the second module. According to formula (E.3.1), for membrane potentials we obtain $U_3(p+2) = qW_{3,1}X_1^1 + W_{3,2}X_2^1$. This is done by the choice of synaptic matrices $X_3(p+3) = \theta\left(qW_{3,1}X_1^1 + W_{3,2}X_2^1 - U_*\right)$ (Proposition 3). These arguments can be repeated for all subsequent modules.

Let remain in force all of the remarks made in the previous paragraph. In particular, for the storage of sequences one can use other ways to bypass modules (for example, in the order that is the reverse of the initial order). Directing the appropriate synaptic matrices, we can record a new sequence without changing the previous records. Using any path of conduction of excitation, one can store several short sequences instead of one long sequence.

The obtained results are easily generalized to the case $r_0 = p - s$, $p - 1 \geq s \geq 2$. For the existence of the above-described stable periodic mode of functioning of the network, we must control the synaptic matrices $W_{k,k-1}, W_{k,k-2}, \ldots, W_{k,k-s+1}$.

E.5 Organization of Mode of Burst Wave Activity in Networks of W-Neurons

As we noted in the description of W-neuron, if in formula (E.1.1) we omit factor $\delta_0(s(t - 1))$, then the membrane potential of the W-neuron will not be "dumped" after its transfer to the state of excitement. Let at some moment of time the membrane potential of a neuron be larger than the threshold value, and the element generates a single output signal. At the next clock cycle, the W-neuron goes into the refractory state, i.e., the neuron becomes refractory to external action. However, the value of the membrane potential is greater than the threshold, and the neuron generates one more pulse. Thus, the W-neuron will frequently generate single-output signals, i.e., bursts. By choosing the strength of synaptic connections for a single neuron, one can control the length of bursts (i.e., the number of pulses in the burst). By choosing the force of synaptic connections for a single neuron, one can control the length of burst (i.e., the number of pulses in the burst). Denote the length of burst by γ.

The ability of the W-neuron to generate burst action in case $\gamma = 2$ was used in (Mayorov and Shabarshina 1999) to solve the problem of storage and reproduction of a sequence of binary vectors of the space R^{2N}.

For pulse generation by the W-neuron for exactly two consecutive clock cycles t and $t + 1$, it is necessary and sufficient that the value of the membrane potential satisfies inequality:

$$q^{-1}u_0 < u(t) < q^{-2}u_0, \tag{E.5.1}$$

where u_0 is the threshold value of the membrane potential.

Note the importance of the next statement: By virtue of the point set out in Sect. E.2 on the rules of training, the contribution of summed synapses to the dynamics of the neuron is nonnegative, and, therefore, the membrane potential itself is nonnegative.

Let the W-neuron at time $t_0 + 1$ be passed to the refractory state. The transition could be caused either by the pulse generation at the previous clock cycle or the arrival of a nonzero signal on an absolutely inhibitory synapse at moment of time t_0. Let nonzero input signals on its synapses arrive after the release of the neuron from the refractory state. Then the value of the membrane potential is

$$u(t_0 + r_0 + 1) = u(t_0)q^{r_0+1} + \sum_{i=1}^{N} w_i x_i(t_0 + r_0 + 1).$$

Due to the fact that $u(t_0) \geq 0$, the values of the membrane potential satisfy inequality

$$u(t_0 + r_0 + 1) \geq \sum_{i=1}^{N} w_i x_i(t_0 + r_0 + 1).$$

If at moments of time t_0 and $t_0 + 1$, the W-neuron generates pulses, then by (E.5.1) we obtain the limit on the values of the membrane potential from the top:

$$u(t_0 + r_0 + 1) < u_0(t_0)q^{r_0-1} + \sum_{i=1}^{N} w_i x_i(t_0 + r_0 + 1).$$

Thus, regardless of whether or not pulse generation occurred at moments of time t_0 and $t_0 + 1$, one could estimate the value of the membrane potential at time

$$t_0 + r_0 + 1.$$

By virtue of the two preceding inequalities and (E.5.1), the condition of double-pulse generation by the W-neuron at moments of time $t_0 + r_0 + 2$ and $t_0 + r_0 + 3$ requires the holding of inequalities:

$$q^{-1}u_0 < \sum_{i=1}^{N} w_i x_i(t_0 + r_0 + 1). \tag{E.5.2}$$

$$q^{r_0-1}u_0 + \sum_{i=1}^{N} w_i x_i(t_0 + r_0 + 1) < q^{-2}u_0. \tag{E.5.3}$$

Recall that at training of the perceptron (see Sect. E.2), the value

$$\sum_{i=1}^{N} w_i x_i = \|w\|^2 = M > 0$$

for all $X = (x_1, x_2, \ldots, x_N)$, for which

$$\sum_{i=1}^{N} w_i x_i > u_0.$$

For all X such that

$$\sum_{i=1}^{N} w_i x_i < u_0,$$

is the value

$$\sum_{i=1}^{N} w_i x_i = 0.$$

Using inequalities (E.5.2) and (E.5.3), we find that the threshold value of the membrane potential should be chosen satisfying inequality:

$$\frac{Mq^2}{1 - q^{r_0 + 1}} < u_0 < Mq. \tag{E.5.4}$$

Inequality (E.5.4) requires certain restrictions on the damping factor of the membrane potential q:

$$\frac{q}{1 - q^{r_0 + 1}} < 1. \tag{E.5.5}$$

Normalizing the synaptic weights, we can assume that the constant M satisfies inequality

$$\frac{Mq^2}{1 - q^{r_0 + 1}} < 1 < Mq. \tag{E.5.6}$$

By virtue of (E.5.4) we can write $u_0 = 1$.

We describe the oscillatory mode that interests us. Let at zero time part of W-neurons of one of the modules pass to the excited state and generate single-output signals. In the next moment, there are nonzero output signals in neurons of another module and secondary pulsation in neurons of the first module. These signals induce the generation of pulses by the elements of the third module. The wave of excitation consistently bypasses all of the modules. The next clock cycle of the passage of the excitation wave through the network is opened by the generation of single-output signals by part of the neurons of a source module (i.e., not coinciding with the set of neurons excited at the first clock cycle). After a series of clock cycles of the wave's passage, only neurons of the first and second modules pass into the state of excitation, and from them begins the process of the wave's propagation. In the future, this process is periodically repeated. Each time the elements of the two modules are in the excited state.

Let the duration of the refractory state of W-neurons be equal to $p - 3$. We discuss the problem of training the network, i.e., the choice of synaptic weights. Consider the sets of binary vectors $X_1^i, X_2^i, \ldots, X_p^i (i = 1, \ldots, m; \ m \le 2N + 1)$. We assume that the vectors $(X_{k-1}^i, X_k^i) \in R^{2N}$ are linearly independent. Choose the synaptic matrices $W_{k,k-2}, W_{k,k-1} (k = 3, \ldots, p)$ according to the rules of training the single-layer perceptron so that neurons of the kth module would respond by the output vector X_k^i to the input vectors $X_{k-2}^i, X_{k-1}^i (i = 1, \ldots, m)$ (formed by the $k - 2$th and $k - 1$th module respectively), i.e.,

$$X_k^i = \theta\left(W_{k,k-2}X_{k-2}^i + W_{k,k-1}X_{k-1}^i - U_0\right).$$

The coordinates of the vector of threshold values U_0 for all neurons in the network are assumed to be equal to one.

Choose matrices $W_{1,p-1}, W_{1,p}$ so that the vectors X_{p-1}^i, X_p^i, as the output vectors of the first module, would serve vectors $X_1^{i+1}(i = 1,\ldots,m-1)$, and for the vectors X_{p-1}^m, X_p^m would serve vector X_1^1. For the neurons of the second module matrices $W_{2,p}, W_{2,1}$ are chosen so that

$$X_2^{i+1} = \theta\left(W_{2,p}X_p^i + W_{2,1}X_1^{i+1} - U_0\right),$$
$$X_2^1 = \theta\left(W_{2,p}X_p^m + W_{2,1}X_1^1 - U_0\right).$$

We choose the matrices of synaptic weights using inequality (E.5.4) to ensure the generation of neuron pulses during two consecutive cycles.

Statement E.5.1 *At the indicated choice of synaptic weights (if the duration of the refractory state is $r_0 = p - 3$), there exists a periodic mode of functioning of the network, in which at output signals are generated at consecutive times:*

$$\left(X_1^1, X_2^1\right), \left(X_2^1, X_3^1\right), \ldots, \left(X_p^1, X_1^2\right)\left(X_1^2, X_2^2\right), \ldots, \left(X_p^2, X_1^3\right),$$
$$\ldots, \left(X_1^m, X_2^m\right), \ldots, \left(X_p^m, X_1^1\right), \left(X_1^1, X_2^1\right)\ldots$$

Pairs of vectors $\left(X_{k-1}^i, X_k^i\right)(i = 1,\ldots,m)$ are formed by the $k-1$th and kth modules, respectively.

We explain how to initialize the oscillatory mode. Let us assume that the signal $x_{\text{ext}}(t) = 1$, which is transmitted to the external input, transfers the W-neuron to the state of excitation at the next time moment and thus essentially sets the value of the membrane potential $u(t+1)$ in the range (E.5.1) (this latter point is important). Let at time zero all W-neurons be in a wait state. Using external inputs consistently at moments of time $t = k(k = 1,\ldots,p)$, transmit absolutely inhibitory signals to neurons of the kth module. At time $t = p$, by way of external inputs, we act on the elements of the first module so that at $t = p+1$ its output vector would be the vector $X_1(p+1) = X_1^1$. In the same way, acting through the external inputs at time $t = p+1$, transfer the outputs of W-neurons of the second module into the state $X_2(p+2) = X_2^1$.

After initialization, the network will function in the planned oscillatory mode. Let us illustrate it. At moment of time $t = p+2$, neurons of the third module passed into a wait state and are under the influence of the secondary pulses of the elements of the first module and the pulses of neurons of the second module. By the choice of the matrices of synaptic weights, it follows that the vector of output signals of the third module is $X_3(p+3) = X_3^1$.

It is clear that this construction can be continued. It is important that at any moment of time, nonzero vectors of output signals form W-neurons of the kth and $k+1$th modules (after the pth follows the first module) and the $k+2$th module is in a wait state. Elements of the other modules are in a refractory state. At moment of time $t = p+1+mp$, the output vector of the first module is $X_1(p+1+mp) = X_1^1$. From this moment onward, the process is periodically repeated.

Consider generalization of the problem. Let the value of the membrane potential of the W-neuron at time t satisfy the inequality:

$$q^{-(k-1)}u_0 < u(t) < q^{-k}u_0,$$

where u_0 is the threshold value of the membrane potential. Then the neuron generates, starting from moment t, a burst of length $\gamma = k$.

Let hold inequality (E.5.5). Using this fact, we can prove the following. Let the vectors $X_i^j (i = 1,\ldots,m, j = 1,\ldots,p)$ be such that the kNth dimensional vectors $X_{j-k}^i, X_{j-k+1}^i, \ldots, X_j^i$ are linearly independent.

Statement   **E.5.2** *We   choose   synaptic   matrices $W_{j,j-1}, W_{j,j-2}, \ldots, X_{j,j-k}$ ($j = 1,\ldots,p$) so that there exists a periodic mode of operation of the network in which output signals are generated in consecutive moments of time:*

$$\left(X_1^1, X_2^1, X_3^1, \ldots, X_k^1\right); \left(X_2^1, X_3^1, \ldots, X_k^1, X_{k+1}^1\right); \ldots \left(X_{p-k+1}^1, \ldots, X_p^1, X_1^2\right);$$

$$\ldots \left(X_1^m, X_2^m, X_3^m, \ldots, x_k^m\right); \left(X_{p-k+1}^m, \ldots, X_p^m, X_1^1\right); \ldots$$

Vectors $\left(X_{j-k}^i, X_{j-k+1}^i, \ldots, X_j^i\right)$ are formed by modules with numbers $j-k$, $j-k+1, j-1, j$, respectively.

Note that the above Remarks 1 through 3 are true.

E.6   Networks of W-Neurons in the Problem of Planning Optimal Pathways for Point Robots

We consider the problem of planning of optimal pathways for two point robots on a regular triangular lattice. Some of the lattice vertices act as obstacles on the plane. While moving, the robot must bend around obstacles and not have positions for collision. To solve problem, we use a neural network, the elements of which are W-neurons. The problem under consideration in planning the pathway serves a broad range of applied problems, e.g., generating the path of autonomous mobile equipment, navigation problems, etc.

The solution of the problem of planning of optimal pathways for two robots is based on applying the idea of effective replacement of their movement to the movement of one generalized robot. The idea was developed in (Mayorov et al. 2005); the implementation of it, other than the given idea (based on the neural network of wave propagation), is presented in (Lebedev 2000).

E.6.1 Formulation of the Problem

Consider on the plane of a regular triangular lattice, the length of an edge of which can be considered equal to one. Denote the vertices by $V_{i,j} (1 \leq i \leq N, 1 \leq j \leq M)$. We call the neighborhood of vertex $V_{i,j}$ "set $U_{i,j}$," which consists of six adjacent vertices. Denote by $e_{i,j}^{k,l}$ edge the connecting point $V_{i,j}$ with the point $V_{k,l}$, where $V_{k,l} \in V_{i,j}$.

We call a path (trajectory) connecting two fixed vertices $V_{i,j}$ and $V_{r,s}$, an "alternating sequence of vertices" and edges $V_{i,j}, e_{i,j}^{k,l}, V_{k,l}, \ldots, e_{p,q}^{r,s}, V_{r,s}$. The length of the path is the sum of the lengths of its edges (actually it is the number of edges).

Note two vertices on the lattice where we put point robots. The robots function in discrete time. Per one clock cycle, each robot must move along the edge to one of the adjacent vertices, or it can remain in place. We assume that some of the vertices of the lattice act as obstacles on the plane, i.e., the trajectory of robot motion cannot contain these points. In addition, the robots cannot simultaneously be at the same vertex at any time.

The task is to find the shortest path for each robot connecting the initial and end points given for it. In this case, the paths must bend around obstacles, and to be safe it should have possible positions of collision. Note that on regular lattice, the solution of the problem is not unique in the general case.

The proposed solution is based on the considered auxiliary problem of finding the shortest path for the robot, which is presented in the next paragraph. For its solution, we use the neural network of W-neurons.

To solve the main problem, we introduce on the plane a coordinate system, choosing one of the lattice vertices with axes directions in accordance with lattice generators as the origin. Then the position of the robots on the plane will be characterized by integer coordinates $(a_1(t), b_1(t)), (a_2(t), b_2(t))$, respectively. Here t is the current moment of time.

In four-dimensional space, we consider the set Ω of points of the form a_1, b_1, a_2, b_2, where a_1, b_1 and a_2, b_2 are coordinates of the vertices of the lattice. Then the movement of the robot on the plane corresponds with the displacement of point $(a_1(t), b_1(t)), (a_2(t), b_2(t))$ on the set Ω. Here the first two coordinates describe the movement of the first robot, and the third and fourth coordinates describe the movement of the second robot. We obtain a complete analogy with the problem of planning the robots' path on the plane.

E.6.2   Finding the Shortest Pathway of the Point Robot on a Regular Lattice

W-neurons, each of which can act on six adjacent elements, are located in the nodes of regular lattice. We assume that the neurons corresponding to the obstacles have definitely inhibitory synaptic inputs and that the rest of the neurons have summing inputs. Through absolutely inhibitory synapses, the neurons are constantly kept in a refractory state at the nodes, which are obstacles. Denote by A and B the neurons of the network corresponding to the initial and final positions of the robot.

Let $w_{i,j}^{k,l}$ be the synaptic coefficient of action of the k, lth neuron located at the vertex $V_{k,l}$ on the (i,j)th neuron, which is located at the vertex $V_{i,j}$, where $V_{i,j} \in U_{k,l}$. We assume that the connections between the neurons of the network are symmetrical, i.e., $w_{i,j}^{k,l} = w_{k,l}^{i,j}$. Initially, we set all of the synaptic weights to equal one.

We equip each W-neuron with external input. If, at the moment of time, t the neuron is in a wait state ($s(t) = 0$), then the external signal $x_{\text{ext}}(t) = -1$ (inhibitory) transfers it to the next moment into the state of refractoriness, which lasts for one cycle. In turn, the signal $x_{\text{ext}}(t) = 1$ (excitatory) transfers the W-neuron at the next moment of one clock cycle to the excited state (i.e., it generates a single output signal). Then follows the state of refractoriness, which lasts for one clock cycle. The external zero signal does not act on the neuron.

Suppose that at time zero all W-neurons are at rest. Transmit external signal $x_{\text{ext}}(t) = 1$ to neuron A, corresponding to the initial position of the robot. At time $t = 1$, the neuron generates a pulse. We assume that the threshold values of the membrane potentials of each element of the network are identical and equal to $u_0 = 1$. This means that at time $t = 2$, W-neurons in the neighborhood of neuron A— except those which correspond to the obstacles and are in a refractory state—pass one clock cycle to the excited state. Next the wave propagates through the network. Neuron A passes to the refractory state.

The process of generating pulses is accompanied by a modification of the existing synaptic connections. If a neuron receives a signal from another neuron and therefore becomes active, then the weight of connection between these neurons, according to Hebb's rule (Hodgkin and Huxley 1939), should be increased. In view of symmetry, make such coefficients equal to two: $w_{i,j}^{k,l} = w_{k,l}^{i,j} = 2$. If the neurons generate pulses simultaneously, then a cause-and-effect relationship between them is absent. We assume that synaptic coefficients for them are equal to zero. We also make equal to zero the weights of the connections, the activation of which has not led to the generation of the pulse.

The process of the sequential generation of pulses by neurons of the network ends as soon as the excitation wave reaches point B. We call this the "first cycle" of the passage of the excitation wave through the network. Note that here is determined the length of the minimal path AB. At the end of the first cycle, all of the neurons are at rest. It is sufficient to apply to all external inhibitory neurons the signal $x(t) = -1$.

After two clock cycles, all neurons will be at rest. At the second cycle, a nonzero signal is transmitted to the external input of neuron B. We assume the threshold values of the membrane potentials to be equal to $u_0 = 2$. This means that at the second cycle, only those neurons whose synaptic inputs have sufficient weight pass to the excited state. A change of the synaptic coefficients is similar to the changes at the first cycle. The second cycle ends as soon as the wave of excitation, generated by the neuron at point B, reaches the neuron at point A. After the second cycle follows the third cycle: A nonzero signal is transmitted to the external input of neuron A, etc.

Each cycle of excitation wave propagation through the network is accompanied by an increase of the threshold value u_0 by one as well as a change of the synaptic weights. List the rules of the change of synaptic weights.

1. If for two neighboring neurons $s_{k,l}(t)s_{i,j}(t+1) = 1$ where $s_{k,l}(t)$ and $s_{i,j}(t+1)$, are their states at moments of time t and $t+1$, respectively, then we increase by one the synaptic weights $w_{k,l}^{i,j} = w_{i,j}^{k,l}$. If $s_{k,l}(t)s_{i,j}(t+1) = 0$, then the weights $w_{k,l}^{i,j}$ and $w_{i,j}^{k,l}$ are equal to zero.

2. If for two neighboring neurons i, j and k, l $s_{k,l}(t)s_{i,j}(t+1) = 1$, then the synaptic weights of the connections $w_{i,j}^{k,l}$ and $w_{k,l}^{i,j}$ are equal to zero.

3. If in the neighborhood of the $U_{i,j}$ (i, j)th neuron generating pulse at a given time t, there is no element with index (k, l) such that $s_{k,l}(t+1)s_{i,j}(t) = 1$, then we assume that the synaptic coefficients $w_{i,j}^{m,n}$ and $w_{m,n}^{i,j}$ for any pair of indices (m, n), where $V_{m,n} \in U_{i,j}$, are equal to zero.

The first two rules are consistent with Hebb's rule (Hodgkin and Huxley 1939). The third rule is in "support" of effective paths leading from A to B.

Recall that the cycle is the process of propagation of the pulse wave from node A to node B or vice versa. Denote by A_k the set of W-neurons generating pulses at the kth cycle. Here we make an obvious remark. If the neuron does not generate a pulse at the kth cycle, then it cannot pass to the state of excitation at the $k + 1$th cycle. Therefore, $A_{k+1} \subseteq A_k$.

If $A_{k+1} \subset A_k$, then we should pass to new $k + 2$th a cycle of excitation wave propagation through the network. If $A_k = A_{k+1}$, then the set of vertices corresponding to these neurons is the sought set of vertices belonging to the optimal routes. There are no extraneous vertices here. Otherwise, the wave of pulses generated by neuron A, in passing through these vertices, reaches neuron B when the latter has already passed to the refractory state.

The algorithm converges for a finite number of steps k_0. At the same time, the threshold takes the value $u_0 = k_0 + 1$. The set A_{k_0} is the union of all vertices of nonintersecting paths from A to B. Path lengths are identical and minimal. Along any path the synaptic weights at adjacent neurons are $k_0 + 1$. If neurons are in the adjacent nodes but belong to different paths, the synaptic weights of the connections between them are equal to zero.

We explain how to restore the path when knowing the synaptic weights of the connections. If $w_{i,j}^{k,l} = k_0 + 1$, then $V_{k,l}, e_{i,j}^{k,l}, V_{i,j}$ is the element of optimal path (i.e., the excitation between neurons (k, l) and (i, j) can be transmitted). First, the network is in a state of rest (i.e., sleep). To he neuron located at node A is transmitted the external signal $x(t) = 1$. Pulses will propagate along paths from A to B. In the general case on a regular triangular lattice, this is not the only way. As a rule, we have a series of equivalent paths, i.e., paths that are the same length. We can leave only one way. To do this, in the neighborhood of the neuron located at node A, it is sufficient to block (i.e., transmit external signal $x(t) = -1$) all but one neuron. Naturally, the nonblocking neuron is connected with neuron A by a nonzero synaptic connection, i.e., it is in one of the selected paths from A to B.

E.6.3   Finding the Shortest Pathways for Two Point Robots on a Regular Lattice

We return to the problem of the motion of two robots. Consider the set Ω belonging to the four-dimensional space, of points of the form a_1, b_1, a_2, b_2 where a_1, b_1 and a_2, b_2 are coordinates of the vertices of the lattice. Then the movement of the robots on the plane correspond to the displacement of point $(a_1(t), b_1(t), (a_2(t), b_2(t))$ on the set Ω. Here t is current time; the first two coordinates describe the movement of the first robot; and the third and fourth coordinates describe the movement of the second robot.

Denote by $p \subset \Omega$ the domain of obstacles. It is that part of the set Ω, in which there is impossible movement of the corresponding point $(a_1(t), b_1(t), (a_2(t), b_2(t))$.

We give a formal description of set P. Let (c, d) be the coordinates of the vertex belonging to the set of vertices of obstacles on the plane. Then the point a_1, b_1, a_2, b_2 belongs to set P if $a_1 = c, b_1 = d$ or $a_2 = c, b_2 = d$ (or if both of these inequalities are held at the same time). Within the meaning of the problem, this condition means that the trajectories of the movement of robots in R^2 must not intersect with the obstacles.

To set P we must also include the points of set Ω, the coordinates of which satisfy the following conditions: $a_1 = b_1, a_2 = b_2$. The movement trajectories of robots in R^2 should not intersect with each other.

After we have described sets Ω and P, we place W-neurons at points Ω. In this case, neurons placed at points of the set, have absolutely inhibitory synapses. Further to the neural network, the algorithm of solution of the auxiliary problem of finding an optimal path for a single robot is applied.

Note that the basic idea of solving the problem—the replacement of the movement of two robots with the movement of one generalized robot, i.e., the transition to multidimensional space, allows us to extend path planning in the case of k robots.

References

Bakhvalov, N. S. (1973). *Numerical Analysis* (vol. 1). Moscow: Nauka.

Dunin-Barkowski, V. L. (1978). *Information processes in neural structures.* Moscow: Nauka.

Frolov, A. A., & Muravyov, I. P. (1987). *Neural models of associative memory.* Moscow: Nauka.

Hodgkin, A. L., & Huxley, A. F. (1939). Action potentials recorded from inside a nerve fiber. *Nature, 144,* 710–711.

Lebedev, D.V. (2000). Planning of optimal paths of planar robots using the wave neural network architecture. In *Modelling and Analysis of Information Systems.* Yaroslavl, 7(1), 30–38.

Mayorov, V. V., & Shabarshina, G. V. (1997). Report on networks of W-neurons, *Modelling and Analysis of Information Systems, 4,* 37–50. (Yaroslavl).

Mayorov, V. V., & Shabarshina, G. V. (1999). Simplest modes of burst wave activity in the networks of W-neurons. *Modelling and Analysis of Information Systems, 6(1),* 36–39 (Yaroslavl).

Mayorov, V. V., Shabarshina, G. V., & Anisimova, I. M. (2005). Neuronet solution to the problem of planning of optimal paths for point robots. In *Proceedings of the VII All-Russian Conference "Neuroinformatics-2005",* Part I (pp. 197–202). Moscow: Moscow Engineering Physics Institute.

Rabinovich, M. I., Varona, P., Selverston, A. I., & Abarbanel, H. D. I. (2006). Dynamical principles in neuroscience. *Reviews of Modern Physics, 78,* October–December, 2–53.

Wiener, N., & Rosenbluth, A. (1961). Conduction of impulses in the heart muscle. *Cybernetic Collection, 3,* 7–56 (Moscow: IL).

Printed in the United States
By Bookmasters